Advanced Industrial Development

Books from the Lincoln Institute of Land Policy/OG&H

Land Acquisition in Developing Countries
Michael G. Kitay

Introduction to Computer Assisted Valuation
Edited by Arlo Woolery and Sharon Shea

Second World Congress on Land Policy, 1983
Edited by Matthew Cullen and Sharon Woolery

The Zoning Game Revisited
Richard F. Babcock and Charles L. Siemon

Land Supply Monitoring
David R. Godschalk, Scott A. Bollens, John S. Hekman, and Mike E. Miles

Land Markets and Land Policy in a Metropolitan Area: A Case Study of Tokyo
Yuzuru Hanayama

The Urban Caldron
Edited by Joseph DiMento, LeRoy Graymer, and Frank Schnidman

Land Readjustment: The Japanese System
Luciano Minerbi, Peter Nakamura, Kiyoko Nitz, and Jane Yanai

Measuring Fiscal Capacity
Edited by H. Clyde Reeves

Economics and Tax Policy
Karl E. Case

A Redefinition of Real Estate Appraisal Precepts and Processes
Edited by William N. Kinnard, Jr.

Advanced Industrial Development

Restructuring, Relocation, and Renewal

Donald A. Hicks
University of Texas–Dallas

A Lincoln Institute of Land Policy Book

Published by
Oelgeschlager, Gunn & Hain
in association with the
Lincoln Institute of Land Policy

International Standard Book Number: 0-89946-196-4

Library of Congress Catalog Card Number: 85-13876

Printed in the U.S.A.

Oelgeschlager, Gunn & Hain, Publishers, Inc.
131 Clarendon Street
Boston, MA 02116 U.S.A.

Library of Congress Cataloging in Publication Data

Hicks, Donald A.
Advanced industrial development.

"A Lincoln Institute of Land Policy Book."
Includes index.
1. United States—Industries. I. Title.
HC106.8.H53 1985 338.0973 85-13876
ISBN 0-89946-196-4

To my father, Donald E. Hicks
1925–1979

Contents

List of Figures

List of Maps

List of Tables

Foreword

Everything that is written about economics is in one way or another about the process of change. Yet even as they describe how prices rise and fall or how industries expand and contract, few economists have the wit or daring to embrace a truly dynamic view. Each shift in prices, technology, or factors of production often means a widely rippling series of social and political changes that can reach far beyond the realm of conventional economic study, and which in turn alter the conditions for the next round of economic change. In their quite different ways, Joseph Schumpeter and Karl Marx each left an impression upon economic thinking because of their willingness to consider where the long chain of constant economic changes might lead.

The great virtue of Donald Hicks' work is that he is as intent as Schumpeter or Marx on understanding change as the one constant of economic activity. Like them, he sees it not as an irritant or an aberration but as the essential condition of economic life. The circumstances of the material world are constantly in flux: new technologies emerge, resources grow suddenly scarce or cheap, customers alter their demands and tastes, new competitors take their place on stage. Healthy economies are distinguished not by their ability to make any one thing in particular but by the suppleness with which they respond to their constantly changing environments.

The steps that add up to such an ability to adapt are the subject of *Advanced Industrial Development*. Every chapter in this book reflects Hicks' willingness to think seriously about the causes and consequences of economic dynamism, not simply to repeat the familiar platitudes about the inevitability of change. Not everyone will agree with his perspective, even after considering his evidence, but I cannot think of anyone seriously interested in modern political economics who will not profit from reading this book.

The themes that Hicks discusses will be useful for citizens of any modern industrial economy, but I think they will seem particularly vital and enlightening to readers in the United States. In the U.S., the latest round of economic transformation and dislocation, which Hicks focuses on in this book, represents the collision of several deeply held American beliefs. The resulting tension and uncertainty help explain several of the fundamental struggles in American politics of the 1980s, and Hicks' analysis is a powerful tool for understanding their implications.

One of these traditions is the long-held American faith that "change" will eventually prove synonymous with "opportunity" and "progress." The nation was, after all, populated by those arriving from somewhere else, and the very industrial regions now most discomfited by structural change are themselves the products of the dislocations and transformations of earlier years. The entire national mythology, from the seventeenth-century settlements in New England to the twentieth-century boom in California, from the completion of the transcontinental railroad to the development of new technologies in space, has been based on the faith that the nation's distinctive advantage was its embrace of the new. So today, as certain industries emerge and others decline, one familiar strain in American thinking can interpret the transition as nothing more than the twentieth century's version of the process that took Americans from the farms to the factories and thereby made the country the most productive on earth.

But there is another prominent theme in American culture that leads to a far less cheerful view of today's economic uncertainties. Practically as soon as the continent was colonized by Europeans, the settlers began worrying that what they had achieved could easily be undone. Through the nineteenth century, the standard work of Fourth of July oratory would alternate a celebration of the young republic's virtues with a warning of how quickly they might be lost. Nearly every European visitor, from Alexis de Tocqueville to Charles Dickens to Sigmund Freud, noticed how confident Americans were of their potential as a fresh, new continental power. But most visitors also noticed the Americans' anxiety about the threat of decline. *The Decline and Fall of the Roman Empire* was popular in the United States because it was considered an explicitly cautionary tale. In the post–World War II era, the decline and fall of the British imperial economy has served the same admonitory function. First the republican virtues—savings, hard work, discipline—were squandered, and then, according to the standard formulation, so was the republic's international dominance. In the 1950s and 1960s, the Soviet Union made a brief appearance as the upstart power whose future potential (mainly military) would eclipse that of the United States. Since the early 1970s, Japan has ably filled that role.

Out of its preoccupation with the possibility of decline, the United States has developed a deeply ambivalent view of changes like those that Donald Hicks describes. What if the death of the old is merely that, a death, and not simultaneously the birth of something new? What if the loss of familiar habits and occupations and ways of life is a loss plain and simple, rather than a painful but ultimately beneficial transition? Certainly the world has seen such unredeemed declines before: who now fears the economic might of such once-potent societies as Portugal, Spain, and the Venetian city state? Why should the American economy be exempt from the forces that have brought ruin to many others?

This fear cuts with an especially sharp edge in the 1980s because of one other traditional American belief. Among the reasons that previous waves of change seem in retrospect to have been so beneficial is that they created the broad, democratic opportunity that is fundamental to the American ideal. From the villages of Sicily, from the farms of Indiana, from the sharecroppers' shacks of Mississippi and Georgia, Americans could come to the cities and make new lives for themselves, through the opportunities created by new industrial technologies. Anything that endangered that process would at the same time endanger American democracy, since it would weaken the glue of shared opportunity that helped a nation of different races, classes, and beliefs to cohere.

The most familiar analyses of today's "deindustrializing" economy suggest that the fears of decline and lost opportunity are being borne out. According to this interpretation, the American economy is losing its productive powers, and it can no longer open the avenues of opportunity that have been so important to the nation's political stability as well as to its economic strength. If this analysis is correct, then the proper response to economic change is to resist it. It is painful for the people and communities immediately affected, and it bodes no one any good.

Through his case studies and his overall surveys, Donald Hicks provides an alternative to this despairing view. With an analysis that is careful, compassionate, and—to me—persuasive, he suggests that today's economic dislocations do not constitute a break from the long pattern of American expansion and opportunity but rather are a clear continuation of a long-established national trend. Some businesses are troubled, he says, but not all of American enterprise; some locations are declining, but not whole regions; some people are undoubtedly in distress, but the whole economy is not dying, nor should they and their children be without hope.

"Hope" may seem too subjective a word to apply to a study as carefully documented and as dense with statistics as this, but to me it is the most important implication of Hicks' message. Throughout economic history, change has produced discomfort; farmers have left the land, cobblers have been denied their traditional means of support. Yet these cruelties of economic life look much different, depending on whether they ultimately help create more opportunity and greater wealth or are merely symptoms of stagnation and decline. There is cruelty too in today's economic transformation, but the United States can better manage to cushion the blows, while letting the transformation go on, if the changes seem to be part of the long process that created the nation's opportunity and wealth.

In one other way, *Advanced Industrial Development* highlights a crucial analytic distinction. Through the 1980s, the American economy has been affected by two forces that differ completely in their origins and implications, but which share a common set of symptoms. The symptoms in each case are the factory closures, regional shifts, and other changes that together create a sense of economic and social dislocation. One of the causes of this change is the ultimately hopeful process that Donald Hicks describes. The other, of course, is the destructive impact of the overvalued dollar, born of the unprecedented federal deficits of the early 1980s, which in effect saddled American manufacturers with an artificial export tax. The first process promises to create as well as destroy; the second represents a few years of subsidized consumption at the cost of a burden of debt that impedes the prospect for future growth.

The difference between these processes—both disruptive, but one in a promising and the other in an ominous way—again underscores the value of Donald Hicks' analysis. He has not written to endorse economic disturbance for its own sake, but only as a consequence of the ferment that typifies a healthy and adaptable economy. But disturbance can be a sign of disease as well as of health. The more clearly we understand the difference, the easier it will be to encourage one kind of change and retard the other. By teaching us how to tell the difference, Donald Hicks has with *Advanced Industrial Development* performed a tremendous service.

James Fallows

James Fallows is the Washington editor of The Atlantic; *he has also been an editor of* The Washington Monthly *and of* Texas Monthly. *From 1977 to 1979 he served as President Jimmy Carter's chief speechwriter. Twice the winner of the Champion Media Award for Economic Understanding, Fallows is the author of* National Defense, *which won an American Book Award in 1981, and of* Human Capital *(forthcoming from Random House).*

Lincoln Institute Foreword

The Lincoln Institute of Land Policy is an educational institute dedicated to the development and exchange of ideas and information pertaining to land policy and property taxation. It is a school offering opportunities for instruction and research. It welcomes government officials, working practitioners, and students to the pursuit of advanced studies.

The Lincoln Institute is also a center for linking the university and the practice of government; for bringing together scholars, professionals, and officials; and for blending the theory and practice of land policy. Professor Donald A. Hicks has helped us fulfill these roles. He was the author of *Urban America in the Eighties* (1981) of the President's Commission for a National Agenda for the Eighties, while on leave from the University of Texas at Dallas. Because of this work, we invited him to present a paper at our February 1982 joint conference with the University of Southern California Law Center, "Urban Land Policy for the 1980s: The Message for State and Local Government." His paper, "National Urban Land Policy: Facing the Inevitability of City and Regional Evolution," became the basis for further research during the following two years. We are pleased to have been able to assist him in this endeavor and to publish the final result of his research and writing.

This book critically examines the evolution of industrial development. It provides policymakers and practitioners with basic background information and carefully researched detail, all to assist them in addressing current ideas concerning the restructuring, relocation and renewal of America's industrial base.

Frank Schnidman
Senior Fellow
December 1985

Preface

As this manuscript leaves my desk for the last time prior to publication, I am reminded once again of the reasons why I undertook this project in the first place. I began the research and planning portions of this book more than three years ago at a time when the United States and most of the other First World nations were in the grips of a severe and extended recession. Talk of an industrial "crisis" was widespread. As a consequence of what was widely pronounced and thus widely perceived to be a "deindustrialization" of the United States, fear was developing that this nation's industrial status was being jeopardized and the prospects for sustaining high living standards were being diminished. From this perspective a new chapter in the continuous clash of "capital" and "community" had commenced. Inevitably, much debate over the wisdom of inducing a "reindustrialization" via explicit industrial policies ensued. An extended, uneven, and seemingly incomplete recovery has only dispelled some of these concerns while deferring others, no doubt, just until the next business downturn arrives. And so go our continuing efforts to comprehend and differentiate the substance from the illusions, the realities from the metaphors, of industrial change.

This book presents an alternative interpretation. It is intended to be neither optimistic nor pessimistic, liberal nor conservative. Rather, it is meant to offer a broad—if necessarily incomplete—and coherent perspective on a wide range of issues that are implicated in a society whose industrial economy is undergoing continuous demographic, technological, and cultural changes. It offers guidance on how to locate and interpret the resulting evidence of industrial change in a more even-handed fashion. It does not attempt to discount the wrenching changes in personal lives that accompany industrial change, but it does attempt to help us appreciate the implications of new developments and be less apprehensive about the more familiar features of an industrial world that we may be leaving behind.

I am indebted to many present and former students and colleagues in and beyond the university for both the explicit and subtle ways in which they helped me proceed with this task. High on that list are the Lincoln Institute of Land Policy (and especially Arlo Woolery), which supported this project through a postdoctoral fellowship during 1982–83. I extend special thanks to my former graduate research assistant, Bill Stolberg, for his friendship and tireless and conscientious contributions to the research projects that provide a foundation for Chapter 5. Also I wish to thank Czi-C. Pann, Joel James, Mel Letteer, Phil Berry, Florence Cohen, Dorothy Luttrell, and Cynthia Keheley for their assistance in preparing portions of the research materials that are reported elsewhere in the book. For the editorial skills and calming temperament of Penny Stratton, my Senior Editor at OG&H, I am also very grateful.

Finally, I am reminded of my debt to my father, a man whose very life provided me with my earliest and most lasting impressions of the complex ways in which the abstract process of industrial change can be registered on the private lives of us all. A proud union man whose entire work life was spent in mills, garages, and factories, the full span of his adult life exhibited the range of adjustments, including retraining and relocation, by means of which individuals can and do create and respond to new opportunities and possibilities for community. Both he and the factory job he held during his last work days are gone now. But for those lessons, and the quiet and dignified ways in which they were taught, I am eternally grateful.

D.A.H.
Dallas, Texas
December 1985

1

Advanced Industrial Development: An Introduction

By many accounts we are closing the books on an older industrial era and opening a new set for the next one. A so-called second industrial revolution is underway, accompanied by wrenching adjustments facing older basic industries and localities, new technologies that permit production activity to disperse to new settings while much corporate control continues to concentrate in older ones, and the rise of new jobs in new industries defined by new production arrangements. There is much talk about this being a "time of transition," even though all times tend to take on the protective coloration of a time of transition. Nonetheless, events of recent years have conspired to reinforce the idea that we have reached some sort of industrial threshold. While there has been a tendency to exaggerate the extent to which the present and future represent a clean break with the past, it is clear that our industrial economy is experiencing profound changes.

The implications of our exit from an older industrial era and our transit into whatever follows have become the subject of considerable speculation during the 1980s. Increasingly, the debate surrounding these implications has acquired a sense of urgency as signs of what is viewed as industrial distress have become more visible and seemingly more resistant to traditional policy prescriptions. For many, the prospect of alterations in older industrial arrangements is unsettling and threatening. While industrial change is acceptable in the abstract, they contend that the process should somehow be steered clear of major features of our industrial past such as traditional basic industries, as well as factories, farms, offices, cities, neighborhoods, regions, and households. From this perspective, industrial change should be a bounded and directed process, as though we could ignore somehow the tie between economic growth and industrial change. As a result, the lexicon of industrial change has become increasingly confused and inadequate. Moreover, the full range of industrial adjustments we are experiencing seems to require a reconceptualization of both the nature of economic change and the policies that predictably have been proposed in response.

Growth, Development, and Industrial Change

Our understandings of growth and development have failed to keep pace with the processes themselves. In an earlier era—especially prior to the 1950s—a simple conception of economic development could not easily be disentangled from economic growth. Growth understood in a quantitative sense was seen as leading inevitably to development in a qualitative sense. The linkage between growth and development was established during a period when the latter implied net increases across a wide range of economic aggregates such as total employment, housing starts, industrial output, household and per capita income, consumption, and capital investment.[1] And the bulk of the increases derived from the expansion of the industries at the heart of an industrial economy—durable and nondurable manufacturing, construction, and the energy sector (mining).

This aggregate growth was registered *physically* in ever-larger settlements that offered more jobs and housing to residents; fostered larger and more numerous factories, farms, offices, and shops as settings for production and consumption; evolved more elaborate infrastructures to bind together separate places of work, residence, and recreation; and especially, created the need for more land and space development to accommodate all this growth. Simultaneously, this growth was registered *socially* and *politically* in higher incomes and standards of living, more leisure time, and beginning in the late 1930s, supportive political arrangements including retirement and unemployment benefit systems and local economic development strategies for securing greater economic security for people and places alike. Concurrently, macroeconomic policies, keyed as they are to the behavior of selected aggregate indictors of economic functioning, were viewed as reasonable economic development policies for the larger economy.

The economic growth of the past half-century inevitably led to economic development that took many forms. New technologies and institutional arrangements were required to administer economic activity on ever-larger scales. Among these innovations was the rise of new multilocational corporate forms, including branch plants and other satellite production settings spun off from headquarters offices and regional distribution centers. Entire regions, especially the South, were slowly brought into the economic mainstream of the nation. Continued settlement growth led to the spread of population within and across regions and of jobs to suburbs and beyond.

Economic growth also revealed itself to be capable of unfolding at uneven rates among industrial sectors and regions and of bringing unequal industrial benefits to different subgroups of the population. Within the context of continued growth, emphasis shifted by the 1960s from the *scale* of growth to the *structure* of growth. As a result, economic development translated into disparate employment rates and income levels across labor markets subtly structured as to occupation, race-ethnicity, gender, and age. Inevitably, as the industrial economy continued to develop, primary concern for the continuation of that growth began to yield to a building concern for the distribution of the costs and benefits of that growth. The scope of economic development gradually broadened to include notions of access, distribution, and fairness.

Consequently, macroeconomic policies came to be viewed as necessary, but insufficient substitutes for explicit economic development policies and microeconomic strategies, because the policies aimed at aggregate growth could neither ensure growth nor avoid the inequalities accompanying uneven industrial development. By the 1970s, the United States had begun a quest for "balanced" growth implemented by "targeted" economic development policies designed with specific geographical, sociodemographic, and industry-sectoral constituencies in mind. No longer could economic growth be counted on to lubricate social adjustments and provide a constant margin for handling politically difficult distributional tasks through a wide range of policies with social welfare goals. And our federal system of government, whose primary dynamic for organizing and responding is territorial, took on that task.

Decoupling Growth and Development in Advanced Economies

By the late 1970s and early 1980s, economic circumstances began to change. There is, of course, nothing new about firms, industries, and even entire sectors not expanding for extended periods. However, by this time, slow or no growth was a common feature across several industries whose very expansion had defined the development of our industrial economy. Textiles, with the nation's largest share of manufacturing employment, machine tools, and of course, the highly visible automobile and steel industries ceased registering growth on major indicators. Over the course of several cyclical shifts, the services industries were looked to for nearly all new employment growth. While the larger economy continued to grow, it began to do so slowly and with greater difficulty. Nonetheless, such high rates persisted in labor-force expansion, business formation, population redistribution, and other forms of capital mobility that it became apparent that considerable industrial development could continue even in the absence of traditionally high rates of output and productivity growth.

That so much else was changing so rapidly in the economy at the same time that the industrial performance of many of our basic industries was so sluggish suggests that the industrial development of the nation had reached a new stage. Continued restructuring and redistribution within industries and regions, if not of the overall economy itself, became the basis for distinguishing between industrial development and growth. It could now be anticipated that several of our most basic industries would continue to evolve in the future without much, if any, new net growth. The nation's industrial core, in particular, would increasingly be characterized by stable output shares and slowly growing or declining employment shares registered both by basic industries and by the regions that they dominated. Continued, and even significant industrial development in the absence of growth across a range of traditional indicators, then, is the basis for what is referred to in this book as *advanced industrial development.*

Furthermore, industrial development could now be viewed as a process fully capable of continuing—even accelerating—during recessions. An industrial economy may undergo major shifts in the structure and location of industrial production, investment, residence, and employment as it adjusts to a changing global economy. This conclusion applies even, if not especially, during periods of slow or no growth, and it draws attention to the need for a reconceptualization of what is meant by economic growth and development, as well as of the policies designed and implemented to promote them.

Key Features of Advanced Industrial Development

I use the term *advanced industrial development* to underscore the idea that what we are experiencing represents a *continuity* with what has gone before rather than a break with it; that is, what we do not have is a *crisis.* To the extent that *industrial* connotes an enduring commitment to goods production, I argue that we continue to have an industrial economy. However, this choice of terms is not meant to preclude the use of *post-industrial* as a legitimate adjective. Indeed, I will also argue that the process of advanced industrial development has created the first truly post-industrial places—a reality reflected in the new land uses, architectural forms, and economic functions of many of our older central cities.

Advanced industrial development has been shown to have registered its impacts broadly throughout the larger economy and society. What appears to be a series of conflicts and collisions between industrial sectors (e.g., goods vs. services), major industries (e.g., cars vs. computers), regional economies (e.g., Snow Belt vs. Sun Belt), urban economies (e.g., central city vs. suburb), occupations (e.g., primary vs. secondary labor markets), and even land uses (e.g., urbanization vs. agriculture) can be deceiving. A major effort of this book will be to show that selected dominant features of an earlier industrial era have not changed, while other more subtle ones have changed dramatically. In the end, the more consequential impacts of advanced industrial development are shown to be located *within* sectors, industries, regions, cities, occupations, and patterns of land use, rather than in conflicts between them.

Economic Growth and Perspectives on Change

It has long been acceptable to speak of the "economy" as a thing apart. As an entity, the economy is judged to have evolved to the point that it exerts a set of powerful influences on everything around it. There are dangers in this conceptual shortcut if we lose sight of the fact that the economy—and the apparent order it offers—is ever the abstract consequence of individuals responding to the circumstances they find themselves in by exercising the choices they perceive available to them. What appears to be a situation wherein firms, governments, households, and other corporate actors are responding to the economy as something external and uncontrollable is ultimately nothing more than individuals who are responding to their perceptions of what they sense is going on around them.

The notion of an embodied economy generally causes few problems. However, in a discussion of industrial change—how to define, locate, interpret, and evaluate it—the consequences of such a notion may lead to problems. The economy, like the society that it reflects, is better viewed as a result, than as a cluster, of original causes. Shifting patterns of economic activity are rooted in how individuals view and ultimately put to use so-called "natural" resources, space, and their time and effort (and those of others), as well as in how new ideas make their way from invention and innovation to eventual application.

As a result, two very compelling, if somewhat contradictory, images of economic change in the United States have developed. The first suggests that our industrial economy has been undergoing a monumental restructuring or transformation, and that to that extent we are leaving behind a sort of industrial "golden age" of the third quarter of the twentieth century. The focus is on the fact that a transition between industrial arrangements is underway; the implication is that insofar as the new arrangements serve us poorly, trouble may lie ahead. A second image is related but is tied less to the existence of change than to the capacity to adjust to it. This image emphasizes the perception that the critical capacity of our economy and larger society to adjust to new economic realities is deficient. The prospect for trouble looms as a result of well-intentioned attempts to protect existing industrial and social arrangements from developments that may have beneficial overall effects but that do not yet enjoy the advocacy of well-developed constituencies.

Industrial "Decline" as a World View

During the past decade there has arisen in this nation and elsewhere around the world a belief that the United States has commenced a slow industrial retreat. Since World War II, we have seen a quickening succession of recessions, recoveries that fail to restore either employment in major industrial sectors or overall levels of unemployment to prerecession levels, long periods of dampened productivity growth, a deepening dependence on the services sector for net employment growth, and increased vulnerability to competition from equally and less developed nations. These events are commonly seized upon as evidence that the U.S. industrial hegemony is slowly ebbing to the point that our future prosperity may be in jeopardy. While, as of this writing, speculation that this trend constitutes a long-term secular decline has subsided somewhat in recent months, there is every reason to expect that it will flare up again as the business cycle reasserts itself and ingredients of the next economic downturn fall into place.

Evidence of the "costs" of industrial change is highly visible and all around us. Flagship industries lose exports; major industrial settlements lose jobs, the opportunities for replacement jobs, and therefore residents; established neighborhoods lose investment; and households lose their ties through traditional employment to the local economies around them. In contrast, recent patterns of growth and development in the U.S. economy often are acknowledged with considerable misgivings, largely because any gains are considered insufficient to compensate for what we appear to be losing. New jobs in both new and renewed industries often have sprung up in new locations and reflect dependence on new forms and mixes of production inputs—land, energy, technology, human skills, and other capital. This situation is viewed as evidence of the gradual unraveling of the traditional combinations that for decades have been perceived as essential to maintaining leadership in an industrial world.

Is the United States experiencing an industrial decline? Is the economy severing its ties to its basic industrial moorings and slowly drifting toward something new? Are goods production, in general, and basic manufacturing industries, in particular, losing their capacities to serve as the spine of our advanced industrial economy? Or, putting aside for the moment whether such changes constitute an industrial retreat or advance, is the core of the economy evolving into something so historically unfamiliar that in one way or another we are ceasing to be an industrial society? These are complex and complicated economic questions that in turn give rise to equally thorny social and political issues.

If industrial decline is perceived as an eclipse of the industrial core of our economy either by an expanding domestic services sector or by able and aggressive foreign competitors, then the available "hard" evidence indicates that talk of a "deindustrialization" is most probably not justified. The transformative industries of manufacturing, and to a lesser extent construction, constitute the key sectors in this debate. And while the construction share of the GNP has declined slowly across the past quarter-century, the manufacturing share has been relatively stable at slightly less than a quarter of total output since 1975. The manufacturing share of total employment has shown a modest decline during the 1970s, with the construction share remaining relatively stable. At the level of aggregate shifts, at least, there is little evidence that our industrial sector is receding against the backdrop of the larger economy.

This surface calm, however, conceals a series of major trends that may be just as consequential for this nation as any presumed contraction of our major transformative industries. If industrial decline is perceived more as an abandonment of familiar landmarks, technologies, and organizational features of business and older forms of work itself, then a stronger case for industrial decline, understood in the sense of a departure from the past, can be made. Our major industries reveal ample evidence of changing mixes of production and nonproduction workers; they also show an increased variety of relationships between goods production and both in-house and outside-contracted services, including product research, financing, marketing, legal, and other producer services. Beneath surface indicators of net change, one can find evidence of a changing structure of the labor force, of the group that has left the labor force, and of the unemployed. Rising net employment conceals rising levels of both business formation and business failure, with differences in the volatility of these underlying processes particularly evident across regional and local economies. Also concealed are redefinitions of traditional features of employment, as illustrated in the rise of part-time and seasonal employment and self-employment, the deskilling of some work activities, and the rising skill requirements of others. Home-workplace distinctions are blurring, and even the substance of "work" itself and how it is perceived as an activity is being redefined.

Now, it is quite another matter whether or not these changes constitute processes of "decline" and should be construed as undesirable or be regarded as symbolic of trends that must certainly lead to diminished standards of living. My view is that they do not, even though honest differences of opinion currently exist. Defining and interpreting evidence of industrial decline has become yet another treacherous proving ground for the social sciences. Indeed, the very limits of science are laid bare as analyses evolve into advocacy of one or another set of incompatible or contradictory policy options. Increasingly, it is clear that comprehending and responding to these questions, either for their own sake or in an effort to formulate policy options, rests as much on interpretations rooted in normative analyses as on attempts at rigorous scientific measurement. In the end, the final answers hinge both on initial assumptions concerning what should properly constitute evidence of industrial "decline" and "dominance," and on what conceptions of desirable industrial arrangements appear to be most threatened by ongoing industrial change.

In this book I offer an alternative interpretation of selected features of the larger process of industrial change. This interpretation is markedly different from the conventional view of industrial decline, even though it acknowledges many of the same shifts and trends on which that view rests. It also acknowledges a major transition from one set of production arrangements to another without necessarily assuming that this constitutes a net loss or decline. Such change does, however, involve a reworking of the nation's industrial landscape in both major and minor respects. The production arrangements of an industrial society are assumed to be continuously in transit, and this process inevitably involves departures from traditional patterns of industrial structure, investment, production, employment, and consumption. In addition, this same landscape is being continuously resettled as firms and households come to define and use land and space in new and historically unfamiliar ways. Ultimately, at any one point in time, the industry-sectoral, employment and occupational structures of an advanced industrial economy and society are viewed as less compelling or important than is the success with which the economy responds and adjusts continuously to countless technological, sociodemographic, and political factors. A capacity for industrial adjustment to new production arrangements, rather than the ability to retain and defend any fixed set of existing ones, is viewed as the premier feature of an advanced industrial economy and society.

Summaries of the Remaining Chapters

Chapter 2: Locating Industrial Change

Economic growth and its structure provide a backdrop against which to view industrial change. Chapters 2 and 3 offer descriptive overviews of key features of existing industrial arrangements and whether, and if so how, they appear to be changing. In Chapter 2 the central focus is on the difficulties of interpreting what some would construe to be evidence of industrial decline. The most visible indicators include the long-term slowdown in productivity growth of the past decade and a half, as well as an even longer sequence of business recession-recovery cycles.

I turn to productivity trends first. In the most general sense, productivity indicates the efficiency of production and is therefore essential to the goal of rising living standards. While overall, the United States remains the most productive economy in the world, this is no longer true for all industries. Through the 1970s and early 1980s, and even now as we enter the fourth year of an extended economic recovery, the U.S. rate of productivity growth is the lowest of all major industrial nations. Most notably, Japan, West Germany, and France have been successful in nearly closing the gap with us after three decades of relatively more rapid productivity growth. Several partial explanations for this slowdown are examined, although none appears to suggest that the current productivity stagnation need be considered more than a temporary and relative phenomenon. These explanations offer little support for the view that even an extended period of slow productivity growth warrants concern that a process of industrial decline is underway.

Next, I look at recessions and movement through the business cycle and examine their direct and indirect influences not only on employment and output trends but also on our interpretations of those trends. On the surface, a recession is understandably viewed as an "event" or a constellation of essentially negative circumstances including a slackening in demand and industrial production that leads to employment loss and new rounds of reduced demand and production.

A major impediment to disentangling industrial "decline" from a more benign process of industrial transformation can be attributed to the necessity to search for evidence of structural change amid continuous and simultaneous movement through business cycles. A persistent susceptibility to recessions is judged by many to be consistent with longer-term industrial decline. Taken together, therefore, the recession-recovery sequences of recent decades are troublesome not only because they have occurred but also because they can be interpreted as the chronic sputtering of the production apparatus of an entire nation. The same evidence, however, is also largely consistent with the conclusion that a new and more advanced industrial economy is gradually taking shape. From this alternative view, our industrial economy is not so much receding as it is being gradually restructured and continuously renewed. Cycles of recession and recovery may obscure certain structural adjustments while making others more visible. Therefore, cyclical fluctuations in a recession-prone economy are wisely not equated with industrial decline.

An argument is offered centering on the fact that recessions have far from uniform impacts—either sectoral, spatial, or sociological—across the economy. Moreover, recessions, far from being the isolated crises that industrial society must endure, can actually be times of great creativity and flexibility that serve as bridges between stages of industrial development. As the momentum of older industrial activities slows, new directions are more easily chosen. A preoccupation with stagnation and contraction in highly visible economic indicators may blind us to important changes taking place beneath the surface. Recessions can accelerate the transit of a firm or even an entire industry from older and less efficient production arrangements to newer and more efficient ones. In the long run, these adjustments are necessary to preserve or restore an industry's competitiveness. Admittedly, such a detached and mechanistic perspective is difficult to maintain when the effects of these adjustments include older jobs that do not come back or the growth of new jobs tied to skills that are in short supply. Invariably, recoveries do not retrace the steps that led to them in the first place. However, a recovered economy relies on restoration of the capacity to produce, not the replication of the production arrangements that existed prior to a recession. The sequence of recoveries, then, no less than the sequence of recessions since World War II, illustrate important features of advanced industrial development.

What evidence substantiates the claim that our industrial economy is evolving into something new? The answer offered in this chapter is that the resulting restructuring involves subtle but consequential shifts that indicate continued, rather than curtailed, industrial development. I reject the argument that the centrality of goods production to our industrial economy is being threatened by rapid growth in the services. The manufacturing share of total output has remained relatively stable during the post–World War II era, while the decline in manufacturing's employment share has unfolded far too slowly to constitute an industrial retreat or decline. Since 1970, levels of personal consumption expenditures for services have edged out first those for nondurable goods and eventually those for all manufactured goods. However, this trend is important largely because it reflects the dramatic ways in which the services sector itself has restructured and the more complex ways in which it contributes to an advanced goods-production sector. Even though the traditional goods orientation appears to be intact, goods production has evolved to the point that production sequences are increasingly complex and extended and therefore require a much more developed services sector than ever before.

Is industrial society changing in ways that suggest that socioeconomic life will be more severely stratified and tend toward greater inequality than has been the case in the society to which we have grown accustomed during our industrial past? Probably not. I consider the possibility that industrial restructuring involving occupations and income distributions, rather than industry sectors, may provide the missing evidence of alleged industrial decline. While there has been a marked shift since World War II toward both high- and low-paying white-collar occupations, and a shift away from all blue-collar occupations, especially those that in the past helped create and sustain a blue-collar middle class, there is little evidence that this has had—or necessarily will have—a profound effect on the income distribution traditionally associated with contemporary industrial society. While it may yet be too early to be confident in this view, there appears to be no compelling reason to anticipate that the changing ways in which the work of an advanced industrial nation gets done somehow must result in lowered living standards or greater polarities among social groups. Inequality and disadvantage can be expected to filter into new sociodemographic and geographic locations, although without necessarily becoming more prevalent.

Finally, I explore an even more subtle shift that might account for the rise of what some have called the post-industrial society. This shift involves the transformation of "work" itself in an advanced industrial society. I find that there has indeed been a steady and rapid decline in the proportion of workers who actually make material things and a corresponding increase in the proportion whose work consists of creating, communicating, and applying information in its myriad forms. What people do, rather than the sectoral and occupational locations of their activities, does appear to constitute evidence of a major revolution underway in our industrial economy and society. The result may be a departure from an older conception of industrial society and may be sufficient in certain respects to justify the term *post-industrial.* In the end, though, there appears to be no restructuring, reorientation, or even redefinition underway that can unambiguously be interpreted as evidence of industrial decline.

Finally, it is unwise to assign too great a significance to slow sectoral

is faced with intensifying global competition.

Chapter 3 briefly describes the way in which the U.S. metalworking sector, in aggregate and through a succession of new plant cohorts, has slowly made the transit from older to newer physical settings, from unionized to nonunionized labor environments, and from lesser to greater technological sophistication, all without abandoning its historical regional locations. The purpose is to showcase the ways in which countless investment decisions at the level of individual small plants and shops are even more constrained than previously thought by an unplanned process of industrial metabolism. This process illustrates the way in which even the unintended aggregate consequences of calculated investment decisions at the plant level are shaped by a simultaneous and continuous "turnover" of plants within and between plant cohorts.

Just as a specific industry may experience a restructuring over time through intraindustry cohort succession and firm-level investment decisions, a nation's industrial base may likewise evolve and thereby be repositioned in a global economy as a result. Development cycles in two "marker" industries—steel and computer software—are sketched out to illustrate how the contraction of older industries and the expansion of newer industries are integral to advanced industrial development and thereby contribute to the restructuring of the industrial base of an entire nation.

Chapter 4: Resettlement and New Industrial Location Trends

A principal legacy of the industrial era has been a network of major urban settlements. The spatial and locational features of advanced industrial development in and between these settlements—that is, the impacts registered in specific places—are the explicit focus of Chapter 4. The locational counterpart of deindustrialization is urban and regional "decline." Settlements and entire regions that historically have been heavily dependent on manufacturing are unavoidably implicated in this discussion, if for no other reason than because one of the defining features of advanced industrial development, like earlier industrial development, has been the continuous dispersal of manufacturing from once core regions and central locations. As a result, it would appear that a new urban-industrial landscape has taken shape. This conclusion would be only partly correct. In actuality, a key message of this chapter is that location is ever a major dimension of a ceaseless process of industrial settlement and resettlement by households and industrial organization by firms. Patterns of industrial location and specific industrial places, like industries themselves, are the targets of continuous rearrangement.

Chapter 4 begins by describing, at three spatial scales, the patterns of deconcentration and dispersal of both population and manufacturing jobs throughout the past century. The purpose is to highlight those spatial features of industrial organization that have *not* changed, as well as those that have. Over time, the nation's four major regions have come to resemble each other industrially. While the South has emerged as the major source of the nation's total output and the location of the majority of the nation's population and manufacturing employment, manufacturing continues to account for higher employment shares in the Midwest and Northeast than in the South or the West.

At the next smaller scale, even though the negative relationship between manufacturing employment growth and place size has grown stronger, especially since the 1960s, the overall distribution of manufacturing employment by place size has not appreciably changed. It is true that manufacturing employment has been dispersing to newly attractive locations outside metropolitan areas to the point that manufacturing now constitutes nearly equal shares of total metropolitan and nonmetropolitan employment; but the bulk of U.S. manufacturing employment remains located in the largest metropolitan places. The capacity for advanced industrial development, then, to move multistate regions and metropolitan and nonmetropolitan economies toward increased convergence in their industrial structures has not appreciably lessened the affinity of manufacturing for major population centers. This is true even though much manufacturing has simultaneously exchanged locations in selected older regions for locations in selected newer ones.

At the third spatial scale, the focus is on central cities and local urban economies. Large dense industrial-era cities with heavy commitments to basic manufacturing industries, as much as the mass production arrangements of the factories they hosted, are the physical expressions of an older industrial era. It might be expected that advanced industrial development would reserve for these older cities especially dramatic impacts. The century-long process of residential and manufacturing suburbanization is recounted in Chapter 4. While the advanced industrial economy has indeed expanded beyond the urban-local to the metropolitan-regional scale, in light of the patterns already discussed at larger spatial scales, it is unlikely that advanced industrial development is capable of unleashing forces of dispersal powerful enough to erode the metropolitan scale of industrial organization as dramatically as happened at the smaller urban-local scale. While the observation may be more abstract than practical, it does suggest that for the time being at least, space and location will retain their powerful joint influence over patterns of advanced industrial development. It is inconceivable that either population or manufacturing employment could be distributed uniformly across the landscape at any time in the forseeable future.

I next turn to consider what Wilbur Thompson has called the "recycling of the industrial-era city." Levels of population and employment long have been regarded as the most important indicators of traditional industrial development. A major reason for this view has been that the industrial-era cities were political as well as social and economic entities, and expanding tax bases in support of a fixed jurisdictional base permitted attractive economies-of-scale in the delivery of urban services. Predictably, in recent decades as our largest industrial-era cities lost first people, then jobs, and in many cases both, deep concerns were expressed that our cities were in decline. A succession of urban policies dedicated to reversing these trends, and then later to resisting them where possible, have been offered in response since the late 1950s.

In Chapter 4, I urge a reorientation of the key indicators by which we measure the health of our cities from levels of population and employment to their structure or composition. Population and employment contraction and redistribution, no less than their growth and expansion, are viewed as integral features of advanced industrial development. Many industrial-era cities have been able to achieve at lower population and employment levels the "health" that they were unable to achieve decades ago at much higher levels.

The "recycling" of older cities—especially central cities—can teach us many new lessons about what cities are in the first place. It has long been commonplace to view cities as freestanding entities with the autonomy to adapt to only those technologies that reflected collective choices. Over time the image of the city beset by unwanted technologies—the automobile remains the classic case—supported the view that urban life and a wide range of new technologies were naturally antagonistic. Even today new technologies are commonly cited as the principal factors accounting for—or at least permitting—city-leaving behavior and overall urban decline. However, the city does not simply orchestrate the social uses of assorted technologies in order to create wealth. It, too, is a production tool. And to the extent that advanced industrial development proceeds by replacing older tools with new more productive ones, cities can be expected—even encouraged—to be part of that process.

For the past quarter-century the "city" has been widely viewed as being on a collision course with successive waves of new transportation, communication, and production technologies. I strongly disagree with both the orientation and the implications of such a view. In Chapter 4,

I seek to reorient the "city versus technology" view to a "city *as* technology" view. A city may come to be many things to us, and many parts of it may be assigned primarily historical, cultural, and related derivative functions. But the first test of a city as a whole is its ability to combine resources to create wealth. No urban settlement limits its repertoire of functions to that of a production platform, but should that primary task begin to fail, the city's secondary roles cannot be expected to insulate it indefinitely from the inevitable consequences.

Chapter 4 presents a discussion aimed at reinterpreting both traditional indicators of "urban decline" and the goals of conventional urban economic development strategies. To illustrate the "place-sculpting" aspects of advanced industrial development, I have chosen to examine recent stages of the development of Dallas, Texas. Commonly viewed as one of the nation's relatively new and prosperous cities, Dallas's economic base and physical capital are perceived to reflect major national industrial and economic trends more closely than most older industrial cities. But it would be a mistake to assume that Dallas, as a place that has hosted a central city for nearly a century and a half, is somehow exempt from continuous urban restructuring and redistribution. While Dallas continues to experience population and employment growth, the vast majority of that growth is confined to only a few residential and business "microclimates." Most parts of the city have been "thinning out" or growing more slowly than the overall city on several key indicators including population, housing, and business activity. How are we to interpret this redistribution of activities and people *within* the city; are these trends indicative of incipient urban decline? Again, probably not.

Finally, then, urban decline, no less than general deindustrialization, is a notion that depends heavily on the premise that mobile capital and "community" in its several senses are increasingly antagonists in a capitalist economy. In response, Chapter 4 concludes with an attempt to suggest the ways in which new forms of "community," beyond the narrowly territorial, may be emerging as increasingly important and valued. In the end, the impacts of advanced industrial development depend heavily on the mix of endowments of individual cities; the outcomes of their economic and technological transitions will be far from uniform. Nonetheless, many of our central cities—transformed by either rapid expansion, contraction, or sometimes both—have emerged as the clearest examples of "post-industrial" places that have yet been produced.

Chapter 5: High-Technology Industrial Development Features

While industrial development is a continuous process, there is great danger in assuming that it is somehow self-initiating. An industrial economy is forever making internal adjustments to itself, to changing demography, new technologies, and public policies, as well as international developments. However, the introduction of new elements into an economy is wisely viewed as something separate from the rearrangement of existing elements. Chapter 5 is devoted less to the momentum of advanced industrial development and more to the factors responsible for its new directions. Here the focus is on innovation, entrepreneurship, business formation, risk-taking, and the industrial settings that best nurture these things. The chapter explores the local-regional features of high-technology industrial development by bringing the technology and industry life-cycle dynamics of Chapter 3 together with the spatial and locational dynamics of Chapter 4. High-technology industrial development within Texas is used to illustrate the key points of this chapter. Three research projects I conducted are used as sources for the data reported.

The first study examines the development features of a relatively new cluster of high-technology services industries—computer software and data processing services. The computer services industry has experienced explosive growth over a relatively short time. The evidence suggests that while this new industry has relatively few locational constraints imposed by its essential production activities, its heavy reliance on pools of appropriately skilled workers has caused it to be distributed across economic geography in particular ways. At the largest scale, the computer services industry is nearly exclusively a metropolitan phenomenon in Texas with the bulk of the industry located in only the very largest metropolitan areas. Moreover, as the components of the larger computer-equipment manufacturing industry have positioned themselves during its recent development, a spatial division of labor has emerged among Texas's largest metropolitan areas. Computer services appears to be gravitating toward certain metropolitan areas while computer manufacturing gravitates toward others.

At the next smaller scale, the development features of the computer services industry are explored within the Dallas Primary Metropolitan Statistical Area (PMSA). The vast majority of both computer services establishments and employment in the PMSA are discovered to be located within the city of Dallas. At an even smaller scale, the distribution of computer services establishments across business microclimates within the city reveals the way in which the industry has been swept along by Dallas's real estate, residential, and commercial development in recent years.

A second study supplements secondary analysis of data on business formation within Texas with primary data from a Texas subset of a nationwide survey of firms in the U.S. computer software industry. The focus is on the location factors governing the early development stage of this new industry. Of particular importance was how software start-ups, generally small ones, perceived and oriented themselves to a sequence of "nested" locations defined at ever smaller scales down to the level of the actual physical structures in which the firms set up business.

The remainder of Chapter 5 examines the origins and developmental anatomy of a regional high-technology sector. The data are drawn from a third study that explored the emergence of the high-technology sector over the quarter-century since 1960 within the Dallas-Fort Worth regional economy. While the D-FW region's high-technology sector is currently the nation's third largest and fastest growing, a major message of this section is how complex and extended are the roots of this sector within the seedbed that nurtured it. By examining a succession of high-technology birth cohorts, it becomes apparent that the shift to services taking place within the larger regional economy is reflected clearly in this sector as well. Furthermore, despite the increased fascination with the dynamics of small business formation, the bulk of existing employment and new employment is tied to the growth dynamics of large established firms within the sector rather than to small business start-ups.

The D-FW region's high-technology sector provides evidence of the widespread expansion and contraction that accompany advanced industrial development. Many high technology firms and the industries of which they are a part have been losing employment even as the larger sector has expanded. Moreover, in an analysis of concentration and dispersion trends, there appears to have been a long-term slow dispersion of high-technology employment throughout the eleven-county region. The same dynamic is apparent within the city of Dallas itself. Nonetheless, the central city's historical role as an incubator of new technologies and the industrial development triggered by them likewise appears to be supported by the data. That the continued development of today's high-technology sector will likely be accompanied by gradual dispersion from its central city origins also appears to be supported.

Chapter 6: The Land-Based Economy Revisited

Another dramatic agricultural revolution appears to be accompanying the so-called second industrial revolution. The original agricultural revolution signaled the expanding ability to exploit through increased mechanization the productive capacity of the land and land-based resources. The current revolution in agriculture goes well beyond this to include the capacity not only to liberate agricultural production from traditional conceptions of "land," but even to bypass land altogether.

The industrial adjustments evident in agriculture—the original land-based economy—are the focus of Chapter 6. This chapter is included in a book on advanced industrial development so as to underscore the point that agriculture is first and foremost an economic activity capable of making a wide variety of substitutions of nonland for land inputs in the production process. At a time when food production in advanced nations is being relied upon to meet the demand of rapidly growing Third and Fourth World nations, it may be dangerous to cling to an unnecessarily rigid view of agriculture that envisions it as inherently dependent on fixed combinations of so-called natural inputs of soil, water, and sunlight, as well as on traditionally defined labor and organizational inputs. I find no compelling evidence to indicate that agriculture as a whole is suddenly encountering insurmountable "natural" limits to its continued development. In recent decades agriculture, understood as an economic activity, has been able to make dramatic adjustments, thereby demonstrating remarkable flexibility in creating and combining the resources necessary to high-productivity food, fiber, and livestock production.

The decoupling of employment growth from increases in productivity and total output, the "mobility" expressed by agriculture as it repositioned itself from the South to the Midwest and as the corporate organization of farming and land holdings has restructured, and the diffusion of new production technologies throughout the sector offer an easily overlooked model for understanding and interpreting trends that are restructuring modern manufacturing. The chapter also explores land policy controversies generated by an overly rigid conception of an inevitable collision between traditionally urban and traditionally rural activities. Finally, I briefly discuss the wisdom of an industrial policy for U.S. agriculture in light of the difficulty of reaching consensus on what constitutes a healthy industry in the first place.

Chapter 7: The Decline of "Localism" and A New Scale of Governance

A continuously renewed economy can be expected to require sufficiently flexible public institutions to administer it. Chapter 7 offers a brief overview of the way in which traditional institutions and tools of public administration have gradually fallen out of step with the more fluid and flexible processes of social and economic development that easily sweep over political and institutional boundaries. The resulting mismatches are illustrated by examining shifts within systems of public finance. The local property tax, in particular, is identified as a fiscal tool whose use is more compatible with the traditional land uses and conceptions of "localism" that are rapidly being left behind. Declining local autonomy in the context of increasing interdependence among local, regional, national, and international economies reduces the likelihood that a mosaic of local political jurisdictions can continue to rely on locally generated property tax revenues to fund substantial public sector functions. The expanding scale of economic and social organization is unlikely to abide for long the fiction that traditional political jurisdictions can be even substantially self-sustaining.

Chapter 7 examines patterns of dependence on the property tax among advanced industrial nations as well as trends in tax capacity and tax effort at the state and regional levels. The influence of state-local tax policies on advanced industrial development is also explored. The chapter looks at series of physical capital and demographic trends that place constraints on the extent to which local property tax systems can be expected to generate sufficient revenues at a politically acceptable rate. Finally, property tax reform efforts and the extent to which they reflect a continued preoccupation with land and its uses are considered.

2

Industrial Dominance and "Decline": Illusion and Reality

The United States has long been the world's most productive economy, and it continues to retain that distinction even after more than a decade of sluggish growth. Yet productivity growth in the United States has lagged behind that of a major international competitor, Japan, all through the 1970s, and behind that of major Western European economies, including West Germany and France, since 1973. A more immediate concern stems from the fact that despite the recent rapid expansion of the economy, labor productivity growth has not rebounded as robustly following the recessions of the early 1980s as it did following those in previous decades.[1] Moreover, productivity growth in manufacturing has surged ahead of that in nonmanufacturing industries during the 1980s.[2] Any loosening of the tie between output growth and productivity growth could be taken as evidence of an economy more on a cyclical rebound than in the midst of a systemic recovery that is sustainable through the emergence of more efficient production arrangements. The deeper concern over extended slow productivity growth is related in part to the speculation that the frequency and resulting impacts of recessions are symptoms of slow, but profound, structural changes in the U.S. economy, and that these changes threaten to diminish the economy's capacity for efficient production and the prospects for rising living standards in the future.

Productivity Trends and Industrial Change

Productivity is a dimension of economic functioning that is as complex to analyze as it is challenging to measure.[3] Its level and rate of change respond not only to a changing mix of production factors, as when new technologies or altered assumptions about the price and availability of energy resources influence labor requirements or capital investment decisions. They also respond to the upgrading of labor skills through increased education, training, and on-the-job work experience. Productivity also reflects the changing scale of production as an industry experiences increased concentration among larger firms[4] or as an increased proportion of output comes to be accounted for by rapid small business growth. These and other more subtle managerial and organizational factors are all routinely implicated in explanations of productivity trends.[5]

During the 1972–1982 decade, labor productivity for nonfarm business rose at an average annual rate of only 0.7 percent, compared to the 2.5-percent average annual increase between 1947 and 1972. Moreover, there were absolute declines recorded in 1974, 1979, 1980, and 1982. Multifactor productivity, which registers the relationship of output to tangible capital inputs combined with labor, mirrored the steady relative decline of labor productivity from 1948 to 1978 and its absolute decline between 1978 and 1980. Data for the 1980–1983 period reveal that while labor productivity increases averaged 1.6 percent, eclipsing those (1.2 percent) of the 1973–1978 period, increases in multifactor productivity since 1980 (0.7 percent) have failed to match those (0.8 percent) of the 1973–1978 period.[6] Beneath these trends for the total private nonagricultural economy, Bureau of Labor Statistics (BLS) data indicate that certain industries, especially those that have upgraded their production technologies and face rising demand, have increased their productivity at faster rates. Nevertheless, the slowdown in productivity growth has been broad based. Industry-specific measures reveal a slowdown during the 1973–1979 period for three-fourths of the industries for which productivity measures are available.[7]

Accounting for Reduced Productivity Growth

The dampening of productivity growth is not easily explained. Both informed speculation and rigorous analysis have drawn attention to a wide variety of potential influences, including the shifting age-, experience-, and gender-based demography of a rapidly growing labor force; a slowdown in the rate of capital investment; the dislocations associated with the accommodation of higher energy costs after the 1973 oil embargo; rising labor costs and the constraints on flexible firm-level adjustments imposed by unionization; the burden of government regulations in such areas as health and safety and pollution control; a deteriorating work ethic throughout the labor force; and conflicting signals from the federal government concerning its commitment to the task of taming inflation, lowering marginal tax rates, or reducing deficits. These factors vary considerably in their sensitivity and responsiveness to specific public policies and business management strategies. A seeming blizzard of inconsistent and often contradictory incentives facing business, labor, and state and local jurisdictions have been identified that make policies difficult to implement and evaluate. Management orientation, employee morale, product mix, organizational flexibility, quality control, work rules defining the labor environment, labor costs, accessibility to upgraded technologies, and expectations regarding investment and inflation are but a few of the pressure points identified for policy consideration. Yet the impacts of explicit initiatives are often slow to register their effects which, in turn, are difficult to disentangle given the complex interactions that are doubtless at work. While several of the proposed explanations of low productivity growth are amenable to policy responses, some—especially those related to demographic and cultural shifts in U.S. society—appear to be relatively immune from the effects of policy intervention.

And what are the policies most often suggested? Variations of taxation, regulatory, and research and development policy options are those most frequently proposed. In a 1983 report, the Committee for Economic Development urged that as cornerstones of a national productivity policy, the United States should simplify its tax code, develop programs for technology incubation and commercialization, and continue to deregulate business.[8] Indicative of the importance assigned to the roles of research and development in productivity growth, a 1984 Congressional Budget Office report called for greater tax incentives, a liberalized regulatory environment, and increased funding for cooperative efforts with other nations.[9] In addition, there have been calls for greater taxation of consumption to stimulate business investment, abolition of the corporate income tax, and the revision of antitrust laws so that more intercorporate and industry-government cooperation can be encouraged. Finally, since high real interest rates are commonly (if erroneously) believed to discourage capital spending, slowing the growth of government spending that piles up huge deficits—unless relieved by broadening tax bases and higher tax rates—is suggested to place downward pressure on interest rates. Few policy tools, it would appear, have failed to attract the interest of those seeking increased productivity growth. Let us take a closer look at some of the partial explanations with an eye toward their potential responsiveness to public policy prescriptions.

Labor-Force Growth Rate and Composition Shifts: The key dynamic here is as much cultural as demographic. This explanation centers on the unprecedented absorption during the 1970s of successive waves of young entrants from the 1947–1956 baby boom birth cohorts and of the adult women rapidly reentering the labor force. The growth rate of the labor force nearly doubled from a 1.3-percent average annual increase in the early 1960s to a 2.7-percent average annual increase between 1976 and 1980. (Annual increases after 1980 dropped to approximately 1.5 percent, with continued slowing expected for the rest of the decade and on into the next century.) Consequently, the labor force was inundated with relatively well-educated but inexperienced newcomers, a development that probably had the immediate effect of stunting productivity growth, especially in the services, which absorbed the bulk of these new entrants.

As the labor force expanded, it also began to restructure. Throughout the 1970s, the labor force grew by nearly 3 million workers annually, and its changing composition reflected this patterned growth. For example, males aged 25–54 comprised 40 percent of the labor force in 1970; by 1980, that proportion had dropped to 36 percent. From the perspective of labor-force expansion and restructuring, reduced productivity growth is a temporary effect whereby a nation's economy adjusts to its changing demography. It simply takes time, while new entrants acquire work experience and skills, before productivity will begin to rise at a faster rate.[10]

Indirectly, all that relatively plentiful labor functioned to reverse years of labor scarcity and thereby to hold down wages in crowded occupational categories and labor-surplus regions. In turn, this created the incentives to slow the substitution of relatively expensive capital—including productivity-enhancing plant and automation technologies—for relatively cheap labor. These incentives, too, could be expected to dissipate once the larger economy accommodated these demographic surges. As we look ahead, the continuing effect of declining birth rates and the shrinkage of succeeding birth cohorts can be expected to have largely opposite effects on the labor force beginning in the late 1980s. Just as the composition of the work force slowly shifted in the past ten years to reflect the rapid expansion of entry–level employment cohorts aged 25 to 34, a gradual shrinkage of this same group will occur by the end of this decade. As these same cohorts acquire job experience and training, the background conditions supporting accelerated diffusion of new technologies through factories and offices will improve. Thus, an increase in the productivity of this cohort of workers can be expected in the future.

Labor Costs and Unionization: It is widely held that wage and benefit levels and their rates of increase, especially in industries beset by slackening demand and intense foreign competition, have likewise helped account for dampened productivity. The automobile and steel industries are commonly offered as clear illustrations. Labor costs are estimated to account for approximately 70 percent of production costs, although these estimates reflect considerable variation in labor intensity across specific industries. The services sector has long been regarded as especially, if not inherently, labor-intensive and not amenable to significant productivity increases. Continuing structural shifts toward more services employment can be expected to exacerbate the labor-cost issue.

Trade unionism throughout a wide range of industries that appear to have the greatest need for, as well as the greatest difficulty in adjusting to, new competitive pressures is commonly cited as a chief impediment to industrial adjustment.[11] The implication is that unionization somehow stifles or impedes productivity growth. Strong and successful unions, those in the railway industry, for example, in seeking to protect jobs and improve working conditions have long functioned to preserve traditional and increasingly counter-productive roles for labor in industries whose competitiveness is thereby stymied. The dampening of productivity growth has been viewed as a regrettable consequence of the power and influence of unions.

Frequently obscured, however, is the possibility that unionization, for essentially the same reasons, may well have the opposite effect of spurring increased capital investment by firms attempting to accommodate higher labor costs. There is evidence that indicates that productivity in unionized plants can exceed that in nonunionized plants in the same industry.[12] Beneficial effects on productivity growth may also derive from higher skill levels and shop-floor stability that may be associated with unionization. Yet, while the relationship between unionization and productivity growth is not clear-cut, it is probable that even though productivity growth may be substantial in an industry, its effect may be diluted by collective bargaining agreements that generate even higher wage-benefit increases. It is increasingly difficult to offset rapidly rising wage-benefit packages with productivity increases alone.

Capital Investment Trends: While capital spending may take many forms, its role in productivity growth is well established. Capital spending for new and replacement plant and equipment is commonly regarded as essential for ensuring productivity increases. All through the post–World War II period until 1973, increasing labor productivity reflected the upgrading of capital stock available to workers, especially in the form of new machinery and related equipment but also in structures like new factories and shops. A 1980 Joint Economic Committee report noted that while 77 percent of the nation's growth since World War II is attributable to investment in plant and equipment, the proportion of growth attributable to new capital stock was approximately twice that derived from replacement stock and upgraded skills.[13]

Since 1948, and across successive periods of reduced productivity growth, there is evidence that the importance of changes in capital-labor ratios has increased relative to other factors. During the 1948–1973 period, capital-labor–ratio changes accounted for approximately one-fourth of the change in productivity; in contrast, during the 1973–1980 period, even though the productivity increase was far smaller, changes in capital-labor ratios accounted for half of all productivity changes. This pattern shows evidence of the potential, if not always realized, role of technology and physical-capital upgrading in spurring productivity growth.

Since 1960, real fixed investment by private business has expanded so dramatically that the temporary dampening of capital spending during business-cycle downturns seems to pale in comparison. From 9 percent of real GNP in 1960 to 11 percent in 1982, the outlays after inflation in the past quarter-century have increased by more than 160 percent. Furthermore, there has been a detectable shift in the composition of those capital investments from production machinery to high technology equipment such as computers throughout manufacturing, trade, and producer services. Evidence of productivity gains in the wake of technology innovation and subsequent adoption exists for a wide variety of industries.[14]

Still, there is reason to question whether productivity might today be somewhat uncoupled from patterns of traditional capital spending. Support for this conclusion can be found in BLS data on the development and testing of a new multifactor productivity statistic.[15] This analysis isolated the contributions to productivity of capital growth and so-called "efficiency" factors, which include technological innovation, management changes, economies of scale, and educational level of workers. The distinction is of considerable importance because these efficiency factors permit the establishment of new arrangements for production, rather than a simple increase in production via existing industrial arrangements. Their measurement permits us to distinguish an economy undergoing structural change and industrial reorganization from one that is simply discovered once again to be growing in a more conventional manner.

The results indicate that during the 1948–1973 period, the 3.0-percent average annual rate of productivity growth for all private business and the 2.9-percent rate for manufacturing separately, were largely attributable to efficiency as opposed to capital growth factors. In contrast, during the 1973–1981 period, the meager 0.8-percent average annual productivity growth rate for all private business and the 1.5-percent growth rate for manufacturing were far more attributable to capital growth factors. However, an analysis of the productivity slowdown for all private business across the two periods indicates that the bulk of the decline (– 2.2 percent) was attributable to efficiency factors (1.9 percent) rather than capital growth (0.3 percent). Manufacturing is widely assumed to be at the center of the most turbulent changes in a restructuring or recession-plagued industrial economy, and these efficiency factors include technological innovations and organizational "investments" so important to the incubation of new industries and renewed production arrangements in older mature ones. Thus it is especially noteworthy that for manufacturing these efficiency factors emerged as even more decisive relative to traditional capital growth—1.9 percent versus 0.4 percent, respectively, in accounting for the – 1.4 percent change across the two periods since 1948.

Capital spending does not take place in a vacuum. In both dramatic and subtle ways the taxing and spending decisions of government influence investment patterns by channeling investment among and within industrial sectors. Over a hundred categories of tax expenditures, including a variety of deductions, credits, and exemptions, cause the U.S. tax code to function as a source of complex, and often contradictory, incentives and disincentives that shape the spending behaviors of firms. The Congressional Budget Office estimates that these expenditures totaled $71 billion in FY 1984 with another $22.5 billion in direct spending and business credit programs also provided for.[16] This pattern is replicated at the level of state and local governments. At any one time, the pace and even direction of advanced industrial development is influenced by the federal income tax system alone. Depreciation schedules and investment credits influence the turnover rate of capital assets within older traditional industries, as well as the success with which the capital assets of new information and knowledge-based industries compete with plant and equipment for public favor.[17]

Heavy or rapidly increasing government spending, rising deficits, and high real interest rates are generally thought to bode ill for capital investment planning designed to boost industrial productivity. Yet this logic turns on whether or not interest rates are sensitive to high or rising deficit spending and resulting federal-debt levels. The debate over this point is considerable, and evidence is far from conclusive, especially for the decade just past.[18] Nonetheless, such adjustments do not occur in a closed system, as was commonly assumed when the domestic economy was more insulated from other national economies. When real interest rates remain high, large amounts of foreign investment become available to lessen pressure on domestic money markets. While a strong dollar can hobble the ability of U.S. firms to export their products, budget deficits in and of themselves need not necessarily imply reduced capital investment in the domestic economy. Indeed, not only has capital investment continued to increase as deficits have mounted, but an inability to sell products abroad may pose a greater immediate problem than the inability to produce efficiently at home.

Research and Development Activity: Cross-investment by government, business, and universities in each others' programs of systematic research and development are generally considered essential to boosting productivity in existing industries and ensuring the long-term health of the economy through the nurture of entirely new industries. Total R&D spending by government, industry, and universities reached an estimated $100 billion in 1984. While the U.S. federal government has been exemplary in this regard throughout the post-World War II era, in recent years this activity has been in relative decline. Annual government outlays for research and development were stable at approximately $27 billion over the 1970–1975 period. Since then, they have risen steadily, topping $40 billion as the recession began to wane in early 1983 and nearly $50 billion in 1984. Despite the fact that R&D support in the United States has generally exceeded that of other industrial nations, and that the commitment of scientific manpower to R&D is, as a percent of GNP, higher in the United States than elsewhere, both total and industrial R&D expenditures declined through most of the period since the mid-1960s. Total spending fell from a peak of 2.9 percent in 1963 down to 2.2 percent in 1977, although it has been slowly rising recently, reaching 2.6 percent by 1984.[19] Corporate R&D commitments account for more than half of total expenditures. While corporate R&D budget increases may be essential for achieving productivity increases, such spending may follow rather than lead business-cycle upturns. According to National Science Foundation data, expanded corporate revenues have enabled double-digit increases in corporate R&D expenditures since 1983. Rising along with these R&D outlays has been the speculation that the synergy among science, information, research, and technology can provide the basis for more rapid and efficient economic growth.

A "Shift-to-Services" Effect? A final factor likewise has been offered as a partial explanation for why productivity growth has stagnated since the 1970s. It has been suggested that the increased dependence of the economy on the output and employment contributions of the services sector has resulted in diminished productivity growth. The precedent for exploring the role of sectoral shifts in productivity change is found in earlier research which showed that since World War II, productivity levels have generally risen in response to the increased capital intensity of farming and the shift of redundant labor from the farm to nonfarm sectors.[20]

Despite the general accuracy of predictions regarding the continued tendency for employment expansion in the services sector to eclipse that in goods production, during the 1970s, the United States was alone among advanced industrial economies—Canada, Japan, France, Germany, Italy, and the United Kingdom—in experiencing growth in manufacturing jobs. Moreover, as we shall see below, any intersectoral shift has been a slow one indeed. Nevertheless, speculation persists that the shift of employment among nonfarm sectors—especially from goods to services production—is responsible for the decline of productivity growth. In a strict sense, productivity growth is thought to be jeopardized not from any sectoral shift per se, but from the presumed imperviousness of services sector production to productivity increases tied to such things as capital investment and the adoption of new production technology. Also, the traditionally high labor intensity of service industries is cited as evidence of the modest returns to productivity of a wide variety of capital investments.

However, impressions drawn from a traditional view of the services sector as being static and either amorphous or homogeneous lag behind new realities. BLS data estimate that no more than 12 percent of the increase in total output between 1947 and 1980 was accounted for by such a shift of employment from one sector to another. Rather, increased output was attributable largely to increased productivity.[21] Furthermore, a 1983 report by the Office of the U.S. Trade Representative concluded that "None of the productivity slowdown experienced in the United States in the 1970s . . . was caused by the shifting of U.S. employment from goods to services industries."[22] Other studies report only slightly negative effects of sectoral shifts on productivity trends.[23]

It appears, moreover, that the conventional view of the service sector is significantly outdated. In a study ranking 145 industries by capital intensity, it was discovered that nearly half of the top fifth were service industries.[24] The service economy—with respect both to the production and distribution of producer, consumer, and other services—has become increasingly "industrialized."[25] Between 1975 and 1982, Bureau of Economic Analysis data indicate that the investment in new technology per service worker soared 96.6 percent.[26] The increase in capital intensity has spawned a broad cluster of service-producing high-technology industries as well as upgraded the potential for productivity gains and higher-quality services at lower prices from less technology-intensive services. Therefore, the argument that major structural shifts in the economy—even if those shifts were as rapid and massive as they are widely assumed to have been—account for a substantial portion of the slowing of overall productivity growth of the past quarter-century seems not to be particularly convincing.

Industrial Restructuring and the Mask of Recession

A recession provides a context as well as a source for more direct influences on industrial adjustment throughout the business cycle. There have been eight official business recessions since World War II—in 1948–'49, 1953–'54, 1957–'58, 1960–'61, 1969–'70, 1973–'75, 1980, and 1981–'82. These recessions have varied both in intensity and duration, and together they have registered their collective impact on the national and subnational economies. Increasingly evident have been the disparate effects on regional and local economies, since cyclical impacts are typically distributed unevenly across the economic landscape. The worst off are those economies whose prosperity is tied to the production of goods for which demand has slackened, for which increased market share has been captured by foreign producers or investment has lagged, or for which physical plant and equipment and human-skill mixes reflect outmoded and inefficient production arrangements. Such economies are usually the first to recede and the last to recover—to the extent that they do—across the business cycle. Throughout the United States, regional economies anchored to auto, steel, rubber, and oil-gas refining currently illustrate this phenomenon. The greater the dependence of a region—its workforce and localities—on such production arrangements, the greater the impact of a recession and the wider its ripple effects throughout other parts of the economy composed of suppliers of parts and other intermediate inputs, home and business consumers, and assorted distributors.

The temporal patterning of recessions has prompted considerable interest in the study of business cycles per se—how to anticipate and influence their courses, and how to lessen their impacts by inducing a policy-led recovery. A mix of macroeconomic and microeconomic policy tools designed to stimulate general, but balanced, growth has commonly been resorted to during a recession, just as it has been to achieve economic stabilization during a recovery. So accustomed have we become to viewing economic growth as an essentially volatile, yet manipulable, process.

Interpreting Recession: Continuity Vs. Crisis: An individual recession viewed as an event does have a distinct structure, trajectory, and impact worthy of careful examination, but it has been the staccato pattern that has characterized the repeated occurrence of recessions through the post–World War II era that has been the more compelling reason for assigning them significance in this discussion of industrial restructuring. The U.S. economy has been in the grip of an official recession nearly one-fourth of the time since the onset of the Depression of 1929.[27] While few would suggest that recessions alone constitute evidence of industrial decline, for some observers their spacing and timing across the past four decades signal our exit from an historically familiar industrial setting and the growing pains accompanying the emergence of something new and potentially threatening. This is especially so as recoveries—particularly recent ones—do not so much restore those productive arrangements that existed prior to a recession as they function to renew or replace them.

This chapter portrays a recession as a vehicle for a larger process of industrial adjustment. Like economic growth and expansion, it can accelerate some changes and retard others, even though it may not necessarily inaugurate any wholly new ones.[28] The 1980–1982 period of economic stagnation that twice deteriorated into an official recession was the culmination of a wide variety of disparate trends. It may be viewed as having provided an especially potent stimulus for the rationalization and reorganization, and frequently the relocation, of the components of the nation's industrial economy. Yet, this pair of recessions, like the extended sequence over the past four decades, may well mask the details of the longer-term development that has been unfolding throughout the century and more of the evolving industrial era.

Recession-recovery sequences assume major significance in discussions of economic development and industrial restructuring, since they define variable growth in the larger economy as well as across and within individual sectors and industries. A recession is in many ways an arbitrarily demarcated set of circumstances that may obscure certain trends even as it highlights others. By bracketing consecutive quarters in which key economic indicators such as output, productivity, wages, profits, hours worked, or employment may have either stalled or slipped, there is a tendency to view such periods as somehow detached from longer sweeps of economic change. It may be true that by viewing a recession as an event, we are once again impressed by the special potency of the interaction of such conditions as sluggish capital investment, increased unemployment, and dampened consumer demand in selected industrial sectors. Yet, viewing a recession as such a crossroads or concatenation temporarily obscures the equally important perspective from which the separate parts of a national, regional, or local economy are seen to be constantly and necessarily adjusting to each other, as well as to a changing demography, the diffusion of scientific developments and technological innovations, the implementation of government policies, cultural and related value shifts, and global competition in foreign and domestic markets.

Economic growth or its absence—being in the recession or recovery phase of the business cycle—probably has more to do with the visibility of industrial adjustment than with whether or not that adjustment constitutes evidence of major restructuring or industrial decline. The occasion of a recession tends to provide an important "window" through which to view a wide variety of events. Unfortunately, when framed in this way against a backdrop of economic stagnation, trends are in danger of being misinterpreted or misunderstood. A recession is an event only to the extent that a number of factors that influence the structure of an economy relate to one another in specific ways. Declining industrial output and profits; stagnant productivity; building proportions of unused and often obsolete industrial capacity; fierce foreign competition resulting from comparative advantages being eroded, created, or sustained with government subsidies; unstable or unbalanced exchange rates; and the decoupling of employment from unemployment as the labor force contracts are all now familiar ingredients in the chemistry of a recession. Yet, more importantly, these same factors all figure in a much larger-scale process of advanced industrial development as well. It is wise, but difficult, to avoid confusing the two processes.

A recession is especially noteworthy because with little or no aggregate growth, the friction of economic change increases considerably. Expansion in an economy can lubricate even the most dramatic restructuring, as has generally been the case with the remarkable shift of employment and other resources out of the land-based economy, including agriculture and energy, and into the factory-based economy during the century of initial industrialization. However, when profits decline, demand recedes, and employment contracts, the productive apparatus of a nation loses the growth margin that has at other times been used to compensate for the absolute and relative losses invariably associated with slowly growing or even contracting sectors and industries. When new jobs are created faster than old ones disappear, and when new markets expand rapidly enough to accommodate the entry of new competitors, the resulting restructuring is often relatively painless and may go largely unnoticed. A less detached and charitable view ensues when such changes wend their way through an economy that has temporarily ceased to grow, as does one that is recession prone. Not the changes per se, then, but the context in which they take place, inevitably invites special scrutiny. Recessions have served as contexts that have heightened the visibility of what has often been assumed to be major intersectoral restructuring. But the real realignments in the economy may be better viewed as taking place *within,* rather than between, major industries, sectors, and subsectors. It is to this possibility, then, that I turn in the following section.

For these reasons, it may be a mistake to assign too much importance to the occurrence and pattern of recessions. Movement through business cycles does not so much leave a special imprint on a stable economy in mythical equilibrium as it functions to accelerate and make more visible those ceaseless adjustments that characterize an economy in continuous and dynamic disequilibrium.[29] Despite the ease with which the longer-term industrial restructuring accompanying advanced industrial development can be confused with shorter-term business-cycle fluctuations, the former is better understood as systemic than as a crisis occurrence—as evolutionary rather than episodic.

Recession: Exploring Beneath the Surface: It is apparent that patterns of structural change are easily confounded by patterns of cyclical change at all levels of economic analysis. A restructuring economy defined by differential growth rates across its several sectors, in revenues, profits, employment, wages, or output, is easily obscured against a background of robust economic growth among all of these indicators. Such was the case during the decades immediately following World War II. However, as selected rates of growth slowed, as they did during the early 1980s, clearer signs of the possible restructuring of the U.S. economy stood out in relief and became the object of considerable optimistic speculation and pessimistic foreboding.

Too often, while focusing on what grows slowly or even ceases to grow during a recession, we overlook other dimensions of an economy that have continued to grow or have even increased their rate of growth. One example is the continued capacity of the U.S. economy to create jobs. During 1960–1984, fully 41 million new jobs were created, 13.4 million of them during 1972–1980 alone.[30]

Reevaluating Recessions: Intensity Versus Impact: It is never easy to evaluate completely the severity or significance of an individual recession. While aggregate changes in output, employment, and related key indicators of economic functioning are used to demarcate a recession, net changes may obscure the degree to which rapid changes in component indicators may simply be canceling out one another. Moreover, certain indicators of economic functioning may continue at very high levels or even increase during a recession.

By most traditional indicators, the second of the back-to-back recessions of the early 1980s was perhaps the most severe—with the possible exception of the 1974 recession—since the Great Depression a half-century ago. Unemployment was deeper, broader, and of longer duration than at any time since the 1930s.[31] In 1982 alone, with an unemployment rate hovering above 10 percent and more than 11 million workers officially counted as unemployed, the median duration of unemployment increased from 8.6 weeks to 12.3 weeks.[32] Reflecting the swelling ranks of the long-term unemployed among all unemployed, by December 1982, the mean length of unemployment rose to 22 weeks.[33] In such figures were found the spurs to a wide variety of economic-adjustment-policy proposals (industrial, trade, sectoral, research and development, export promotion, capital construction and public infrastructure, public employment, retraining and relocation programs, and so on). Their goals were to assist the economy in restoring, through a recovery, those patterns of employment and investment that were undone by a recession.

Yet, when measured from an alternative perspective, the same recession may lead to quite different conclusions. The most recent recession offers evidence of a set of market-derived adjustments that operate to dampen the impacts of business cycles. Delong and Summers have noted that general price stability has been imposed by the increasing industrial concentration and high levels of unionization in major sectors.[34] This appears to be evidence of a set of long-term industrial adjustments, supplementing the social welfare policy interventions following the 1930s Depression, which have restricted the range and severity of cyclical swings and their impact on industries, localities, and workers alike. As employment has expanded faster in nonunion than in unionized plants and industries, and as selected industries are deregulated, these patterns of organization among firms and among workers within firms may erode, as may the forms that have evolved since the 1930s to cushion against cyclical changes.

From a markedly different perspective, Gilder has contended that the recent recession has been the least severe of the postwar era, since key aspects of economic functioning never really stagnated despite the downturn in aggregate growth, productivity, hours worked, and employment.[35] To support this unorthodox claim, he has noted that throughout the same decade of stagnant growth, the birth rate for new business reached a historical high. In the decade since 1974, new business starts rose from 319,000 to 581,000, with a slight downturn in 1982 only. Accompanying these high birth rates were high death rates, and together the two led to volatile impacts on local and regional economies.[36] During the 1978–1982 period there was a dramatic rise in the rate of business failures from 24 to over 100 per 100,000 companies. That a severe recession can exact a huge toll in the form of business failures may too easily obscure that business start-ups can take place as rapidly as they do even during such inhospitable periods. Such a perspective, then, views the retention of the capacity of an industrial economy to make adjustments as being at least as important as the achievement of any specific levels of aggregate growth.

Consequently, while the rate of business metabolism in recent years has been accelerated, such data may testify to the underlying strength and continuing resilience of the economy, rather than to any temporary stagnancy and vulnerability to cyclical downturns. Such a diagnosis can be easily obscured by a preoccupation with growth per se, as it has traditionally been understood, since such a preoccupation may blind us to the capacity for significant industrial adjustment through restructuring within and across sectors.

Shifting Structure of Employment and Unemployment: Analysts have likewise noted that during the recent recession the employment rate remained relatively high—58 percent; this contrasts with a 49-percent employment rate during the expansive growth period in the 1960s. B.H. Moore, of the Center for International Business Cycle Research, has noted that the extent and duration of unemployment may well obscure the influences of both the business cycle and the prevailing "social wage"—that is, wage and benefit packages for both the employed and unemployed.[37] In the boom year of 1929, 58 percent of the population over 16 were employed; this was the same proportion as in 1982. Yet the structure of unemployment differed radically between the two points in time, as the entry of young workers and the reentry of adult women made the labor force broader and considerably more diverse over the past half-century. In 1929, only 2 percent were unemployed, while 40 percent were not considered to be in the labor force; in 1982, 6 percent were unemployed and 35 percent were not seeking work. Consequently, for reasons relating more to a narrow and outmoded conceptualization inherent in our indicators of economic change than to measurement imprecision per se, we may be hampered in our ability to see in cyclical-change data evidence of more profound structural changes.

Interpreting Patterns of Cyclical Recovery

The structure of a recovery, like that of a recession that precedes it, displays features of cyclical change that may reveal clues about processes of larger-scale industrial adjustment. The key is to look for ways in which the new expansion of the economy fails to retrace the steps of the preceding contraction. Increasingly, there is evidence that as it proceeds, an eventual recovery may bypass certain firms, corporate characteristics, industrial sectors, local and regional economies, and subgroups of the labor force.

The most recent recession bottomed out in December 1982. What followed resembled an economic rebound as much as a classic recovery. In the succeeding fifteen months, employment growth in both construction (10.4 percent) and durable goods manufacturing (9.6 percent) greatly exceeded that in services (5.3 percent), thus confirming that much of the previous employment contraction was cyclical rather than structural. Employment expansion was especially robust in electrical and nonelectrical machinery, transportation equipment, and fabricated metals, with these industries accounting for nearly half of the total increases in manufacturing employment. In the wake of this development, manufacturing has nearly resumed once again the role it has held consistently across nonrecessionary periods since World War II, accounting for slightly less than a quarter of the GNP. This rebound in output has been accompanied throughout both durable and nondurable goods manufacturing by expanding use of capacity in factories and other structures. Between late 1982 and early 1984, utilization rates surged in the automobile industry from 35.8 percent to 83.2 percent, in electrical machinery from 74.7 percent to 92.1 percent, in rubber and plastics from 74.0 percent to 92.5 percent, and in paper and paper products from 84.2 percent to 96.8 percent. By mid-1984, even the steel industry managed to increase its rate from 38.8 percent to more than 80 percent.[38]

The data on spending for new plant and equipment and on small business growth hint that the structure of a recovery may offer evidence of the economy regaining its ability to expand through a process of restructuring, as opposed to continuing to derive growth from its prerecession industrial arrangements. U.S. Department of Commerce data show that annual spending levels more than doubled in the decade since 1975, from $158 billion to nearly $350 billion. Slight declines were experienced only in 1982 and 1983. Yet patterns of capital investment during the recovery continued the quarter-century shift to new equipment and away from new physical structures like new factories and shops.[39] Similarly, Small Business Administration data indicate that between 1980 and 1982, virtually all the new job growth in the United States came from independently owned and operated small business establishments with fewer than 100 employees. And finally, Labor Department analyses indicate that there has been a substantial rise in the number of self-employed in recent years, from 7.0 million to 9.1 million during 1970–1983 alone. Taken together, these developments constitute evidence of the resilience of existing industrial arrangements, as well as of the patterns by which an advanced industrial economy seeks to restore and rejuvenate itself: it seeks increased output and higher productivity by first retrofitting older industrial spaces with new industrial machinery and organization and only secondarily by replacing older physical plant through relocation or secondary expansion to new locations. In addition to these physical capital adjustments, the patterns of employment gains made during a cyclical recovery reveal the special flexibility of small businesses and individuals in taking advantage, through entrepreneurship and self-employment, of opportunities for economic growth.

One implication of these patterns of output growth, capacity utilization, and business investment in a recovery is that any productivity gains are often registered as lagging employment growth in key sectors such as manufacturing. While employment in selected manufacturing industries may illustrate the fastest rates of rebound, in aggregate the growth may fail to restore prerecession employment levels, especially among production workers. Consequently, even if aggregate manufacturing employment levels are restored, beneath the surface the character of manufacturing will probably continue its shift away from reliance on production workers and traditional man-machine combinations within increasingly automatable factories that represent outmoded and uncompetitive production arrangements. Both recessions and recoveries, then, provide opportunities for the continued shift within goods-producing industries toward occupations and human-capital combinations that we have come to associate with the more advanced segments of the services sector.

In this way, manufacturing continues to reveal the same gradual decoupling of both employment and its composition from output during a recovery that may have been visible during the preceding recession. As this process continues, a new structure of unemployment is likewise revealed. Older workers with outdated skills, displaced from industries in which productivity increases are achieved through new capital investment, frequently constitute the long-term unemployed or even the permanently dislocated both during a recession and in the recovery that follows.[40] As a result, we see in recovery data, as in recession data, patterns of industrial change wherein the transition from older production arrangements to new or upgraded ones proceeds through the gradual elimination of redundant and/or obsolete physical and human capital.

Searching for the Post-Industrial Society

> What is clear is that if an industrial society is defined as a goods-producing society . . . then the United States is no longer an industrial society.[41]

The term *industrial society* typically connotes a characteristic set of relationships among a full range of industry sectors, and especially between the sectors devoted to goods production and services provision. The core of an industrial economy generally has been regarded as the capacity for goods production, or more specifically, for manufacturing itself—the fabrication of finished products from raw materials. In recent years, it has become common for discussions of industrial restructuring to view goods production and services provision as two distinct, largely independent, and somewhat adversarial subeconomies, particularly as regards technology diffusion, productivity, and employment creation. From this perspective, the services sector is frequently viewed as ancillary and subordinate to the goods sector.[42] Understandably, then, the perception of a building dependence on services has been interpreted by many as being tantamount to a process of industrial decline.

Industrial Retreat or Advance?

Implicit in the argument that the United States is experiencing some sort of industrial decline is the assumption that the economy is being restructured in ways that diminish, rather than enhance, our capacity for producing goods or even creating wealth altogether. From a *growth* perspective, the extent to which expansion occurs in each of these two sectors is important. From a *development* perspective, however, growth in the goods sector relative to that in the services sector generally has been regarded as being of more fundamental importance. Differential rates of employment or output growth between the two sectors are commonly seized upon as evidence of the industrial restructuring of the national economy. The result can be viewed as a retreat from or an advancement beyond traditional industrial circumstances. Extended periods of slow growth or even temporary contraction of employment or output in the goods-producing sector, accompanied by relative expansion in services, may even constitute for some a "deindustrialization" process.[43] Others, such as Bell, perceiving a society whose material progress increasingly depends on intangible factors such as specialized knowledge and information, and whose social stratification arrangements have reconfigured to reflect these new priorities, offer the term *post-industrial society*. It is in this concern for sectoral shifts in the larger economy that discussions of economic development and industrial restructuring converge. Rather than growth per se, the structure and pace of that growth become the paramount concerns in analyses of industrial change.

For those who choose to emphasize the important *continuities* within the process of industrial development, our passage may be viewed as one from an older industrial era into an advanced, even post-industrial era. A key assumption of this view is that while goods production—and especially manufacturing—may experience a decoupling of employment growth from output growth, *the centrality of goods production to the larger economy is not being fundamentally challenged.* Slow rates of employment growth in manufacturing reflect the increasing productivity of the goods-production sector rather than its diminished significance to the larger economy. For those who wish to draw attention to the *discontinuities* of this same process, the terms *industrial* and *post-industrial* are commonly juxtaposed under the assumption that two historical eras are separated by major structural dissimilarities. This perspective suggests that goods production is gradually being dislodged from the central position it has held thus far throughout the industrial era. Taking its place is a new economy wherein goods production is being downgraded from its former position as the dominant contributor to a nation's industrial status.

The interpretation developed in this section is a hybrid of these two views. Our industrial economy *is* experiencing more than simply an extended sequence of effects accompanying movement through the business cycle. Major industrial adjustments *are* indeed taking place. However, any intersectoral realignments that entail a decline in employment, or even output, within the goods-producing sector do not in themselves constitute conclusive evidence of industrial decline. Since such intersectoral adjustments are both small and slow, we are encouraged to search elsewhere for the emerging features of an advanced industrial economy. A main message of this chapter is that developments *within* the respective sectors are viewed as more compelling than any intersectoral shifts between them. To assign great significance to the different trajectories and rates of change of the goods and services sectors can cause us to overlook the more consequential ways in which the two sectors have come to relate to one another. New and renewed industries throughout both sectors reflect new production arrangements wherein the application and dissemination of knowledge and information, science and technology, and research and development have eclipsed the centrality of more familiar inputs of land and other natural resources, experience and manual skills, and traditional forms of physical and financial capital. This has proceeded largely through the redefinition of older inputs, rather than through their elimination. As a result, each kind of factor input contributes to production in both sectors in new and important ways.

The "Shift-to-Services" Reexamined

In past decades, it had been commonplace to note that the long-term employment contraction in the primary sector had been as much a feature of modern industrial society as had the employment expansion in the secondary and tertiary sectors. These sectoral trends have long been associated with basic industrialization and urbanization. More recently, as the growth rate of employment in services continued to eclipse that in goods production in the post–World War II economy, this basic transformation became known as the "shift-to-services."

Even though employment growth continues within both the goods and services sectors, since that growth invariably proceeds at different rates, the mere existence of relative rates of growth has reinforced the assumption that the two sectors grow at the expense of one another, and that the growth of one presumes the decline of the other. Until recently, intellectual interest in a presumed shift-to-services had not developed into a concern that something ought to be done to slow or retard the process. Increasingly, models of sectoral change have become more detailed and elaborate as work has progressed on both the structure of the larger industrial economy and of the services sector in particular. It has been suggested that the shift-to-services is largely an employment phenomenon and that the use of other criteria permit a somewhat different, and decidedly less threatening, perspective on the structural changes taking place in the nation's economy. The addition of many new low-wage, low-productivity jobs in the services results in relatively modest contributions to overall economic growth. We have seen, however, that such an assumption is not entirely warranted.

The relatively rapid growth of services compared to goods production—the so-called shift-to-services—has been referred to frequently thus far to account for features of recent industrial change. Let us now examine the dynamics of this shift in more detail. As a result of the Depression of the 1930s and the new importance of explicit economic policy interventions—including the pivotal role of World War II—in explanations of our emergence from it, the industrial structure of the economy began to be studied with renewed interest. Nonetheless, the industrial taxonomies commonly used have had to contend with considerable conceptual ambiguity reflecting, in large part, the ways in which changing relationships between goods production (especially manufacturing) and services provision have been perceived. In 1940, Clark proposed a three-sector model which held that while both the secondary or industrial (that is, manufacturing and construction) and the tertiary (that is, services) sectors were expanding, the growth in services came at the expense of the primary (that is, agricultural/extractive) sector.[44] Clark's taxonomy has strongly influenced how patterns of industrial change have come to be interpreted and evaluated.

Subsequent sectoral classification schema have offered alternative conceptions of the structure of the services sector and the more complex ways in which goods and services production have begun to relate to one another. Fuchs and Kahn and Bruce-Biggs subdivided the services into two major groupings—the tertiary services, which include those services integral to goods production (for example, transportation, wholesale and retail trade, finance, and selected business services), and the quaternary services, which include services (such as nonprofit and government services) that are integral to increasingly complex service production, as well as those (such as consumer services) that are consumed independent of other production sequences.[45] Finally, Stanback and Stanback et al., building on the work of Singelmann, have likewise drawn attention to the lack of uniformity and the diverse trends within the broad services subeconomy.[46] They offer a services sector subdivided into six main groupings—distributive, retail, nonprofit, producer, consumer, and government services—each experiencing differential rates of growth and relating to the production of goods and other services in new and disparate ways. As a result, the services sector is viewed as undergoing a major transformation prompted by changes in the larger industrial economy.

Locating the Shift-to-Services

On the surface, the growth of the U.S. economy during the post–World War II era has been considerable. While rates of change in output, employment, and income growth are related to outcomes of social and political significance, including increased job creation and living standards, recently the structure or distribution of that growth across industry sectors, occupations, regions, and other social groupings has been attracting considerable academic and policy attention.

Total Output: Relative Stability in Sector Shares: Aggregate output has risen considerably in recent decades. The output of the economy—the Gross National Product—has expanded from less than $1 trillion in 1946 to roughly $4 trillion today. Yet the rate of growth across this period has been quite uneven. During the 1973–1982 period, the economy expanded at an average annual rate of 1.8 percent. Compared to the 4.2-percent average annual rate for the 1960–1973 period, the growth of the past decade appears anemic indeed. Yet, countering the building interest in the possibility that the economy is rapidly restructuring, the sectoral patterns illustrate relative stability over the same period. Between 1950 and 1983, the proportion of total output contributed by manufacturing declined only slightly, from 24.5 percent to 23.1 percent (Table 2.1). Between 1977 and 1982, cyclical downturns stunted manufacturing to the point that its output share was estimated to have

Table 2.1

Gross National Product Share by Industry Sector, 1950–1983

(billions of 1972 dollars)

Industry sector	*1950*	*1955*	*1960*	*1965*	*1970*	*1975*	*1980*	*1983*
Gross national product	$534.8	$657.5	$737.2	$929.3	$1,085.6	$1,233.9	$1,475.0	$1,534.7
Total private industry	84.7%	83.8%	84.7%	85.3%	84.9%	85.4%	86.4%	86.9%
Agriculture	5.5	4.9	4.4	3.5	3.2	3.0	2.7	2.5
Nonagriculture	80.4	80.5	81.0	82.7	82.7	83.7	85.3	85.9
Mining	2.1	2.0	1.8	1.7	1.7	1.5	1.5	1.4
Contract construction	5.5	5.8	6.3	5.9	4.9	3.9	3.5	3.3
Manufacturing	24.5	25.2	23.3	25.5	24.1	23.5	23.8	23.1
Durable goods	14.6	15.7	13.7	15.6	14.3	13.7	14.3	13.6
Nondurable goods	9.9	9.5	9.6	9.9	9.8	9.8	9.5	9.5
Transportation, communication, and public utilities	7.7	7.6	7.8	7.9	8.7	9.2	9.5	9.3
Transportation	5.2	4.5	4.1	4.0	3.9	3.7	3.5	3.1
Communication	1.2	1.4	1.6	1.8	2.4	2.9	3.6	3.8
Public utilities	1.3	1.6	2.0	2.1	2.4	2.5	2.4	2.4
Wholesale and retail trade	16.4	15.7	15.9	15.9	16.3	17.0	16.7	17.4
Wholesale	5.5	5.4	5.8	6.1	6.6	7.1	7.1	7.5
Retail	10.9	10.3	10.1	9.9	9.6	9.9	9.6	9.9
Finance, insurance, and real estate	11.9	12.4	13.9	13.8	14.4	15.2	16.0	16.5
Other services	11.2	10.3	11.3	11.1	11.7	12.0	12.8	13.5
Government and government enterprises	14.1	14.6	14.6	13.8	14.1	13.3	12.0	11.6

Source: U.S. Department of Commerce, Bureau of Economic Analysis, *The National Income and Product Accounts of the United States, 1929–74, Statistical Tables,* and *National Income and Product Accounts, 1976–79, Survey of Current Business,* respective years. Totals do not add to 100.0% without accounting for official statistical discrepancies.

dropped temporarily to just above 20 percent. This drop however has been largely reversed as the recovery continued with the output share for manufacturing rebounding to 23.8 percent in 1984. Meanwhile, the services sector expanded its output share from 62.7 percent to 66.1 percent, an increase that appears to have come largely at the expense of agriculture, whose contribution declined from 5.8 percent to 2.9 percent.[47] The services share has continued to grow during the recession-dominated period following 1977.[48]

Changing Composition of Employment Growth: While an industrial society pays attention to the impact of the contribution to total output made by each industry and sector, the contributions of each to total employment hold special social significance. As the 1980s began, total civilian employment hovered just above 100 million. Manufacturing was still the nation's largest employer, accounting for 21.1 percent of total civilian employment and 25.0 percent of total private sector employment (Table 2.2). Yet, over the preceding two-decade period, manufacturing employment had expanded only 26.4 percent, while the total labor force had expanded 50.6 percent.

During the 1970s, increasing significance was attached to the realization that it was no longer manufacturing but the diverse services sector that was assuming the principal role of employment generator in the U.S. economy. This was especially the case, as we shall see later, in the restructuring economies of the nation's central cities. Between 1968 and 1977, for example, total employment in manufacturing actually contracted, while employment expanded at a 2–3-percent average annual rate in finance, insurance, real estate, government, wholesale and retail trade, and other services. Of the 19 million new jobs created during the 1972–1982 decade, only 5 percent were in manufacturing. Moreover, BLS projections indicate that between 1982 and 1995, only one in seven—or approximately 14 percent—of the 24.4 million jobs that will be created will be in the manufacturing sector. Herein lies a second trend that has been seized upon as evidence of our retreat from a predominantly goods-oriented economy. On the surface, these differential growth trends in employment, even more clearly than those of output, are consistent with the image of an economy jettisoning its industrial commitment to goods production in favor of a qualitatively new and different services orientation for the future.

Recession and the Decline of Goods-Production Employment: Not only do the composition and pace of any shift-to-services deserve careful scrutiny, but so does the context for that shift. As we have seen, conclusions about the rise of the service economy generally have been drawn

Table 2.2
Shifting Shares of Total Employment by Major Sector, 1959–1990

Industry sector	*1959*	*1968*	*1972*	*1977*	*1980*	*1985*	*1990*
Total U.S. employment (000s)	(67,563)	(79,836)	(88,408)	(93,715)	(101,761)	(111,851)	(118,615)
Total private employment	88.0%	85.2%	84.5%	83.8%	84.4%	84.9%	85.2%
Agriculture	8.1	4.6	3.6	3.1	2.9	2.6	2.2
Nonagriculture	79.9	80.6	80.8	80.7	81.5	82.3	81.3
Mining	1.1	0.8	0.8	0.9	1.0	0.9	0.9
Contract construction	5.4	4.9	5.4	5.0	5.0	5.0	4.8
Manufacturing	25.2	25.1	23.0	21.2	21.1	20.6	20.1
Durable goods	14.2	14.8	13.6	12.5	12.7	12.6	12.4
Nondurable goods	11.0	10.3	9.3	8.7	8.4	8.0	7.7
Transportation, communication, and public utilities	6.3	5.7	5.5	5.2	5.1	4.9	4.8
Transportation	4.1	3.6	3.3	3.1	3.0	2.9	2.8
Communication	1.3	1.3	1.4	1.3	1.3	1.2	1.2
Public utilities	0.9	0.8	0.8	0.8	0.8	0.8	0.7
Wholesale and retail trade	20.4	20.5	21.7	22.3	22.9	23.2	23.1
Wholesale	5.2	5.2	5.3	5.3	5.4	5.2	4.9
Retail	15.1	15.3	16.2	17.0	17.5	17.9	18.1
Finance, insurance, and real estate	4.3	4.6	5.0	5.2	5.2	5.5	5.6
Other services	13.5	16.0	17.3	18.9	19.5	21.0	22.5
Private households	3.8	3.1	2.4	2.0	1.6	1.3	1.1
Government	11.9	14.8	15.5	16.2	15.6	15.1	15.1
Total civilian employment	100.0%	100.0%	100.0%	100.0%	100.0%	100.0%	100.0%

Source: This table is adapted from actual and projected figures reported by Valerie A. Personick, "Industry Output and Employment: BLS Projections to 1990," *Monthly Labor Review,* BLS #6, Bulletin 2030, p. 32 (April, 1979).

from observation of differential rates at which the goods-production and services-production sectors create employment. For the most part, inferences concerning the building dominance of the services sector have not been qualified by the increasing frequency and severity of downturns in the business cycle. Yet it is known that services generally gain ground during recessions, as cyclical downturns reserve for goods production their most severe impacts. The rapid succession of recessions in recent years has thus created an unusually hospitable environment for services sector expansion.

Interpretations of both general and sector-specific growth rates over relatively long periods are vulnerable to the complications that inevitably arise when cyclical and structural trends in an industrial economy are not carefully disentangled. In the twelve-month period between spring 1981 and spring 1982, 1.3 million jobs were lost in the goods-production sector, while a half-million jobs were gained in services and finance. By April 1982, employment in consumer, financial, and service industries (24.3 million) had eclipsed for the first time the job total (24.0 million) in goods production—a composite of manufacturing, mining, and construction. Whether this is viewed as a benchmark in a relatively short cyclical process or as the crossing of a major threshold in an inevitable process of industrial senescence and decay does much to determine its significance as a milestone.

Recent evidence of the gradual decoupling of employment from unemployment levels makes employment growth increasingly difficult to interpret. Between December 1982 and March 1983, for example, the unemployment rate dropped precipitously from 10.8 percent to 10.3 percent. While this was welcomed as a further sign of continued economic recovery, the exit of 655,000 workers from the labor force, rather than the addition of only 10,000 new jobs, appears to have accounted for the bulk of that decline.[49] However, as with any extended period of recovery, especially following a deep and extended recession, goods production can be expected to rebound so robustly that the net gains made by services in the short run will be blunted considerably. Still, increases in productivity can also be expected to hold down employment growth in manufacturing even during the recovery phase of a business cycle. Thus, any conclusions about the so-called shift-to-services must look beyond evidence that may simply be depicting the special scourge of recession on manufacturing and construction.

The Slow Pace of the Shift-to-Services: A focus on rates of change in both output and employment share, beyond the observation simply of the differential rates of growth between the goods and services sectors, dispels somewhat the sense of urgency that has heretofore characterized many discussions of intersectoral shifts. While the shift in employment shares is more pronounced than that in output shares, both are taking place relatively slowly. Over the longer sweep of time between the two most severe cyclical downturns of the century—1929 to 1977—the share of employment in the extractive (mining) and transformative (manufacturing and construction) industries has only declined from 36.5 percent to 29.7 percent, while the employment share in services of all kinds has increased from 55.1 percent to 68.4 percent. Consequently, we might justifiably ask whether or not such a plodding and complex development is sufficient to justify pronouncements concerning the demise of the U.S. industrial economy.[50]

Reconciling the Advanced Industrial Economy with a Post-Industrial Society

Daniel Bell introduced the term *post-industrial* first in a series of lectures in Salzburg, Austria, in 1959 and later in print in 1962. He used the term to account for the influence of technology on the structures and substance of work and lifestyles in an evolving industrial era. Bell was well aware of the slipperiness of language and the potential tyranny of labels used to describe important socieconomic and cultural trends. His goal was to trace the evolution of industrial society on a social-technical dimension, while tracing capitalism as a set of administering structures on a socio-economic dimension. His discussions of the transforming economic and occupational structures of society were essentially the means to, rather than the end of, his larger goal of better understanding the new roles of science, knowledge, and technology in civilization.

Thus Bell compared the internal dynamics of the goods and services sectors and the changing nature of workers' ties to them through changing labor markets and skill requirements, and many of his readers drew a casual, if unwarranted, equation between continued industrial evolution—and with it the arrival of a post-industrial society—and the rise and hegemony of an economy dominated by services production. For Bell, the shift from a goods-production orientation to a services-provision orientation was meant to herald a far more subtle, yet substantial shift in the larger orientation of industrial society. Any intersectoral shift-to-services was meant to serve as simply a proximate indicator of that larger reorientation.

Since output shares are shifting so slowly between sectors, and since shifting employment shares, while more visible and pronounced, are exaggerated by business-cycle movements, neither appears to provide compelling evidence of a major restructuring that warrants concern over the pending demise of our industrial economy. Neither, in these trends, is there justification for a view that sees a fundamentally new economic structure being shaped by processes of advanced industrial development. Therefore, we must look elsewhere if the claim of profound change is to be substantiated. For the remainder of this chapter, I will direct attention to several more subtle industrial trends including the illusion of a weakening or even disappearing overall goods orientation in the larger economy, the restructuring of the services economy itself, shifts in occupational or income distributions, and even a redefinition of the notion of work itself.

An Enduring Goods Orientation: An economy's goods orientation can be tied to patterns of consumption as well as to patterns of production. It is true that in terms of production—and differential rates of employment and output—we appear to be very slowly transforming into a services economy (Figure 2.1). However, from the perspective of personal consumption, at least, the indications suggest that we are still very much a goods-oriented economy. Personal consumption expenditures for services surpassed those for nondurable goods for the first time in 1970 and those for durable and nondurable goods combined for the first time in the early 1980s. This development is not surprising given the rising levels of disposable income and standards of living that major segments of U.S. households have come to enjoy in recent years. This is particularly true since services price rises have typically continued apace during the recent recovery period, while the price trends for many manufactured goods have moderated given the low levels of inflation.[51]

These observations reinforce a more compelling conclusion. If we wish to understand how our advanced industrial economy is changing, one key to that understanding rests in the relationship between goods and services and how that relationship is changing. The development of our advanced industrial economy is to be discovered less in major intersectoral shifts or in a diminishing goods orientation and more in the changing patterns of alignment between goods and services production. Both goods and services are registering the impacts of new technologies, new forms of work in the production/provision process, an increased sensitivity to and dependence on knowledge brought to the workplace by appropriate skilled workers, and the location preferences tied to changing demography and cultural influences. While it is increasingly challenged in a global marketplace, the United States is not losing its industrial moorings. Rather, it appears to be building on them. In increasingly complex and extended sequences of production and consumption, the boundary line between goods and services is becoming blurred. Goods that are produced for eventual consumption

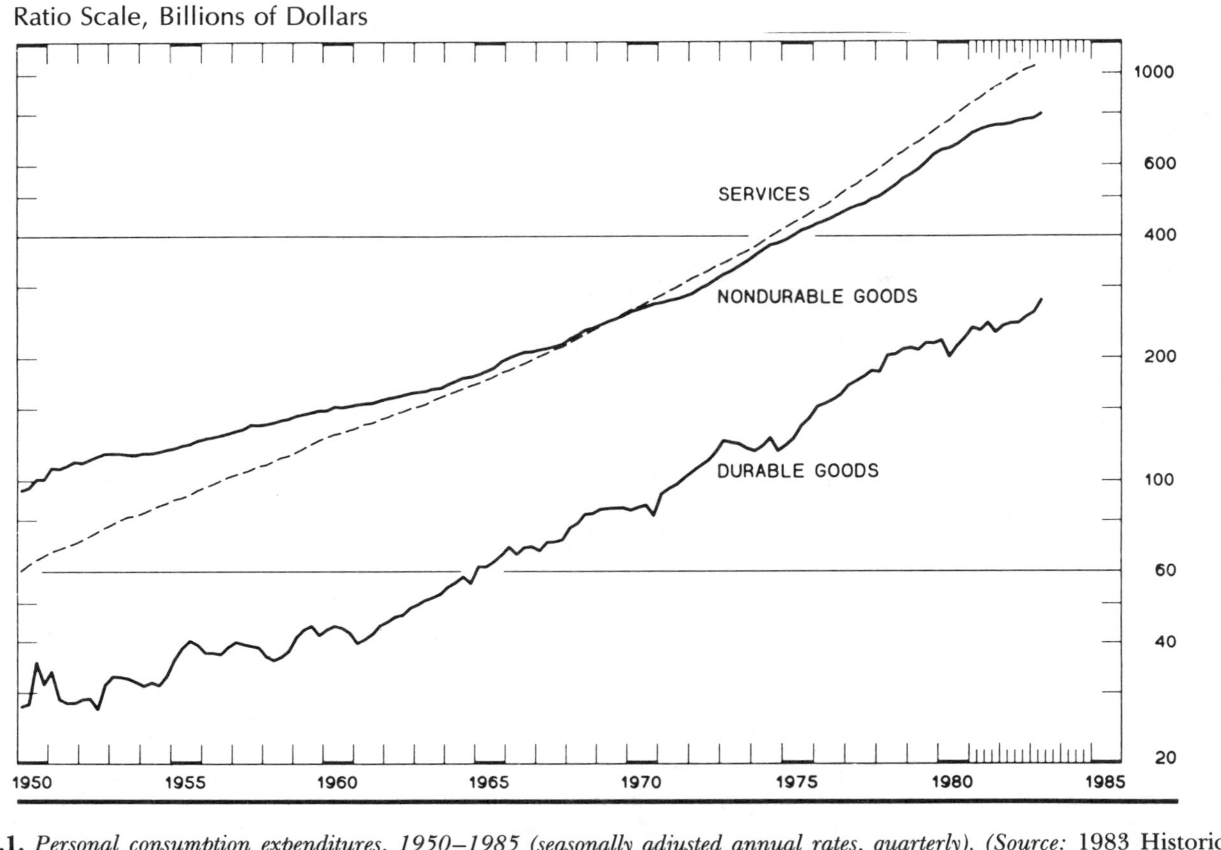

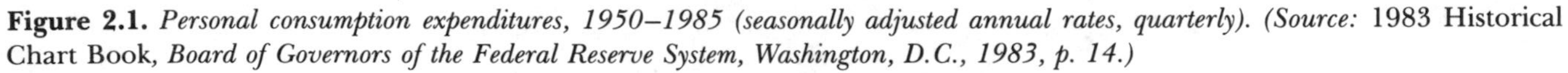

Figure 2.1. *Personal consumption expenditures, 1950–1985 (seasonally adjusted annual rates, quarterly). (Source:* 1983 Historical Chart Book, *Board of Governors of the Federal Reserve System, Washington, D.C., 1983, p. 14.)*

tied to final demand increasingly reflect the value-added of a widening range of services. We require greater understanding of the increasingly complex and diverse processes of production and the new relationships among factors of production that have emerged.

Producer Services and the Restructuring Service Sector: Even though seven jobs in ten and the bulk of new employment growth are located in the services sector, neither our orientation toward goods production nor our status as an advanced industrial economy should be in doubt. The transition from one stage to another renders traditional forms of sectoral analysis inadequate and requires new assumptions about how to analyze and evaluate economic activity and industrial change. What appears on the surface to be the growing dominance of the services sector may be better understood as the continued expansion of the roles played by services in the production both of goods and of other services.

As Stanback et al. have recently demonstrated, of all sectors, the services sector is often mistakenly associated with a set of characteristics that best describes consumer services. Yet this category is becoming the least important of all the services groupings. Not only is it the smallest grouping within the services, but it has shrunk substantially in the past half-century in both employment and output. Other service sectors reveal themselves to be far more important.[52]

In the context of such diversity within the services economy, it is the category of "producer services" to which Stanback et al. and Ginsberg and Vojta draw particular attention.[53] The producer services include the FIRE (finance, insurance, and real estate) industries, as well as business and legal services, among others. Their principal function is to provide intermediate inputs to other productive activities. This sector is growing at a faster rate than any other in the economy, and it currently accounts for a quarter of total GNP—a share that exceeds even the 24.2 percent for manufacturing.[54]

What are the implications of this new diversity within the services sector for the transition into an advanced industrial era? In the past, the growing capacity of the services to create jobs faster than the goods sector raised questions about the long-term prospects of an industrial economy. In an advanced industrial economy, could the expanding services sector serve as the productive industrial core just as goods production did throughout the earlier industrial era? Debate about this point remains considerable.[55] Yet, as services have industrialized, they have come to constitute a more attractive economic base for local and regional economies, and even the national economy, than was thought possible at an earlier time.

But can services be said to dominate the economy now in any meaningful sense? Perhaps more important than any presumed dominance of one sector by another is the expanded and intensified interdependencies that have developed to bind them to each other. The rapid expansion of producer services, and therefore the restructuring of the services sector itself, tells us as much about goods production as about services themselves. The expanding services sector may better be conceptualized as an integral part of a transforming capacity for advanced goods and services production than as a mere appendage or system of supports. The gradual transformation in the way goods production takes place has been matched by the way in which services—and especially producer services—have transformed to stimulate and structure their growth. The goods-production orientation of an advanced industrial economy is not so much being discarded in favor of services, as the services sector itself is being restructured to allow it to play a newly updated and critical role in advanced goods production. To focus on the slowly shifting relative shares of employment or output between the two major sectors tends to obscure the emergence of this far more important relationship between the two.

Occupation-Income Packages and the "Missing Middle": Accompanying the view of industrial change as being dominated by a basic shift-to-services and a retreat from our historical goods-production orientation has been a concern that a highly undesirable social development is underway as well.[56] Acknowledgment of the role of services as a principal employment generator has led to frequent discussions of the inability of services employment to compensate socially for jobs eliminated or foregone in goods production.[57] Since widespread understanding of the services sector has largely been dominated by the perception of the consumer services, jobs created in the services are commonly assumed to be invariably tied to secondary labor markets; associated with low pay and minimal benefits or security provisions; and characterized by an absence of career ladders that offer prospects for advancement.

The issue is an important one. There is a growing fear that occupation and income shifts accompanying continued sectoral shifts will rework the income distribution or even the class structure of industrial society here and abroad. Bluestone and Harrison and Stanback et al. have suggested that the services subeconomy is characterized by a two-tier distribution of employment with high and increasing proportions of jobs of above-average and below-average income. This is in contrast to employment in manufacturing and construction, which yield more of a bell-shaped distribution and therefore greater income equality. Any continued sectoral shift-to-services, then, is expected to reinforce income inequality and even lead eventually to the decline of what we have traditionally defined as the middle class.

Were the labor force to evolve into two distinct job-class groupings, the employment-earnings packages would ensure continued middle- and upper-middle-class status for those with appropriate skills. But those whose skills did not permit them to make a successful transition to new industrial arrangments would fall from their former middle-class status or fail in efforts to be upwardly mobile. The blue-collar middle class would be especially vulnerable; their job prospects in the restructuring basic industries would not only diminish, but their job tasks would also be significantly "deskilled" and thereby would merit substantial cuts in wages and benefits.[58]

In the decades following World War II, there have indeed been differential growth rates among occupational groupings. As Table 2.3 indicates, between 1950 and 1982 the labor force changed from being only one-third white-collar to being predominantly white collar.[59] Moreover, BLS projections suggest that this long-term shift to a white-collar work force will continue through the rest of the century. Importantly, the white-collar work force is far from homogeneous in its composition. Female-dominated occupations, like secretarial and other clerical office jobs, have long represented the largest single subgroup of white-collar workers and are "crowded" into job categories that are often characterized by relatively low wage-benefit packages, little opportunity for advancement, and involuntary seasonal or part-time working arrangements. The combined categories of professionals, technical

Table 2.3
Shifting Occupational Shares in the U.S. Labor Force, 1950–1995

Occupational category	*1950*	*1960*	*1982*	*1995*
White collar workers	36.6%	43.4%	51.6%	52.7%
Professional and technical workers	8.6	11.4	16.4	17.1
Managers and administrators	8.7	10.7	8.4	9.0
Sales workers	7.0	6.4	6.4	6.5
Clerical workers	12.3	14.8	20.4	20.1
Blue-collar workers	41.2	36.6	30.7	29.6
Craft and kindred workers	14.2	13.0	11.0	11.2
Operatives	20.4	18.2	13.6	12.6
Nonfarm laborers	6.6	5.4	6.1	5.8
Service workers	10.5	12.2	16.7	16.8
Farm workers	11.8	7.9	2.7	1.9

Source: Productivity and the Economy: A Chartbook U.S. Department of Labor, Bureau of Labor Statistics, October, 1981, Bulletin 2084, Chart 35, pp. 74–5. Data for 1982 and projections for 1995 are from Data Resources, Incorporated.

workers, managers, and administrators whose jobs are associated not only with higher educational and skill entry requirements, but also with higher compensation packages and better-defined advancement ladders, constitute a far larger group.

In contrast, over the same period all major blue-collar occupational groupings have constituted—and will continue to do so—declining proportions of the labor force. The restriction of employment opportunities in these categories, together with the increase in services employment, is widely believed to have set the stage for what has been termed "the missing middle" phenomenon, wherein the once broad ranks of the middle classes are diminished through downward mobility within and across generations. Not only will those who once held well-paying manufacturing jobs find it difficult to make the transition to new jobs requiring new skills, but young labor-force entrants in the future will discover middle-class jobs to be far more difficult to find. As the ranks of these middle-class occupational categories contract, it is assumed that income distributions will tend toward greater inequality as a result. From this perspective, industrial change appears to provide a recipe for a substantially diminished and undesirable future.

Despite the tempting logic in the notion that industrial restructuring and related occupational shifts lead to an income distribution characterized by increasing inequality, there is remarkably little empirical or theoretical support for such an outcome. Lawrence has demonstrated that within manufacturing, any sectoral shifts of employment to high-technology industries and any shifts in employment between the goods-production and services-provision sectors can be linked to only the very slightest impacts on middle-class earnings.[60] Moreover, the middle class has been far from the enduring monolith that is commonly called to mind. Rather, the United States is characterized by considerable income and occupation "churning," and movements of households between income strata in both directions are both commonplace and continuous. The middle class is, therefore, constantly being rearranged and reconstituted. Duncan, analyzing data from the University of Michigan Panel Study of Income Dynamics, has documented clearly the high degree of metabolism and social mobility that exists in the United States. That households fall from middle-class grace has been discovered to be far from extraordinary.[61] Rather, the relative fluidity of the status-forming process in the United States deserves greater attention. Moreover, Benz has recently shown that within the rapidly growing services sector, high wage-skill occupations, including for example, accountants, architects, computer specialists, and engineers, are leading U.S. employment growth.[62] Employment in the low wage-skill occupations is either slowing or declining. The moderate occupational shifts since 1950 that have led to a change in the predominantly white collar labor force have involved the relative expansion of low-pay sales and clerical jobs and the relative contraction of high-pay factory jobs; despite this shift, the income distribution in the United States has not been significantly altered.[63] The restructuring of the occupational distribution during advanced industrial development need not be accompanied by an inevitable restructuring of the income distribution or a decline of the middle class.

Samuelson has aptly observed that it is not low-wage jobs per se, but how those jobs are distributed across households that has the potential for altering the distribution of income traditionally associated with an industrial society.[64] And there is no reason to believe that the increase in low-wage jobs must necessarily be absorbed by low-wage households. Many low-wage jobs are held by youth and returning housewives whose earnings may supplement, rather than dominate, the income of a household. Moreover, there has been the surge of household formation in the wake of demographic and cultural changes that created and legitimized the preferences of young adults to live alone and that supported the option for the divorced, separated, widowed, and elderly to maintain their independence in separate households. This phenomenon goes further in explaining any increase in households at opposite ends of the income distribution than the assumption that the transit of an industrial economy must necessarily mean greater income inequality and the erosion of the middle class.[65]

During earlier decades of this century, households were more commonly sustained by a sole wage earner and revealed consumption patterns more predictably associated with a widespread child-rearing orientation and common housing preferences. More recent declining birth rates, increasing diversity in household form and living arrangements, and other aspects of a changing demography and culture may be altering the assumptions under which jobs produced by an advanced industrial economy have come to be judged as relatively undesirable and inferior.

As a result, there is the potential for confusion that surrounds both the relationship between any sectoral shift-to-services and the basic goods orientation of an advanced industrial society and the relationship between any occupational shifts and resultant income distributions. This situation testifies to the need for reconceptualizing industrial development more in terms of human capital—in terms of education and skill acquisition throughout the life-cycle. New products for new markets, produced using new production arrangements, with the aid of new technologies, new occupational specialities, and new organizational/management practices, can be expected to generate a range of employment-earnings packages that differ from those of the past. Yet there is no compelling reason to assume that these accompaniments of industrial transition must necessarily have negative net effects.

Disadvantage in the form of income-determined poverty that derives from low-wage employment sectors will not be eliminated with the arrival of new and renewed industrial arrangements—and may only be redistributed, as we are seeing especially with the rise of a central city minority underclass and the feminization of poverty. Still, there is no compelling reason to expect that overall disadvantage will be increased by continued industrial change. The reality that poverty derives from the employment people do have, as well as from the employment they do not have, is not likely to be altered much in the future. It is at least as likely that such poverty and disadvantage can be diminished through the opportunities provided by an advanced industrial economy that has continued to develop and thereby to remain competitive in domestic and foreign markets. For those for whom labor-force participation is largely irrelevant, such as those whose age, lifestage, or health status makes them dependent, the possibility of new, reworked social welfare arrangements to provide a floor beneath which they are not allowed to fall increases in a society whose industrial capabilities continue to evolve. While disadvantage will ever continue to be relocated to newly defined sociodemographic subgroups and geographical locations, there is no compelling reason to believe that the overall standard of living for society at large must decline as industrial development evolves.

Post-Industrial Society and the Redefinition of "Work"

Finally, let us now turn to the one kind of restructuring that most clearly signals that we are taking leave of an older industrial era and entering into a new and different one. A case has been made that we have been experiencing neither a subtantial shift-to-services measured in rapidly changing shares of sectoral employment nor a decline in the overall goods orientation of the economy. Moreover, while there has been a moderate occupational restructuring, there is little evidence of a resulting changed income distribution or even the likelihood of one in the future. What then, if anything, remains to constitute support for the view that an advanced or post-industrial society is displacing an older industrial one? If a major feature of industrial society is the continuous adjustment of production and consumption to one another, of the location of work and residence to technology and personal preferences, of the economy to the demography, and of cultural values to all of it, what constitutes the basis for suggesting that the shift to an advanced industrial society involves something different than an extrapolation of earlier industrial trends into the future?

The answer relates to what we now produce and to how and where that production takes place. The "post-industrial" nature of our developing economy and society is only dimly reflected in any sectoral shifts between goods and services or any occupational shift between white- and blue-collar jobs. Without accounting for the fundamental transformation of what people actually do on the job, and the key features of that work activity, an examination of sectoral and occupational shifts does not provide sufficient evidence of the kind of major restructuring that could constitute industrial decline in any meaningful sense.

There is, indeed, a subtle shift that deserves attention. This shift is related to the changing nature of work itself and the way in which this change has implications for how workers are now linked to the labor force and how they will be in the future. Within both the goods and services sectors, production arrangements are changing. *Where* people work for pay—understood either as sectoral, occupational, or geographical location—is arguably less important than *what* actually constitutes the work people perform for pay. As a result, sectoral, occupational, and related indicators are ill suited to clarifying the emergence of post-industrial society. An alternative approach begins by assigning primary significance to the actual activities that are required for production of any kind. The substance of work, rather than its location in any of that term's different senses, offers a more sensitive reading of the extent and nature of industrial change.

Knowledge, Information, and Production Technology Redefine Work: Today, the vast majority of U.S. workers process information; the value-added attributable to specialized knowledge and information in its various guises constitutes an increasingly large and important portion of both goods and services production, both white-collar and blue-collar work, both high-paying and low-paying jobs. That the very notion of "work" is being redefined in a new industrial era is revealed in data that indicate that as long ago as 1967, over half the wages and just less than half the U.S. GNP could be attributed to the production, processing, and dissemination of information.[66] This emergent information sector is comprised not only of the manufacture of equipment that stores, collates, and transmits information, but also of the myriad industries throughout the service sector for whom both the medium of production and the final product are information and knowledge.

Today, it is estimated that approximately 60 percent of all Americans are employed in creating, processing, and/or distributing information.[67] In contrast, only slightly more than 10 percent of all workers are employed as production workers in manufacturing, down from 13 percent in the mid-1970s. That proportion is expected to drop below 10 percent by the end of the century. Moreover, much of this new work is done in new locations. The time-honored distinction between homeplace and workplace, which has made job-related commuting such a dominant influence on the spatial features of an industrialized economy, is fading. As we will see in following chapters, amidst continued high levels of residential mobility the shortening of commuting patterns is increasingly evident, particularly in rapidly growing nonmetropolitan areas. Furthermore, transportation-assisted commuting is being joined by communication-based commuting ("telecommuting") as ever-smaller proportions of workers have separate and fixed work and home locations. It is estimated that 7.2 million U.S. workers use telephones and computers rather than automobiles and public transportation as primary links to their employers. Moreover, the U.S. Bureau of the Census estimates that 26 percent of the civilian labor force in 21 million households worked at home in 1984; 56 percent did job-related work at home while 44 percent actually conducted business activity out of their homes.[68]

Clearly, much of the rapid growth in the services sector reflects the increased importance of knowledge and information to an advanced industrial economy. That goods production is increasingly dependent on combinations of both in myriad ways is less often appreciated. Traditional indicators of structural change in the economy may therefore divert attention away from what work has become and how it is changing to where it is located in an industrial/occupational classification system. This may cause us to overlook or fail to appreciate a fundamental dynamic that in large part accounts for how and why a new industrial era may differ from an older one.

Conclusion

As we shall see in greater detail in the following chapters, the services sector has been industrializing as the larger society evolves, and this process has been accompanied by the same general effects that accompanied the transformation of the agricultural economy and the basic transformative industries of manufacturing and construction in the century and a half of the industrial era. If anything, the search for the post-industrial society has led to the realization that as a new industrial economy continues to take shape in response to the ever-greater diversity in markets and more complex production and distribution arrangements, the nation's economic geography has come to reflect the increased possibilities for dispersion and deconcentration. The burden of bridging the gaps between consumers and industrial producers has stimulated the development of a transformed services sector. Not surprisingly, the new goods-production sector, wherein mass production defers to more specialized and customized production, appears to require a new services sector wherein more tailored and complex consumption patterns displace traditional patterns of mass consumption. History reminds us that as we entered the industrial era more than a century ago, the prospect of exchanging the farm for the factory as a place to work and the countryside for the city as a place to live, and the spread of the free labor-wage labor system itself, were met with great fear and foreboding. Since then, according to most commonly accepted criteria, the quality of life has increased dramatically. The shift to an industrial society was accompanied by improved health, housing, and education and by reworked notions of how to evaluate the objective and subjective conditions of life. We must avoid assuming that the transit into an advanced industrial era will of necessity lead to a future any less desirable than that which is being left behind. Greater demographic and cultural variability will more than likely permit the continued adjustment of sectoral, occupational, and even income relationships without the general welfare, measured at the household level, necessarily suffering a net decline.

3

Industrial Transition: Restructuring, Relocation, and Renewal

> [R]evolutions periodically reshape the existing structure of industry by introducing new methods of production—the mechanized factory, the electrified factory, chemical synthesis and the like; new commodities, such as railroad service, motorcars, electrical appliances; new forms of organization—the merger movement; new sources of supply . . . ; new trade routes and markets to sell in and so on. This process of industrial change provides the ground swell that gives the general tone to business: while these things are being initiated we have brisk expenditure and predominating "prosperity"—interrupted, no doubt, by the negative phases of shorter cycles that are superimposed on that ground swell—and while those things are being completed and their results pour forth we have elimination of antiquated elements of the industrial structure and predominating "depression." Thus there are prolonged periods of rising and of falling prices, interest rates, employment and so on, which phenomena constitute parts of the mechanism of *this process of recurrent rejuventation of the productive apparatus.*[1]

In 1975, and again in late 1982, the Federal Reserve Board announced that factory use had slipped below the 70-percent level. Amid much concern that the productive capacity of the nation was nearly one-third idle, the nation's economy was judged to be facing an industrial crisis. It was not as widely appreciated that the industrial activity of the nation's economy was sorting itself into new labor environments (for instance, nonunion shops) as well as filtering out of an older and into a newer vintage of smaller and more specialized production settings. Empty plants, shops, mills, and factories were as much signs of a healthy process of shedding outmoded physical arrangements as they were signs of economic stagnation. In the face of evidence that we had built up an excess of physical capacity of little use to an advanced industrial economy, trying to bolster usage in these outmoded industrial shells was likely to be difficult in the short run and perhaps futile in the long run.

Varieties of Industrial Adjustment

Structural changes in the economy may be accompanied not only by the net "decline" of certain industrial activities but also by the increased reliance of contracting and expanding industries alike on a variety of adjustment or "mobility" strategies. Industrial mobility may involve shifts into new corporate arrangements—such as when satellite facilities are spun off as new production or distribution branches in new locations. Such "branching" may be cheaper than retrofitting older plants. Schmenner has reported that recently the pace at which physical plants become obsolete has been picking up rapidly. In a study of plant closings, he found that the average age of a plant at closing was 19.3 years, while a third were no more than 6 years old.[2] More has been going on, however, than simply a retreat from older industrial spaces. There has also been a trend toward rearranging and rewiring older industrial plants to accommodate new technologies and production arrangements. This retrofitting of older production settings is a key component of the process of industrial transformation.

New in-plant production processes increasingly integrate computer-assisted and automated production sequences with human workers whose skills have been selected and developed to fit the new industrial arrangements.[3] In-plant space is being reorganized to fit the new relationships between workers and successive waves of advanced production technology. Computer-assisted design (CAD) and manufacturing (CAM) are used to collect and analyze data from the shop floor; reconfigure machinery throughout the plant; program multiple machine tools; monitor production tasks; handle materials requirement planning and timing; tailor tool and die design; and support graphics, design, and operations simulation.[4] Flexible or variable manufacturing systems are increasingly characterized by a lengthening sequence of automated processes from initial raw materials handling, to palletized workpiece loading and transport between work stations, to inspection and quality control, and eventually to warehousing, inventory control, and shipment from the plant. While the details may vary from plant to plant, depending on the industry, the same essential renewal process is underway throughout a wide variety of older, mature industries.[5] These new production arrangements—expressed physically by the increased use of robotics and similar tools, and organizationally by the "just-in-time" concept employed by the Japanese and adopted by American automobile manufacturers to reduce the need for parts storage—require less physical space and fewer employees.[6] New corporate structures among plants, like production arrangements within them, reflect these new industrial possibilities.

Although often obscured by the consolidation trend in some industries whereby more activities may now take place within a single plant, the larger trend is toward more compact factories and offices. The slow movement of industrial activities since the 1950s—especially of manufacturing—into sprawling plants on extensive land parcels, often located in suburbs and beyond, may now well be retarded. New production arrangements—such as in canning and bottling in Hong Kong or textiles in Singapore—can now be accommodated to multistory plants in ways that were unheard-of before. While increasingly intensive land use will probably be reflected as businesses sculpt new plant and office arrangements on new sites, the possibility of retrofitting older sites built to reflect the dispersed industrial arrangements of an earlier era also exists.

Industrial arrangements that facilitate the movement of information, as well as the flow of people and materials, will be increasingly required. Gradually, this will necessitate an expanding array of telephone, video, microwave, cellular, satellite and related telecommunications utilities. Bell noted many years ago the ways in which successive industrial eras require a succession of supportive infrastructures from transportation and energy transmission to telecommunications.[7]

This chapter is about shifting patterns of economic investment and disinvestment that accompany industrial development. I attempt to identify and illustrate the multiple and reinforcing ways in which the base of an industrial economy evolves through time and across space. Throughout this chapter I will continue to examine key tenets of the so-called "deindustrialization" or "decline of manufacturing" argument, giving special attention to the changing ways in which modern manufacturing contributes to the larger economy. A case is made that given the diversity and detailed composition of an advanced industrial base, the occurrence of a recession provides the opportunity, impetus, and rationale for the relatively rapid growth of selected tiers or strata in an industry. These features normally do not receive much attention, but their expansion may well herald the shape of the industrial economy to come. We will see this clearly, for example, in the restructuring of an old-line industry like steel. We will also see that a similar dynamic has been at work since at least the 1950s and has resulted in the emergence of "high-technology" or "technology-intensive" manufacturing industries.[8] This stratum of the larger economy is one whose development was not fully appreciated until the larger and more prominent industrial sectors and subsectors were stalled in a series of increasingly severe and extended cyclical downturns. Points of contrast and comparison between these two examples will be offered below. The tendency for cyclical downturns to trigger growth in certain strata of the larger economy is traced at the detailed level of specific subsectors. The steel industry is selected as a "marker" industry, given its historic role in defining the leading edge of this nation's past industrial development and typifying the corporate and locational structures that came to characterize that development. The transformation of the steel industry, especially its corporate structural, locational, and capital mobility features in a transforming capitalist system, are then examined.

The computer software industry is then proposed as an heir apparent to steel as a marker industry for our near-term industrial future. Increasingly, the focus of advanced industrial production has shifted to production processes in which ever greater amounts of value-added are derived from created, processed, stored, transmitted, and applied information in its myriad forms. Expressed somewhat symbolically, this movement constitutes a shift in the medium of an advanced industrial economy from *metal* to *messages*. Demands for new skills, new forms of supportive infrastructure, and a whole host of institutional changes can be expected to follow. The case is made that older basic industries themselves are not so much threatened with elimination as are obsolete and inefficient production arrangements in whatever industry. Furthermore, industrial obsolescence is an economic phenomenon, not a calendrical one, and its accelerated pace is tied to the competitive environment within a global economy. The relatively distinct structural, locational, and capital features of the software industry—despite its status as a hybrid manufacturing/service industry—are used as a template for a discussion of the rewiring of the nation's industrial base. An effort will be made to show the way in which the leading edge of the steel industry, one of the oldest industrial sectors, is converging with the infant software industry.

Finally, although the changes experienced by an advanced industrial economy have commonly directed attention to the rise of new industries, multiple paths leading to the renewal of older industries in long-established sectors will also be explored. The diffusion of new and sophisticated production technology through the fragmented base of an older and mature industry will be used to illustrate the dynamics of industrial renewal.[9] In particular, the filtering of advanced production technology through thousands of small plants and shops that comprise the U.S. metalworking sector will be used to illustrate both the greater importance of technology *diffusion* relative to innovation and the capacity of an industry to respond to market dynamics that are influenced, but not dominated, by political ones. The way in which the metalworking sector itself has filtered into new plants, new labor environments, and new production arrangements, without abandoning older regional settings, will illustrate the more subtle ways in which the industrial traditions of an older era are being jettisoned. Yet there is little new in this process, and the results hardly constitute a national deindustrialization.

The Shifting Energy Environment

The "crisis" interpretation of recent industrial adjustment has been tied principally to the perception that our industrial economy is no longer capable of making orderly adjustments to changed economic conditions. From this viewpoint, the pace of such change, even more than the direction, poses major threats. The mercurial shifts in the demand and supply features of the energy environment illustrate this idea well. The reigning energy assumptions of the industrial world were suddenly shattered and recast in late 1973. Yet if the strength of an economy is to be judged by its resilience in the face of such shocks, rather than simply by its ability to anticipate and avoid them, then there is abundant evidence that our industrial economy retains a great capacity to make rapid adjustments to changed conditions.

The energy efficiency of the U.S. economy has steadily increased in recent decades, deriving ever-greater output from its energy inputs. During 1970–1984 alone, energy-use-per-GNP ratios have declined 27 percent, even though actual energy use has recently begun to increase again in tandem with the recovery phase of the 1980s recessions. Department of Energy data indicate that energy consumption rose slightly for residential, commercial, and transportation uses between 1973 and 1978, only to decline slightly to 1982; industrial users contributed the greatest energy savings, especially after 1978. Not only has a more efficient energy-mix of industries evolved within the overall economy, but within specific industries substitutions of new, efficient capital and organizational arrangements have been made. Today, only some six hundred weeks after 1973, we are far less vulnerable to future energy shocks, as unlikely as they appear to be at the present. Real oil prices in mid-1985 have declined to only 60 percent of their peak levels in 1981. Electricity consumption has not risen since 1981, with per capita levels declining by 14.4 percent since 1978. And the mobility of Americans has been maintained even as the fleet mix has shifted toward greater energy efficiency. In 1983 nearly 60 percent of the automobiles sold in the United States were small cars, compared to only 42 percent a decade earlier.[10]

Supply conditions have also changed considerably. Since 1976, a significant strategic petroleum reserve has been created, with a fourfold increase to 400 million barrels since 1980 alone. Over the same 1976–1984 period, reliance on OPEC oil has declined from a one-third share to less than 8 percent.[11]

Industrial Metabolism amid an Economic Stall

As the nation emerges from an extended period of economic stagnation, the conventional macroeconomic wisdom, together with political assumptions derivative from earlier periods of steady and predictable economic growth, have undergone a good deal of scrutiny. The structure of economic growth, its locational features, and its pattern of variation across and within industries, as well as its differing implications for employment and output, are likewise being examined closely.

Cyclical downturns culminating in official recessions have done much to obscure what evidence there is of more fundamental structural changes in the economy. While recessions may appear to slow or even stop economic growth, as measured by indicators of aggregate activity, beneath the surface considerable economic activity still exists that is capable of profoundly altering an industrial economy. As a result recessions, as well as fueling growth in newly coherent industries, can function to displace growth by channeling it selectively among dominant industry groups, sectors, subsectors, individual firms, and even divisions within a single firm.

An industrial economy does not completely cease growing and developing even in the midst of the severest recession. What usually recedes during a recession are rates of change in dominant and very visible sectors of the economy. As we have seen in the most recent recessions, in the early 1980s, durable goods manufacturing and interest-rate–sensitive industries such as residential construction are especially susceptible to cyclical declines. Yet, elsewhere, selected industries in selected subsectors often continue to grow and prosper. When a so-called general recovery begins, these subsectors are often found to have gained on the industry at large, and they are often well positioned to retain and expand their relative advantages.[12] As recession-recovery sequences continue, a realignment of industrial sectors may take place. Recessions and recoveries are far from mirror images of one another. As a result of their sequencing, a redefinition of the industrial base is slowly but continuously forced upon us.

While cyclical downturns are the occasion of rapid growth and development in small and often unnoticed subsectors of the industrial base, the process can be traced to ever-smaller scales within specific industries, specific companies and even specific plants. Ultimately, even the grandest shifts in an industrial economy are rooted in decisions by local plant managers or distant corporate officials to introduce a new product or to produce a familiar product in new ways or using new materials. When aggregated, this scattered activity can constitute evidence of a new economic base replacing an older one. Older industries, then, do not necessarily die when they become uncompetitive. They may well gravitate toward increasingly efficient production arrangements in both new and renewed industrial locations.

As before, evidence of industrial change will be sought principally *within* industries and individual plants, rather than between sectors; intraindustry shifts again will be viewed against the background of recession-recovery sequences. While the larger economy reveals continuous adjustments at all levels, as we saw in the last chapter, cyclical patterns can amplify or mute, hasten or retard these adjustments. Branson notes that U.S. trade patterns have reflected two such major adjustments since World War II, the first being a response to the rebuilt economies of former enemies, and the second following abrupt 1973 rises in oil prices.[13] The transformation of an industrial economy is particularly evident when a recovery restores growth to the larger economy without necessarily restoring the corporate forms, sectoral shares, or spatial structure of an industrial base that existed prior to its arrival.

The Segmented Economy and the Realignment of Industrial Sectors

By the 1970s, it was increasingly apparent that a new vintage of productive capacities had emerged to demarcate new frontiers of advanced industrial development. Infant industries tied to expanding technologies like photovoltaics, bioengineering, fiber optics, computer and software design, and robotics made it possible for older industries to exploit the latest technological capabilities, to experiment with new mixes of factor inputs, and to filter into new production arrangements and settings. In the United States, the trek to an advanced industrial status has not only involved the eclipse of older marker industries by newer ones—a process that permits higher-quality and more efficient production that uses less energy, exploits new materials and process technologies, generates less (though perhaps new forms of) pollution, and targets more narrowly defined and widely scattered markets than ever before. In addition, individual industries can be seen to have realigned themselves into a handful of distinct groupings.

In a 1981 *Business Week* cover story, a new realignment of the nation's economy was proposed that represented a distinct break with the past.[14] The assumption of a tightly integrated and unified national economy was directly challenged. An alternative perspective was offered and suggested that not a single, but rather five separate economies—excluding wholesale and retail trade—had evolved, each with its own pattern and pace of development; sensitivity to external factors; and labor, industry, and government constituencies and consumer clienteles. One of the five is the services subeconomy; since I devoted considerable attention to that sector in the previous chapter, I present a brief overview of only the remaining four groupings below.

Agriculture and Energy: In this schema the primary sector composed of agriculture and related industries was essentially retained. While the output still consists of food, fiber, forestry, and fishing products, the underlying production arrangements reveal considerable flexibility and change. This sector has experienced the most dramatic restructuring of all throughout the twentieth century, while greatly expanding its scale. Output has soared over recent decades, while the capacity to generate direct employment has continued its much longer decline, and the capacity to generate job opportunities in allied agricultural services has expanded dramatically. To this day, agriculture continues to be at the center of a seemingly endless succession of technological and organizational changes that are slowly altering the traditional relationships among final production, its location, and even land and related natural resource inputs.

The energy sector is viewed as the joint creation of global resource politics and developments internal to the traditional extractive economy. Low-cost energy assumptions came to be hard-wired into the workings of industrial economies around the world. As these assumptions were forcibly recast throughout the 1970s following the 1973 OPEC oil embargo, industrial economies as well as less well developed ones experienced considerable dislocation and political disruption. As reduced petroleum dependence was actively established as a national policy goal in the United States, the policy spotlight widened to draw attention to the nation's vast coal reserves; in turn, the prospects for incubating synthetic energy alternatives likewise attracted considerable, if premature, attention. While agriculture and energy may be separate subeconomies, certain developments, including the increasing use of agricultural products and byproducts for fuel, illustrate ways in which the two may be converging.

High-Technology Manufacturing: The U.S. economy has always had a high-technology sector. At any one time, those industries whose products and production arrangements have been especially dependent on new technologies and special skills have served as frontier industries: mechanized farm machinery like McCormick's reaper; linking technologies like the telegraph, telephone, automobile, and airplane; advanced processes like gene splicing and organ transplantation; as well as robots, transistors and microprocessors. All have in their own time served as frontier industries in a continuously reconstituted high-technology subeconomy. As the product and/or the production process has standardized, movement through the product cycle or industry life-cycle has tended to erode the comparative advantage that once accrued to the product, the industry, and the regional economy in which the industry is located.[15] Since the mid-1950s—although Branson notes that the evidence is clear as far back as the 1930s—a broad range of industries has been spawned that cuts across numerous sectors and subsectors of both goods and services production.[16] Considered together, these industries generally express their technology-intensiveness in the technical sophistication of either their final output or their production arrangements, as well as in their dependence on pools of special labor skills. Increasingly, contemporary high-technology manufacturing has come to be considered a coherent sector of the economy that appears to be moving in a direction and at a pace different from that of the rest of the economy. A longer-term perspective reveals that this coherence is largely an illusion.

These developments are particularly evident in comparisons with the performance characteristics and low-technology features of many of the larger and older manufacturing industries that dominate the industrial base of the nation. The high-technology sector, as currently constituted, only accounts for approximately 3–4 percent of total employment. However, that proportion may well increase as recent technological innovations diffuse through other industries. Although the high-technology sector's rate of growth is twice that of the larger economy, its labor productivity is much higher than for all business, and prices during the 1970s increased at only a third of the rate experienced by the rest of the economy.[17] For the past quarter century, while the U.S. merchandise trade balance has recorded worrisome deficits, the balance of trade in high-technology products has consistently run a surplus, particularly after 1975.[18]

The "hi-tech" stratum is widely believed to herald the shape and substance toward which advanced industrial economies are thought to be gravitating. The ability of many industries in this sector, although not all, to expand during recessions and to provide competitive advantages in an increasingly interdependent world economy has not escaped the notice of the more mature industrialized economies around the world. Department of Commerce figures note that over the 1975–1980 period alone, high-technology employment increased 26.8 percent—to 3.32 million—by adding approximately 700,000 jobs. Understandably, interest in this new stratum of industries has been building among investors, policymakers, and industrial recruiters for state and local jurisdictions around the nation and throughout the industrialized world.

High-technology manufacturing may be a newly appreciated sector, but by no means can the industries that comprise it be regarded as new. In recent decades, the demands on industrial production for greater efficiency, precision, and quality control have been met continuously by the adaptation of theoretical and science-based knowledge to older industrial processes and by the diffusion of technological innovations through many older industries. As a result, more mature industrial activities have been ceaselessly rejuvenated while entirely new industries have been spawned.

In the early stages of their development, many of the industries in this sector remained relatively labor-intensive and so have held great promise for extensive job generation for the future. Recent analysis, however, reveals that employment growth in this sector is not likely to come close to compensating for the employment contraction that results from the automating and further rationalizing of older basic manufacturing processes.[19] However, it would be a mistake to suggest that the higher productivity realized in this sector must necessarily be associated with net employment contraction.[20] Depending on the demand and supply structures characterizing an industry, it is quite common for productivity increases to so lower unit costs that new markets for the product are created, levels of demand begin to rise, and employment must also expand to meet that demand. This sequence has characterized the air transportation, pharmaceutical, telephone communications, and synthetic fiber industries most clearly in recent decades.[21]

The locational flexibility of many high-technology industries has been quickly recognized as one of their most prominent defining characteristics. No compact "new industrial heartland" tied to traditional natural resources or unskilled labor pools sprang up to mimic the original Industrial Heartland for basic manufacturing that extended from Ohio and Pennsylvania in the East and New England in the Northeast to Illinois and Minnesota in the Midwest. A 1981 congressional study of the location of high-technology firms and regional economic development revealed that those states whose overall manufacturing growth was most heavily tied to expanding high-technology manufacturing industries included Massachusetts, California, and North Carolina.[22] At the regional level, high-technology seedbeds also appear to be scattered in and around major metropolitan areas in all regions. However, rather than being accounted for by natural resources or traditional technology constraints, any overlap between the cradles of new and older manufacturing is probably due more to the building influence of new decisive location factors including the distribution of skilled labor pools, attractive business climates, distinguished research-oriented universities, and lifestyle amenities valued by potential employees.

Indeed, in the 1980s, it can be said that the locational dimension of the high-technology sector has broken the bonds of the planet itself and now extends into space. Among the major goals of NASA's space shuttle program have been the tasks of demonstrating the commercial viability of industrial production in space and of laying the "ground-work" for a range of high-technology industries for research and manufacture in the gravity-free vacuum and low energy-cost environment provided by space. Industry frontiers, therefore, have been extended in yet another direction as a result. In addition to the obvious interest on the part of the communications industries and the military, the "extended workbench" that now reaches into space promises to attract the pharmaceutical, metal processing, and electronic crystals industries, among many others. As the era of space manufacturing begins, it has become increasingly apparent how advanced technology has begun to liberate high-technology manufacturing industries from historical conceptions of "location" just as surely as it has continued to liberate the agricultural sector from historical conceptions of land itself.

Old-Line Manufacturing: A final industry grouping is composed of the remaining manufacturing industries that, for a variety of reasons, have been perceived to be resistant to technological or organizational upgrading. Often these industries are discovered to be caught in a treacherous downward spiral. Declining demand reflects new patterns of consumer preferences. These patterns, in turn, reflect shifting age-lifestyle distributions in the population, new energy assumptions, and altered patterns of household formation, as well as the more mundane substitution of new for older materials (such as plastics and ceramics for sheet metal and glass and adhesives for bolts and screws). All these changes may help set the stage for the contraction of producers of both intermediate inputs and products for final demand. This contraction is often reinforced by the rise of aggressive new suppliers throughout the international community. The automobile, steel, semiconductor, and consumer electronics industries remain good illustrations of these trends.

Such substitutions and contractions can erode profits, which in turn often results in reduced expenditures for research and development that could possibly lead to new products and new markets. Given the expanded role of high-technology and science-based ingredients in industrial economies, this newly demarcated sector pools many of those industries assumed to be "left behind." Yet the diffusion of new and even not-so-new technologies through old-line industries has great potential for redefining, restructuring, and rejuvenating industries in this sector. Just such a process, at work in the U.S. metalworking sector, will be the subject of a discussion of technology upgrading and industrial renewal to follow.

Finally, it would be a mistake to equate distressed older industries with low-technology products and production arrangments. The automobile industry, for example, may manufacture a product that still harnesses the century-old technology of the internal combustion engine, yet even this traditional power plant has been made much lighter and more fuel efficient over the years. Furthermore, from computer-controlled ignition systems and advanced braking and steering systems to mechanical features of the power train, increasingly sophisticated technologies have been embedded in the typical automobile. Not only has the product become more technologically sophisticated; the production sequences in automobile manufacturing plants have as well. The automobile industry, like the allied tire and rubber industry in its shift to radial tires, has invested considerable amounts of capital in new product and process technology. These investments have been registered in labor-productivity levels for the automobile industry that have been consistently higher than those for all U.S. manufacturing.[23] Consequently, recent automobile industry distress may have been related less to the absence of technology-related productivity increases per se and more to the combination of labor costs outpacing productivity gains, demographic shifts resulting in flattened demand, and the inroads made by foreign automobile manufacturers.[24] What has been commonly called industrial "distress" is easily as much a consequence of success as of failure.

Managerial Capitalism Versus Long-Wave Models of Industrial Change

At the heart of the concern over the apparent deterioration of this nation's basic industries and the industrial and social welfare arrangements tied to them are the impacts of investment and disinvestment decisions that are invariably distributed unequally across industrial sectors, regions, communities, and workers. From this viewpoint, as long as investment and disinvestment continue to compensate for one another *between* communities, regions, and occupational and social groups, rather than *within* them, the rapid growth in new and small high-technology sectors and services is judged to be little compensation for the declines experienced in U.S. basic manufacturing.

The perspective offered here is that we are experiencing less of an industrial crisis and more of an inevitable period of industrial adjustment whose long-term benefits are largely obscured by more politically visible welfare concerns. Extended stagnation is viewed as an interval during which investment and disinvestment activities reorganize an industrial economy, older mixes of technologies are exchanged for new ones, and the economic climate becomes more conducive to the birth of new firms in infant industries than simply to the continued expansion of existing activities.

Industrial adjustment harnesses powerful dynamics that tie the industrial present to the past century and a half of the industrial era, as well as to whatever our industrial future may hold. Yet frequently the question is asked: Who or what controls the investment-disinvestment process? Do we look to an omnipotent class of corporate managers with the autonomy to orchestrate the allocation of capital and thereby shape and reshape the structure of an industrial economy at will? Or are investors and investment patterns alike shaped by larger historical dynamics in which the economy is seen as moving through long waves of adjustment and readjustment? This is a key distinction that draws attention to two sharply contrasting views of deindustrialization. For some, the seeming erosion of our goods-producing capacities is a willful process revealing the essentially misguided priorities of a profit-oriented capitalist system; for others the same disinvestment may be understood as a constructive stage wherein outmoded and inefficient production arrangements are gradually eliminated and the overall economy is continuously strengthened. Investment decisions may either reinforce the larger adjustment process or, if they attempt to counter the larger trends, be overrun by it. Let us examine these two views in more detail.

Deindustrialization as a Managerial Strategy: The principal feature of this perspective is that deindustrialization is the largely unintended consequence of conscious decisions to alter traditional investment patterns. In the words of Bluestone and Harrison: "By *deindustrialization* is meant a widespread, systematic disinvestment in the nation's productive capacity.... This does *not* mean that corporate managers are refusing to invest, but only that they are refusing to invest in the basic industries of the country."[25] From this perspective, deindustrialization is viewed as a deliberate strategy by the nation's corporate managers to use their control over the allocation of capital to promote their respective corporate interests above all else. As such, the deindustrialization crisis is viewed as the contemporary expression of a century and a half of class-related industrial conflict wherein capital investment-disinvestment decisions engender socially undesirable consequences in pursuit of economic profits.

A major concern of this perspective is the decoupling of the prospects facing corporations during times of industrial change from those facing workers, their families, and the various social institutions and communities that define their lives. A capital "strike" by older basic industries is seen to have dampened reinvestment in people and the places where they live, while capital "flight" leads to patterns of investment in newer forms of physical and human capital to promote new work activities in newly organized work settings, often in new geographic locations.[26] A widespread unwillingness or inability to reinvest in older industries and the rise of new patterns of capital investment and new forms of industrial organization define this sweeping process of deindustrialization, which is viewed as rapidly destroying jobs, displacing workers, and crippling communities. The culprit is seen to be disembodied capital run amok; jobs, communities, and ultimately the lives of workers are threatened by the dynamics of the capitalist system. These dynamics show no allegiances to traditional employment or settlement patterns—or social contracts—that are widely judged to have served us so well during a half-century–long post-Depression era.

The net result is said to be a process of calculated industrial disinvestment that is both systematic and pernicious. Bluestone and Harrison have estimated that as many as 38 million jobs were lost to such private disinvestment in the 1970s through employment contraction, displaced growth, and plant closures. Such disinvestment has recently been a very visible process, coming as it has during an extended period characterized by often-high inflation, unused or unusable industrial capacity, structural unemployment, and stubbornly high interest rates, as well as by slow rates of productivity growth and dampened capital formation. As new investment has sought out new targets, the industrial base of the nation has been seen as jeopardized. As a result of such rerouted investment, special attention is drawn to the historically unfamiliar circumstance wherein growth in the services and in a thin wedge of technology-intensive ("hi-tech") industries within manufacturing has consistently exceeded that in manufacturing in recent years.

Deindustrialization and Economic Long Waves: In contrast, an alternative perspective suggests that shifting patterns of investment and disinvestment are integral to a larger—and largely inevitable—process of industrial development that must be viewed against decades-long sweeps of history, rather than solely in the context of shorter cyclical trends. Expansion and contraction, with investment and disinvestment within and among industrial sectors and individual industries, are viewed as the essential dynamics of economic change. The capacity for change, not preservation, is the focus of this perspective. Economic order is based on the passage between disequilibria, and at any one time industrial structure is defined by key discontinuities, rather than by their absence or elimination; any presumed economic equilibrium is to be viewed as largely an abstraction, a temporary interlude in a complex and continuous economic adjustment and industrial development process.

The prerequisite for this view is that long waves or cycles—commonly lasting a half-century or longer—be discerned in the movement of key economic indicators. Early in this century, Kondratiev and Schumpeter observed such patterns in the movement of price indices, employment levels, and interest rates for the United States and industrialized nations in Europe for the 1820–1920 period.[27] Since then, Rostow, Forrester, Mensch, and others have sought to harness different versions of the basic wave-cycle metaphor to account for different aspects of industrial development and change.[28] Others, including Mansfield and Rosenberg and Frischtak, find little empirical support for such a grand historical dynamic.[29]

A major implication of this perspective is that patterns of economic change cannot be viewed in isolation from the larger sweep of history. Moreover, the interventions of individual actors—politicians and capitalists alike—are destined to be marginal at best and most likely inconsequential if they run counter to larger historical processes of industrial evolution.[30] Therefore, while investment decisions play a role in industrial change, their eventual impacts are conditioned by larger-scale aspects of industrial development. There is little room for individual or corporate autonomy in a process that has the tug of inevitability. Ultimately, it is the locus of control over capital and over the consequences of rerouted investment that distinguishes this perspective from the previous one.

Technology Innovation-Diffusion and Capital Investment: The industrial base with which the United States has faced building global competition in the past quarter-century has been backed by a battery of technologies that have matured and slowly grown obsolete over time. They are certainly not sufficient to guarantee the indisputable competitive advantages in world markets in the coming decades that we enjoyed in the 1960s. A National Science Foundation study noted that between 1929 and 1969, technological innovation was responsible for as much as 45 percent of the nation's economic growth. Consequently, it is disturbing to note from the same study that while in the 1950s the United States was first to market 82 percent of all major innovations, by the late 1960s, the share had dropped to only 55 percent.[31] In a study by Davidson, of nearly 2,000 innovations introduced in the United States, Europe, and Japan between 1945–1974, a pattern of perceived advantages (such as labor, material or capital savings, safety) developed that gradually failed to favor the United States to the degree that it once had.[32] The United States' technology lead over its more rapidly growing rivals—who were committing greater proportions of their GNP to research and development—was gradually dissipating.

Deindustrialization as a managerial strategy places great importance on the view that disinvestment in basic industries deprives the economy of the possibilities of boosting productivity through infusions of capital and of tailoring new technologies to older industries. In contrast, the wave-cycle perspective acknowledges the importance of technology and innovation without reserving for them the role of industrial cure-alls. New technologies are seen to restructure and reorganize industrial activity, rather than being forcibly grafted onto existing arrangements. The adoption of new technologies by new industries and the commercialization of new technologies to accommodate new markets is thought to lead more quickly and certainly to increased productivity than is technological retrofitting alone. While the first perspective looks to older basic industries as the major source of future growth provided that they can be updated and infused with new technologies, the latter perspective places equal emphasis on the promise of new growth from innovations being allowed to germinate into new industries.

Mass and Senge have argued that continued overinvestment, rather than deliberate underinvestment, in existing industries and productive arrangements makes industrial adjustment a painful and protracted process.[33] While capital-labor ratios have increased since 1969, capacity utilization has declined. There exists the possibility that much of the capital invested has sought to expand the production arrangements that the larger economy is busy withdrawing from, rather than seeking to accommodate the new arrangements that the economy is settling into. Moreover, as noted above, capital intensity—the plant and equipment in use per worker—has consistently outpaced labor productivity throughout the post–World War II era, with a particularly sharp divergence developing as the 1980s began. This indicates that the productivity returns on capital investment have been declining over time. Diminished efficiency in the use of capital in turn indicates that investment may have been channeled into the kinds of capacity that an industrial economy in transition increasingly finds inappropriate or outmoded. The *targets* of capital investment, not the levels, are the crucial thing. The wiser use of capital is to encourage the displacement, not the replacement, of outmoded and inefficient physical plant and production arrangements. Continuous flows of capital into existing industrial arrangements will tend to favor larger and older entities and thereby hinder the technological and organizational innovations that are more likely to flow from new and small firms in all sectors.

Reorienting the Deindustrialization Debate: What, then, shall we understand "deindustrialization" to be? Is it a deliberate retreat from our basic industries through disinvestment? Is it a staging area for a new long wave of extended growth deriving from capital investment in new technologies and new industrial arrangements organized around them? Or could it possibly be both?

The search for an answer involves discovering a basis for compromise between the two perspectives. In one sense, deindustrialization connotes a problem rooted in our capacity to produce. It draws attention to those few industries that are still organized around inefficient and outmoded production technologies and organization. Ironically, however, such industries are not in distress solely because their production processes are obsolete or inefficient; distress does not develop until they must compete with a more efficient producer. While a few industries, like auto and steel, may be in long-term decline, the majority of basic industries are not. It is unfortunate, then, that the term "deindustrialization" has come to be equated with the loss or decline of our basic industries, for there is little evidence that that is taking place. Even auto and steel will not so much die as they will continue to restructure in the face of stiffened competition. Consequently, it is not our basic industries that are vulnerable as much as it is the production arrangements that have long characterized them.

It is difficult to find evidence that confirms that "deindustrialization" is taking place.[34] In recent decades, the contribution of manufacturing to the nation's GNP appears to have remained relatively stable during nonrecessionary periods, and the overall productivity of manufacturing continues to exceed that of the larger economy. What has come to be regarded as the demise of U.S. capacity for basic goods production appears to be more of a cyclical than a structural phenomenon.

More accurately, the deindustrialization debate underscores a problem rooted in our capacity to compete, not to produce per se. Here the weakness of our industry lies in being unable to sell our products in selected foreign and domestic markets. This view draws attention to the increasing share of global markets captured by foreign producers. During the 1962–1982 period, foreign manufactured goods captured a share of the total U.S. market that rose from less than 3 percent to 11 percent. A 10-percent decline in the U.S. share of the world market was widespread, with losses registered in 78 of 92 manufacturing industries.[35] Of the 117 top world companies in 10 industries in 1959, 84 were U.S. firms; by 1978 that figure had declined to only 60.[36] Further, between 1960 and 1979, the U.S. share of world output declined from 52.9 percent to 36.7 percent. In the post–World War II period, the U.S. share of world steel production declined from 47 percent to 29 percent, and its share of world automobile production fell from 76 percent to 29 percent. At home, auto imports increased their share of U.S. markets from 6 percent to 28 percent over the twenty-year period since 1962. And in the past decade alone, imports have increased their share of the U.S. steel market from 13.4 percent to 24.7 percent in 1984. By 1981, the U.S. share in world tire production had dropped to 18.7 percent from 30.7 percent in 1972. Such have been the inroads of foreign producers in the auto-steel-rubber production complex that anchors a major portion of the old industrial heartland. Even the U.S. high-technology sector declined in share of global sales between 1959 and 1978—at a faster rate than did other U.S. firms—with sales plummeting from 79 percent to 47 percent.[37] Since the problems of basic industries relate more to their inability to compete than to their ability to produce, *deindustrialization* is a distinctly inappropriate term around which to organize debate.

There remains the concern over employment, however. While manufacturing employment tends to surge back following a recession, the composition of that employment is changing. The proportion of manufacturing workers engaged in actual production—skilled and semiskilled jobs that traditionally have sustained a blue-collar middle class—has been steadily declining. But that decline has involved relative, not absolute, change rates. Department of Labor data indicate that at the end of World War II the ratio of production to nonproduction workers was five to one; today that ratio is two to one. However, the relatively rapid growth in the ranks of nonproduction workers, rather than any absolute decline in production workers, accounts for this shift.

Thus, after subtracting the effects of recession, any remaining evidence of deindustrialization appears to involve the shift of growth prospects and capital investment to new sectoral, sociodemographic, and geographical locations. Opposition to what has been called deindustrialization turns on perceived and presumed inequities associated with industrial dynamics that allocate capital in ways that do not permit the existing industrial system to be maintained. Mergers and acquisitions, corporate branching, corporate migration to lower labor-cost and nonunion environments, and automation are all interpreted as corporate responses to changing economic conditions—responses whose social costs are presumed to outweigh substantially any economic benefits. Yet what we have been losing are the historically high rates and traditional patterns of job growth in the older basic industries—and the insulation of U.S. markets from foreign competitors—rather than the productive capacities of the industries themselves. And the distinction is an important one. It suggests that the argument over who or what controls capital may be moot. Whether individual capitalists are viewed as making the necessary investment decisions to improve their competitive positions or longer-term adjustments are reorienting manufacturing around new or more widely diffused technologies, the destinations and paths traveled appear to be the same. It may make more sense to hasten the adjustments being brought about through these investment/disinvestment decisions than to seek to halt or hinder them with arguments over what drives them. Only enhanced competitiveness can ensure the health of an industry for whose products there is a demand. While healthy industries may not lead to employment growth, uncompetitive industries will certainly not do so. There are enough examples of industries in which productivity increases tied to improved and increased capital investment led to expanded markets and eventual employment growth to encourage us to avoid seeking to trade off increased competitiveness for employment patterns tied to older industrial arrangements.

If what remains of the "deindustrialization" notion is a justifiable concern about employment and dislocated workers, the terrain of the debate will have shifted considerably. Our basic industries are not so much in the balance as are our traditional employment patterns. Since the mid-1970s, nine of every ten new jobs have been in the services, and today only a fifth of the workforce is still employed in either direct or indirect production in manufacturing industries. Yet manufacturing output has nearly tripled since 1950, and employment in manufacturing has increased by a third over the same period. Furthermore, over the past decade of considerable employment growth, output expanded by 38 percent.[38] While manufacturing will continue to be a source of employment growth in the future, production employment will continue its relative decline. This pacing of employment growth by output growth is a telltale indicator of advanced industrial development that has been accompanied by slower-than-normal rates of productivity increase. Given the growth in employment, value-added, and national income tied to manufacturing, there appears to be little justification for equating these new investment patterns with the loss of our basic industries through a process such as is connoted by the term *deindustrialization.*

Short-Circuiting the Industrial Life-Cycle

It has become commonplace to view contemporary industrial change as a series of massive substitutions at the industry level. The problems besetting the automobile, steel, and other durable goods manufacturing industries are frequently seen as symptoms of older arrangements gradually being eclipsed by the efficiencies inherent in newer technologies, factor mixes, and managerial strategies. While the life-cycle perspective is helpful in highlighting the passage of products and even entire industries through lifestages, the metaphor may become limiting and even tyrannical. The search for lower-cost production environments and new markets is often a response to the erosion of comparative advantage within an industry as well as within the regions to which it is tied. Nevertheless, it is the older productive arrangements that are in effect being abandoned, not necessarily the plant or locational setting per se. Ultimately, the relationship between costs and revenues, not the absolute levels of each, determines the profitability of industrial activity. Certainly the industry itself need not be in jeopardy, even though its capacity to create traditional jobs may be either temporarily or permanently affected. As a result, while much of our older manufacturing arrangements may well be obsolete and uncompetitive, the passage through a life-cycle need not be unidirectional or irrevocable. While demise in some sense may be unavoidable, it is more likely to be the fate of uncompetitive production arrangements, rather than of an industry per se. The process may be short-circuited by efforts to upgrade the productive capacities of an industry. This may involve abandonment of older physical capital and even restrictive labor environments. While the competitive advantage that is restored may be only temporary, the dynamics of industrial development reveal little that remains stationary or permanent in any event.

While debate dwells too much on the dispute about whether patterns of capital investment better reflect perverse incentives among capitalists or longer waves of economic adjustment and change, thinking about the role of new technology in industrial change is vastly underdeveloped. Capital investment in new industries tied to new technologies may well speed the process of industrial adjustment through the creation of industries with higher value-added production arrangements and therefore more competitive products. Yet the diffusion of advanced production technologies through existing industries can offer the same outcome. It was Schumpeter who stressed that capitalism continuously unleashed a "gale of creative destruction;" in so doing he did not specify that this force for industrial renewal operated only between industries rather than within them. Today's older industries probably offer the best source of tomorrow's newer industries.

As new technologies become commercially accessible to older industries, not only the range of products but the basic production arrangements themselves can be expected to change. It is the prospect for renewal of older industries, more than anything else, that should caution us against subscribing to a view of industrial change tied too rigidly to inexorable waves or cycles. While the inevitability of the outcomes suggested by wave-cycle approaches is probably accurate, we should be skeptical about the assumption that investment in older industries must necessarily be less productive than investment in technological and related organizational innovations that can spawn new industries.

Industrial Adjustment Patterns in Basic Industries: Multiple Paths to Restructuring

In this section I briefly examine aspects of the industrial adjustment process revealed in one of the oldest industry groups in the U.S. economy—the metalworking sector. This ad hoc industry grouping is defined informally by its medium of production—metal—and is composed of sectors and subsectors that cut across the manufacturing base of the economy. Of special interest here will be the ways in which the metalworking sector has gradually repositioned itself within and between plants in the post–World War II era in an attempt to adjust to a continuously changing competitive environment.

Three separate adjustment dynamics will be briefly examined. Each dynamic is the consequence of slowly unfolding differentials in growth rates among three key industry characteristics—the age of physical plant, the labor environment as measured by whether or not the plant is unionized, and the sophistication of the production technology. In the long run, these dynamics have markedly altered the structure of the overall sector while leaving other features relatively untouched. Together, these dynamics constitute an industrial adjustment process unfolding *at the plant level,* with the aggregate result being that the larger sector has been gradually repositioned in ways that hold the promise of enhancing its competitiveness. Over time, the sector appears to have retreated from selected features that are judged to have rendered it ever-less competitive.

The seeming order revealed by these industrial adjustments, which propel the sector at large toward greater efficiency, demonstrates the considerable indigenous capacity for upgrading and renewal that appears to be retained by even the oldest and most mature industrial sectors. (The industrial adjustment evidenced here is market based rather than policy led. Nevertheless, there is little in the data below to mollify criticisms of sole reliance by an industrial democracy on such processes.) That older industries may not necessarily be the captives of industrial cycles over which they have no influence may be a welcome message to many.

The first dynamic involves the continuous upgrading of the physical plant of the metalworking sector. Over the years as the sector has grown, there has been a succesion of increasingly large plant "birth" cohorts by which the sector has gradually migrated from older production settings to newer ones, even though the actual relocation of a business from one plant to another has been relatively rare. Patterns of disinvestment from older settings and investment in newer settings have been an outgrowth of a kind of industrial metabolism whereby cohorts of new plants succeed cohorts of older plants. Secondly, as the sector at large has filtered into new physical settings, it has simultaneously exchanged an older labor environment for a new one. Workers in plants and shops of more recent vintage are less likely to be unionized than are those in older settings. These first two "migrations" have allowed the sector to exchange older production settings for newer ones without necessarily requiring the interregional migration of either existing plants or even successive cohorts of new plant start-ups. Finally, the adoption throughout the sector of advanced production technology illustrates the way in which capital upgrading of outmoded and noncompetitive production arrangements can be renewed without necessarily relocating to a new plant in a new location.

The data reported here are from a national study I made of metalworking plants and shops throughout the United States.[39] If we look beyond aggregate-level changes, we can find an active industrial adjustment process underway. This process is largely an abstraction, though, since it is simply the net consequences of isolated decisions made in thousands of individual plants and shops throughout the nation. There is little evidence of coordinated and coherent efforts on the part of a unified industry to orchestrate capital mobility, threaten recalcitrant labor with automation, or in other ways pursue a deliberate strategy of worker exploitation. Rather, the evidence seems to be more consistent with efforts on the part of thousands of small and medium-sized enterprises to take isolated steps to stem their individually deteriorating positions in an increasingly competitive marketplace for metal products.

Regional Concentration and Industry Renewal: The first adjustment process draws our attention to the slow movement of the sector into ever newer physical plants and production settings. The U.S. metalworking sector is as old as the industrial era itself. Its historical development reflects both the gradual domination of the U.S. economy by manufacturing and the increasing scale and capacity of production. Together, these trends increased our dependence on the durability and precision with which metal came to endow our production processes and final products.

Since the days of its origins in the nineteenth century, the growth and development of the sector has largely been confined to the East North Central and Middle Atlantic regions—the seedbed for the nation's earliest industrial development. It was not until after World War

II that secondary concentrations developed in the Far West, with development largely confined to California. Despite the gradual spread of the sector across the nation, the earliest pattern of regional concentration persists to this day.

As Table 3.1 indicates, the vast majority of plants and shops in this study were built after World War II. During the post–World War II era, each successive plant cohort has been larger than the last. More than half the plants were built between 1960 and 1980. This is evidence of a continuous renewal process whereby older physical plants have been gradually abandoned in favor of newer plants. It is significant to note that despite this migration of the sector from older to newer plants, the bulk of the new growth has continuously been captured by the older industrial regions. Moreover, to the extent that corporate "branching" has taken place, the bulk of the branching has been intraregional. There is little evidence of a widespread capital flight between regions—such as from the unionized North to the nonunion South.

Industrial Filtering into New Labor Environments: While there is little evidence of systematic interregional shifts in search of more permissive labor environments, there is clear evidence that the sector has gradually sought out more accommodating labor environments within regions. The workers are unionized in only 22.5 percent of all the plants and shops in this study, and beginning in the 1930s, successive cohorts were increasingly dominated by nonunion plants (Table 3.2). As a result of a half-century of relatively more rapid growth in nonunion plants, the sector can be seen to be gradually abandoning unionized production arrangements. Once again, as with the movement from older to newer plants, the movement from union to nonunion settings has been accommodated without the mass migration of the sector from older industrial regions.

Technology Upgrading and Industry Renewal: The third dynamic involves the use of advanced manufacturing technology to enhance productivity. The technology focussed on here involves newer forms of machine control including numerical control and computerized numerical control (NC/CNC-CAM). Numerical control denotes a lineage of technologies from the primitive punched-tape systems of the 1950s to the automated factory of the future. The essential features include the coding of design specifications and production instructions, which are then used to govern the operation of a single machine tool, such as a lathe or drill, multiple operations clustered at a machining center, or increasingly extended production sequences. This form of process technology enhances productivity in many ways including the reduced

Table 3.1

Date When Production Began at Plant, by Region

Plant birth cohort	*U.S. total*		*New England*		*Middle Atlantic*		*East North Central*		*West North Central*		*South Atlantic*		*East South Central*		*West South Central*		*Mountain*		*Pacific*	
	N	*%*	*N*	*%*	*N*	*%*	*N*	*%*	*N*	*%*	*N*	*%*	*N*	*%*	*N*	*%*	*N*	*%*	*N*	*%*
1800–1849	1	0.1	—	—	1	0.5	—	—	—	—	—	—	—	—	—	—	—	—	—	—
1850–1874	3	0.3	1	0.9	1	0.5	—	—	—	—	1	1.4	—	—	—	—	—	—	—	—
1875–1899	19	1.6	3	2.8	7	3.3	2	0.5	4	3.8	2	2.9	—	—	—	—	—	—	—	—
1900–1909	18	1.5	5	4.6	3	1.4	7	1.8	—	—	1	1.4	—	—	—	—	1	2.6	1	0.7
1910–1919	21	1.8	3	2.8	5	2.4	9	2.3	—	—	1	1.4	—	—	1	1.4	—	—	2	1.4
1920–1929	44	3.8	2	1.9	8	3.8	20	5.2	4	3.8	—	—	1	4.0	—	—	4	10.5	3	2.1
1930–1939	44	3.8	6	5.6	8	3.8	19	4.9	5	4.7	—	—	—	—	2	2.9	1	2.6	2	1.4
1940–1949	119	10.2	7	6.5	26	12.3	44	11.3	14	13.2	6	8.7	3	12.0	8	11.4	1	2.6	8	5.7
1950–1959	210	17.9	15	13.9	37	17.5	80	20.6	17	16.0	8	11.6	2	8.0	13	18.6	5	13.2	30	21.4
1960–1969	332	28.3	36	33.3	65	30.7	87	22.4	26	24.5	28	40.6	10	40.0	20	28.6	7	18.4	47	33.6
1970–1979	294	25.1	26	24.1	46	21.7	91	23.5	28	26.4	18	26.1	9	36.0	21	30.0	15	39.5	36	25.7
1980–present	21	1.8	—	—	2	0.9	9	2.3	4	3.8	2	2.9	—	—	1	1.4	1	2.6	2	1.4
Not reported	46	3.9	4	3.7	3	1.4	20	5.2	4	3.8	2	2.9	—	—	4	5.7	3	7.9	9	6.4
Total	1,172	100.1%*	108	100.1%*	212	100.2%*	388	100.0%	106	100.0%	69	99.9%*	25	100.0%	70	100.0%	38	99.9%*	140	99.8%*

*Total does not add to 100.0% due to rounding error.

Source: Donald A. Hicks, *Automation Technology and Industrial Renewal: Adjustment Dynamics in the U.S. Metalworking Sector* (Washington, D.C.: American Enterprise Institute for Public Policy Research, 1986).

Table 3.2
Plant Birth Cohorts, by Union and State Right-to-Work Status

	Union		*Non-union*		*Right-to-work*		*Non-right-to-work*	
Cohort Range	*N*	*%*	*N*	*%*	*N*	*%*	*N*	*%*
1800–1849	1	100.0	—	—	—	—	1	100.0
1850–1874	2	66.6	1	33.3	—	—	3	100.0
1875–1899	13	72.2	5	27.8	4	22.2	14	77.8
1900–1909	14	77.8	4	22.2	1	5.6	17	94.4
1910–1919	8	38.1	13	61.9	2	9.5	19	90.5
1920–1929	24	55.8	19	44.2	2	4.6	41	95.4
1930–1939	16	40.0	24	60.0	3	7.0	40	93.0
1940–1949	35	30.7	79	69.3	21	18.0	96	82.0
1950–1959	52	25.0	156	75.0	22	10.5	188	89.5
1960–1969	58	17.8	267	82.2	57	17.3	272	82.7
1970–1979	23	8.0	264	92.0	70	24.0	221	76.0
1980 to Present	—	—	19	100.0	6	28.6	15	71.4
Not reported	9	25.0	27	75.0	7	12.5	49	87.0
Total	255		878		195		976	

Source: Donald A. Hicks, *Automation Technology and Industrial Renewal: Adjustment Dynamics in the U.S. Metalworking Sector* (Washington, D.C.: American Enterprise Institute for Public Policy Research, 1986).

waste of materials, higher quality control, and shorter setup times between jobs.

The metalworking sector has always included technology-intensive ("hi-tech") subsectors, relative to the capabilities in different eras; but the larger metalworking sector today remains decidedly "low-tech." A principal reason for this is the fragmented structure of the sector. The difficulty of filtering new technology down through the broad base of the sector, composed of thousands of small plants and shops, is a major hurdle. The diffusion of NC/CNC-CAM technology through the metalworking sector has lagged behind the availability of the technology by as much as a quarter-century, largely because of the absence of "fit" between the scale and cost of available turnkey NC/CNC-CAM systems and the fragmented structure of the sector. However, as the competitiveness of these plants and shops deteriorated and machine-control technology continued to develop, a market was created for scaled-down and less-expensive machine control systems. In effect, the movement of numerical control technology through its product cycle enabled it to diffuse throughout the broad base of the metalworking sector and thereby partially arrest the movement of the entire sector through some fixed and predestined sequence of lifestages that must inevitably lead to its demise.

In Table 3.3 we can see that beginning in the early 1960s, progressively sophisticated forms of numerical control technology began to be

Table 3.3
Date of Adoption of NC/CNC Technology

Year plant began	*Age of plant*	*Before 1950*	*1950–1954*	*1955–1959*	*1960–1964*	*1965–1969*	*1970–1974*	*1975–1979*	*1980–present*	*Total*
1800–1824	158–182 yrs.	—	—	—	1 (100.0%)	—	—	—	—	1 (100.0%)
1825–1849	133–157	—	—	—	—	—	—	—	—	—
1850–1874	108–132	—	—	—	—	2 (67.0)	—	1 (33.0)	—	3 (100.0)
1875–1899	83–107	—	—	—	1 (5.9)	11 (64.7)	1 (5.9)	2 (11.8)	2 (11.8)	17 (100.1)
1900–1904	78–82	—	—	—	—	3 (25.0)	5 (41.7)	3 (25.0)	1 (8.3)	12 (100.0)
1905–1909	73–77	—	—	—	—	1 (20.0)	1 (20.0)	2 (40.0)	1 (20.0)	5 (100.0)
1910–1914	68–72	—	—	—	—	—	1 (20.0)	4 (80.0)	—	5 (100.0)
1915–1919	63–67	—	—	—	2 (13.3)	3 (20.0)	4 (26.7)	4 (26.7)	2 (13.3)	15 (100.0)
1920–1924	58–62	1 (4.5)	—	—	3 (13.6)	3 (13.6)	7 (31.8)	8 (36.4)	—	22 (99.9)
1925–1929	53–57	—	—	—	2 (10.0)	5 (25.0)	3 (15.0)	9 (45.0)	1 (5.0)	20 (100.0)
1930–1934	48–52	—	—	—	3 (18.8)	2 (12.5)	3 (18.8)	5 (31.2)	3 (18.8)	16 (100.1)

1935–1939	43–47	—	—	—	1 (4.0)	7 (28.0)	3 (12.0)	13 (52.0)	1 (4.0)	25 (100.0)
1940–1944	38–42	—	—	—	2 (6.9)	6 (20.7)	7 (24.1)	13 (44.8)	1 (3.4)	29 (99.9)
1945–1949	33–37	—	—	—	3 (3.6)	14 (16.7)	20 (23.8)	41 (48.8)	6 (7.1)	84 (100.0)
1950–1954	28–32	—	1 (1.0)	—	8 (8.2)	21 (21.6)	26 (26.8)	34 (35.1)	7 (7.2)	97 (99.9)
1955–1959	23–27	—	—	2 (2.3)	5 (5.8)	21 (24.4)	20 (23.3)	27 (31.4)	11 (12.8)	86 (100.0)
1960–1964	18–22	—	—	—	10 (8.1)	22 (17.9)	29 (23.6)	47 (38.2)	15 (12.2)	123 (100.0)
1965–1969	13–17	—	—	—	—	47 (23.7)	49 (24.7)	80 (40.4)	22 (11.1)	198 (99.9)
1970–1974	8–12	—	—	—	—	—	53 (35.3)	70 (46.7)	27 (18.0)	150 (100.0)
1975–1979	3–7	—	—	—	—	—	—	77 (77.0)	23 (23.0)	100 (100.0)
1980–1982	0–2	—	—	—	—	—	—	—	7 (100.0)	7 (100.0)
Totals		1 (0.1%)	1 (0.1%)	2 (0.2%)	41 (4.0%)	168 (16.6%)	232 (22.9%)	440 (43.3%)	130 (12.8%)	1,015 (100.0%)

Source: Donald A. Hicks, *Automation Technology and Industrial Renewal: Adjustment Dynamics in the U.S. Metalworking Sector* (Washington, D.C.: American Enterprise Institute for Public Policy Research, 1986).

adopted widely by plants and shops of all ages. The adoption rate increased dramatically until hitting its peak between 1975 and 1979. This pattern indicates that throughout the sector, plants and shops were adopting productivity-enhancing technology following the 1974 recession and prior to the onset of the first of two back-to-back recessions beginning in late 1979. Capital upgrading was clearly in evidence well before the first concerns over U.S. deindustrialization were expressed or the first calls for a policy-led reindustrialization were heard.

The importance of this dynamic is that it dispels the notion that older physical settings must necessarily embody older capital plant. Given the relatively recent vintage of the majority of the plants and shops in the metalworking sector, it is apparent that obsolescence has less to do with the chronological age of the production setting and far more to do with the efficiency of the production arrangements located inside. Older plants in older industrial regions are not precluded from seeking to enhance their productivity by retrofitting themselves with upgraded process technology. The role of technology upgrading may then herald the prospect of a limited rebound of regional economies accompanying the renewal of industrial capacities. Much more needs to be learned about the consequences for industrial competitiveness and employment of such capital upgrading. Still, the more compelling message of this section is that even an old and most basic sector like metalworking clearly retains the capacity to boost its productivity through technology upgrading and thereby to adjust to changing industrial circumstances.

From Steel to Software: A New Industrial Base Takes Shape

Steel—its production and use—provided the scaffolding for First World economies until well into the 1950s. In the three decades since, it has become apparent that advanced industrial development is increasingly being tied to new and different industrial products and production processes derived from new technologies as well as from new physical and human capital. Moreover, all of this development may commonly be sited in new locations or organized in more complex ways than has often been appreciated. Symbolically, if not actually, the computer software industry may be viewed as providing a new scaffolding for our industrial future. As such, software is a likely successor to steel as a key frontier industry. The modern industrial forms of the steel and software industries do differ radically. However, both have played remarkably similar roles in the larger industrial economy during the respective era in which each experienced the bulk of its growth. As a result, in its own time, each can be considered an industry that served to shape the structure and influence the development, to the limits provided by its respective medium, of the larger industrial economy of which it was a part.

The production of steel and products made with it not only greatly influenced the structure of industrial development in the United States and other nations, but it also set physical limits on the scale of industrial-urban activity. Similarly, the emerging software industry, a key element in a larger telecommunications and information-based industrial complex, reflects a new set of influences on industrial organization—the structure and location of industry across the landscape—and a vastly expanded scale for economic activity. Through the way it is organized, stored, accessed, and transmitted and the way it defines work and shapes production settings for workers, information does for an evolving industrial order today what the production of steel has long done throughout the industrial era.

In the end, as new industries emerge and older ones develop, an advanced industrial economy comes to require new combinations of land/location, labor, and capital for which there are few, if any, political and social constituencies in the larger society. The rise of the software industry and others like it illustrate these patterns of advanced industrial–urban development. An older industrial base tied to the manipulation of metals is slowly converging with one tied to the manipulation of messages that control new industrial activities. In this section, the steel and software industries have been juxtaposed to illustrate this shift. We will examine the ways in which both the mature steel and the infant software industries reflect a variety of these reorienting industrial dynamics. A case for the convergence of the two industries on several dimensions in an advanced industrial economy will then be offered. Let us begin by examining the steel industry for its potential as a standard of comparison.

Steel: Development Cycles in an Early Frontier Industry

The durability of steel as a building material belies the flexibility with which it gradually endowed an emerging industrial society. Steel and related iron-ore derivatives served as the media through which a succession of new technologies left their imprints on the era. It was ribbons of steel rail that laced together the continent into a coherent whole in the decades before the Civil War. Soon, both settlements and the increasingly productive spaces between them were being stitched together into an integrated, long-distance domestic economy. Long before the nineteenth century ended, the essential *horizontal* features of a rapidly industrializing economy were in place.

As the twentieth century arrived, steel was being used as well to transform the *vertical* features of the industrial—and now largely urbanized—society. Steel girders and structural supports permitted the high-rise construction necessary for the increasingly intensive use of centrally located land parcels in densely compact industrial settlements. As the industrial era approached the end of its first century, steel gave shape to the new forms and arrangements that had been assumed by a transformed economy and society.

The steel industry can be usefully viewed both as a "marker" industry—demarcating the successive stages of the development of an industrializing nation—and as a frontier industry—providing the outer bound of the technological capabilities of a newly industrialized economy. Over the past century of U.S. industrial development, the use of steel figured prominently not only in the production of a wide variety of capital goods and products destined for mass consumption, but also in the development of a multimodal national transportation and goods distribution system, of commercial and industrial construction tied to expanding local and regional economies, of a public works infrastructure above and below ground that laced it all together, and of a national defense system that permitted the pursuit of foreign policy goals and the protection of vital domestic economic interests. In short, it is difficult to imagine the industrial supremacy gradually achieved and retained by the United States throughout the twentieth century had the domestic steel industry not expanded and developed as it did.

A major part of the continued success of the steel industry was tied to its ability to harness a succession of new technologies reflected not only in an expanding array of products it could offer, but also in increasingly efficient production technologies. In the years following World War II, steel making was significantly recast through the adoption and implementation of these new technologies. During the 1950s, open-hearth production expanded through the industry, to be followed during the 1960s by the diffusion of oxygen-blast furnaces—a European innovation. Continuous casting followed. By the late 1960s, however, the industry showed signs of having sacrificed flexibility for productive efficiency. The new expansive scale of steel production and the heavy burden of these capital investments, together with increasingly costly labor, reduced the industry's ability to respond to competition from low-cost foreign producers in a world economy in which it became increasingly clear that demand levels were being overestimated.[40]

With steel at the center, the basic industries in durable goods manufacturing and in construction had developed to the point that they constituted the essential core of our industrial economy. The steel industry performed the role of producer of the premier building material of an industrialized nation. Through that role, it supported production in a wide variety of other developing industries, including the automobile, machine tool, and consumer appliances industries. For many decades the steel industry was able to benefit from the markets it created and the growth it made possible in the larger economy. As the industry matured and the structure of both supply and demand adjusted to a new global marketplace, steel and several allied industries began a wrenching process of long-term adjustment that continues today. The linkages among these basic industries have long defined the industrial base not only of the national economy but also of the local and regional economies throughout an arc of older settlements from the Northeast to the Midwest (and spot locations elsewhere). These linkages are being recast, just as are the industries' ties to the settlements in which they are located and to the legions of workers who for generations have been dependent on them for employment.

Steel as a "Marker" Industry

The production and rapid diffusion of steel serves as a historical marker by which the maturation of an industrial economy can be monitored and observed. Inevitably, the importance of steel expanded to include the industry that produced it. The growing contribution of the steel industry to the nation's Gross National Product and its role as a major employer came to dominate our consideration of it. So important had steel become by the midpoint of the twentieth century that the transit of nations out of a preindustrial status was marked by their ability to produce and work with steel.

Steel has long served as a frontier industry for industrial economies by defining the outer boundary of industrial development. As such a marker, the development and diffusion of steel-making and steel-using capacity has, over the twentieth century, moved down through the ranks of less-industrialized nations. The gradual development of Second and Third World nations around the world has often been accompanied by the steady movement of steel through a sequence of life-stages. The importance of such movement signifies the gradual ability of an industry to exchange one production environment for another. An initial innovation stage places a premium on a set of support factors that can nurture a new product with sufficient amounts of capital for investment, experimentation, and growth. At a later standardization stage, a product both pushes out of the older environment in search of new markets and is pulled out in search of lower-cost labor and related inputs.

Vernon has traced such a sequence in accounting for patterns of international economic development;[41] others have scaled down the process to account for patterns of interregional development involving the spread of manufacturing throughout the United States.[42] The latter-day application of this perspective has focussed primarily on the spatial and locational implications of the product-cycle model; however, earlier work by Schumpeter on business cycles and the industrial life-cycles overlaid on them challenged a Depression-era preoccupation with specific geographical locations—for new markets, sources of raw materials, or related investment targets—as the key to industrial development. For Schumpeter, the frontiers of industrial growth and development were better understood as new sources of competition, regardless of whether or not they were tied to new physical locations:

> [I]t is not that kind of competition which counts but the competition from the new commodity, the new technology, the new source of supply, the new type of organization...—competition which commands a decisive cost or quality advantage and which strikes not at the margins of the profits and the outputs of the exisiting firms but at their foundations and their very lives.[43]

Technology Life-Cycles and Industrial Fragmentation

The steel industry today is regarded as one of the most beleaguered industries in the economy. Since the mid-1970s, employment, total raw steel output, and capital expenditures have all been declining, with the first of these being the most depressing indicator. Employment in the steel industry stood at 682,000 in 1953. Thirty years later, in 1983, employment had sunk to only 230,000. In the 1974–1984 decade alone, employment was reduced by more than half. Presently, steel employment stands at 245,000, reflecting a small rebound in the past two years.[44] Projections for the year 2000 suggest that employment will dwindle further to 185,000. Ironically, while employment contraction is generally regarded as a sign of industrial malaise, a recent report on industrial competitiveness by the Office of Technology Assessment has noted that the most important cause of declining employment in the steel industry has been rising productivity.[45] Unfortunately, productivity increases have not led to lower prices or expanded markets in a global economy characterized by overcapacity and contracting demand. In such an economic context, the increases in productivity unavoidably are registered in the form of employment contraction. This situation has been exacerbated by other factors as well, not the least of which has been the gradual loss of competitiveness in global markets. Between 1956 and 1978 the U.S. share of world production fell from 37 percent to 17.5 percent. U.S. exports likewise fell from 6.3 percent to less than 3.0 percent of the country's total production, while Japan and Germany directed 36 percent and 54 percent of their production to exports respectively.

Over the quarter-century from 1954 to 1978, the profitability of U.S. steel producers was lower than that of overall manufacturing in all but four years, even though it consistently compared favorably to that of steel producers in other industrialized countries. In the steel industry—and to an even greater degree in the auto industry—value-added productivity has exceeded that of all manufacturing in the post–World War II period until the late 1960s.[46] However, since 1967, labor productivity has generally lagged behind that of all U.S. manufacturing.[47] As a result, while wages in the flagship industries of auto and steel have paced those in manufacturing overall, wage increases in steel were unmatched by productivity increases throughout the 1970s. Clearly, then, the shape and prospects of the steel industry are related to factors that both influence it from within and are external to it. The life-cycle features of steel and the diffusion of the technologies necessary both to produce it and to produce with it are clearer than ever before. Yet, as both the industry and the industrial economy anchored to it have matured, the steel industry has become associated with overcapacity, high labor costs, obsolete physical plant, declining employment, and a whole host of product substitutions such as aluminum, titanium and other metal alloys, ceramics, and plastics, all of which are lighter, more durable, or easier or cheaper to produce and transport. Increasingly, producers of automobiles, home appliances, and other durable goods are substituting these newer materials for steel as both product and process technologies mature and energy assumptions underlying fabrication, distribution, and eventual consumer use slowly shift.

Today, the industry that produces steel reflects changes in the larger economy it helped to create. The industry has experienced not only the changes tied to advancements in the technology used to produce steel, but also those changes in allied industries that have made it possible for them to substitute a range of newer materials for steel. The first series of changes, internal to the industry, functioned to expand markets and increase demand for steel, while other changes external to the industry operated to shift demand away from steel and toward other materials.[48] As a result, during the 1980–1982 period alone, the proportion of total domestic steel production destined for use by the U.S. automobile industry shrank from 24 percent to 15 percent.

Ironically, then, within the basic industries that were historically devoted to the fabrication of metals into finished products, a continuous feature has been this process of materials substitution. Just this process has weakened the close link between the automobile and steel industries. A similar decoupling can be expected between industries that supply and those that use zinc, copper, malleable iron, aluminum, synthetic rubber, and other long-used industrial materials.

A New Steel Industry Takes Shape

The responsiveness of the steel industry to this barrage of internal and external influences has not been limited to employment contraction. Under these pressures, the structure of the larger industry has undergone subtle changes at the subsector and corporate levels. Indeed, so significant has this restructuring been that today it may no longer be accurate to consider the steel industry a large monolithic sector in great distress. In actuality there are today several steel industries, not all of which are equally troubled.

The emergence of distinct segments in the steel industry has been influenced by the rise of alternative production technologies and raw materials options. Today the steel industry can be divided into three distinct and markedly unequal segments. The bulk of the industry is composed of large integrated firms, which are today responsible for 84 percent of domestic shipments. A second stratum is composed of nonintegrated firms that produce from scrap metal and account for 13 percent of output. A third and tiny segment includes the producers of alloy and specialty steel. Increasingly, the competitive core of the industry has gravitated away from expansive production arrangements utilizing basic oxygen-blast furnaces and iron ore and toward smaller and more compact producers using electric furnaces, continuous casting, oxygen processing, and scrap metal. While the number of mini-mills in the U.S. has increased from twelve to sixty since 1964, the capacity of the mini-mill sector is expected to nearly double by the year 2000, while that of larger integrated producers is expected to decline by more than a third (Table 3.4).[49]

Moreover, this shift in industrial structure toward the mini-mill sector has been accompanied by a simultaneous shift to new labor environments—generally the work forces within these sectors are nonunion—and to new geographical locations as well.[50] Yet, continuing what has been a pattern since the 1970s, increased productivity to the year 2000 resulting from heavy investment in new plant and equipment will be accompanied by a loss of more than half the 1980 employment. Therefore, while the decline of the steel industry itself, understood as the

Table 3.4
The Shifting Structure of the U.S. Steel Industry

	1983		1985		1990		2000		1983–2000
Industry indicator	*No.*	*(%)*	*No.*	*(%)*	*No.*	*(%)*	*No.*	*(%)*	*% change*
Capacity[a]									
Minimills	18.0	(13.2%)	20.5	(16.3%)	25.0	(21.4%)	35.5	(32.0%)	+97.2%
Integrated	118.5	(86.8)	105.5	(83.7)	92.0	(78.6)	75.5	(68.0)	−36.3
Raw steel production[b]									
Minimills	14.4	(16.5)	18.5	(18.1)	22.4	(22.5)	31.7	(32.7)	+120.1
Integrated	72.5	(83.4)	83.7	(81.9)	77.0	(77.5)	65.2	(67.3)	−10.1
Shipments[a]									
Minimills	12.4	(18.4)	15.9	(20.0)	20.2	(24.9)	29.8	(34.5)	+140.3
Integrated	55.1	(81.6)	63.6	(80.0)	60.8	(75.1)	56.7	(65.5)	+2.9
Productivity[b]									
Minimills	3.3		3.0		2.3		1.5		−54.5
Integrated	8.4		7.8		6.7		4.8		−42.9

[a]Measured in millions of metric tons.

[b]Measured in man-hours per ton.

Source: Adapted from data from Donald F. Barnett, Testimony before the U.S. Senate Committee on Human Resources, Subcommittee on Employment and Productivity, March 22, 1984 (cited in "Steel Industry: Winners and Losers," *The Socioeconomic Newsletter* 10, No. 2, March–April 1985, pp. 2ff.).

capacity to produce steel, may not be underway, the industry's ability to compete for sales in old and new markets has declined. Consequently, the capacity for the industry to create and retain employment opportunities in the future undoubtedly will be severely curtailed.

The Rise of a New Industrial Frontier: The U.S. Computer Software Industry[51]

The computer software industry, at first glance, appears to have little in common with the steel industry. To begin with, software production cannot be regarded as a manufacturing activity in the strictest sense; rather, it is generally included in the producer services portion of the economy as a business service. Unlike steel, the software industry is in its infancy despite its rapid rate of growth.[52]

Between 1981 and 1984, industry sales exploded from $2.7 billion to over $10 billion, with projections approaching $30 billion by 1988.[53] Recently, moreover, software's growth has begun to surge ahead of that for computer hardware. In the end, not only is this industry characterized by rapid growth and development, but it promises to have major implications for the larger U.S. economy. The software industry is expected to surpass the automobile industry in both sales and employment by 1995.

For our purposes, however, software—it's production and integration into a rapidly expanding array of industrial activities—is one of a wave of new industries that have succeeded steel both as a marker and as a frontier industry. Like steel, software promises to be a new kind of building material for an advanced industrial economy; it may well grow to become a basic industry in that economy. Increasingly, software product development embodies a succession of technological advancements through which the larger service sector can continue to "industrialize" and through which older basic industries can renew themselves via enhanced productivity.

Industry Growth and an Accelerated Life-Cycle

Industrial restructuring is as likely to proceed in industries experiencing overall expansion as in industries experiencing overall contraction. The computer services industry (SIC 737) is already in the midst of a restructuring that is similar in form to what we just reviewed in the steel industry, although it is proceeding in the context of much more rapid industry growth. Thus far in the 1980s, the software products industry (SIC 7372) has been the smallest, yet most rapidly growing, of the three major segments of the computer services industry. With an expected growth rate of 27 percent during the 1980–1985 period, the software products segment may well grow at twice the rate of the other segments. Its share of industry revenues is expected to rise dramatically during the 1980s, from 17 percent in 1980 to 25 percent in 1985. In contrast, the data processing services industry (SIC 7374) has long been the largest segment and now is the slowest growing, with a 16-percent annual growth rate. The professional services industry (SIC 7379), with an annual growth rate of 17 percent, is likewise in relative decline within the larger computer services industry.[54]

As is common with infant industries, estimates of industry-wide levels and composition of employment as well as of establishment characteristics are difficult to make. As the fastest growing high-technology industry in the United States, the software industry has benefited from the explosive growth in the computer equipment industry throughout the 1970s. By mid-1984, industry sources estimated that the full sweep of the industry consisted of approximately 10,000 firms producing more than 40,000 software programs. These firm counts include both incorporated and nonincorporated establishments and range from large computer equipment manufacturers that produce software for their own products down to the broad stratum of independent and often invisible entrepreneurs who work at night and on weekends at their kitchen tables. The latter represent the smallest of small business and consequently are not easily captured in either official government or proprietary censuses. Despite the presence in the industry of very large software manufacturers, the role played by the new and small firms has been significant indeed. For example, during a six month period in 1982–1983, the top fifty entertainment software products came from forty-nine different publishers![55]

The Vanishing Distinction Between Hardware and Software

The computer equipment industry experienced its infancy during the 1950s.[56] The software industry, by comparison, is a decade or more closer to its origins as a cottage industry. The explosive development of the software industry has been compressed into a remarkably short period of time. Nonetheless, clear signs of some of the industry's "smokestack" features are already apparent. As software production has moved between generations of technological capacity, increasingly the distinction between hardware and software has become blurred. Part of this rapid movement through the product cycle has been prompted by the need to compensate for the severe shortage of computer programmers and language assembly skills. So-called fourth-generation software language, wherein specific commands are embedded directly onto the microprocessor chip, thereby eliminating ad hoc programming tasks, accommodates the rapid decline in the computer literacy of users that accompanies the diffusion of the technology from concentrated computer complexes to deconcentrated home, factory, and office mass markets. Moreover, software development, like production, is increasingly being automated—software begets software.

The Characteristics of Software Industry Development in the United States

Relatively little information on the development of the computer software industry is available; therefore, I conducted a nationwide survey as background for this section.[57] The data analyzed and reported on here are based on completed surveys from a subsample of software firms located in the core counties of major metropolitan areas around the country.

Internal Environment and Product Development: The establishment dates for companies in this sample range from 1956 to 1982 (Figure 3.1). The rapid and recent expansion of the industry is reflected in the distribution of company formations across this period. In all, 59.7 percent of the firms were begun in the 1970s and another 19.2 percent during 1980–1982. Traditionally, service industries have been considered largely secondary to manufacturing and more export-oriented components of an economic base. As a business/producer service, the software industry reveals a considerable export orientation. Despite the small size of the establishments—62.0 percent have ten or fewer employees and one in twelve is a single-person operation—37.2 percent market their major product nationally and 13.7 market internationally, while 23.5 percent market within the multistate region and 25.5 percent restrict their marketing to their respective states.

Even though the industry is bottom heavy with tiny companies to the point of being atomized, there is evidence of cooperative strategies in product development both within and among software companies. While 64.0 percent of the firms reported their principal activity to be producing and publishing their own software products, industry networks appear to exist among even these small companies, since 41.2 percent indicated that they participated in joint development efforts with other firms. Within these companies, the team approach to product development is used by the majority (68.1 percent), while solo development is reported by only 25.5 percent.

Software products fall into two principal categories—operating sys-

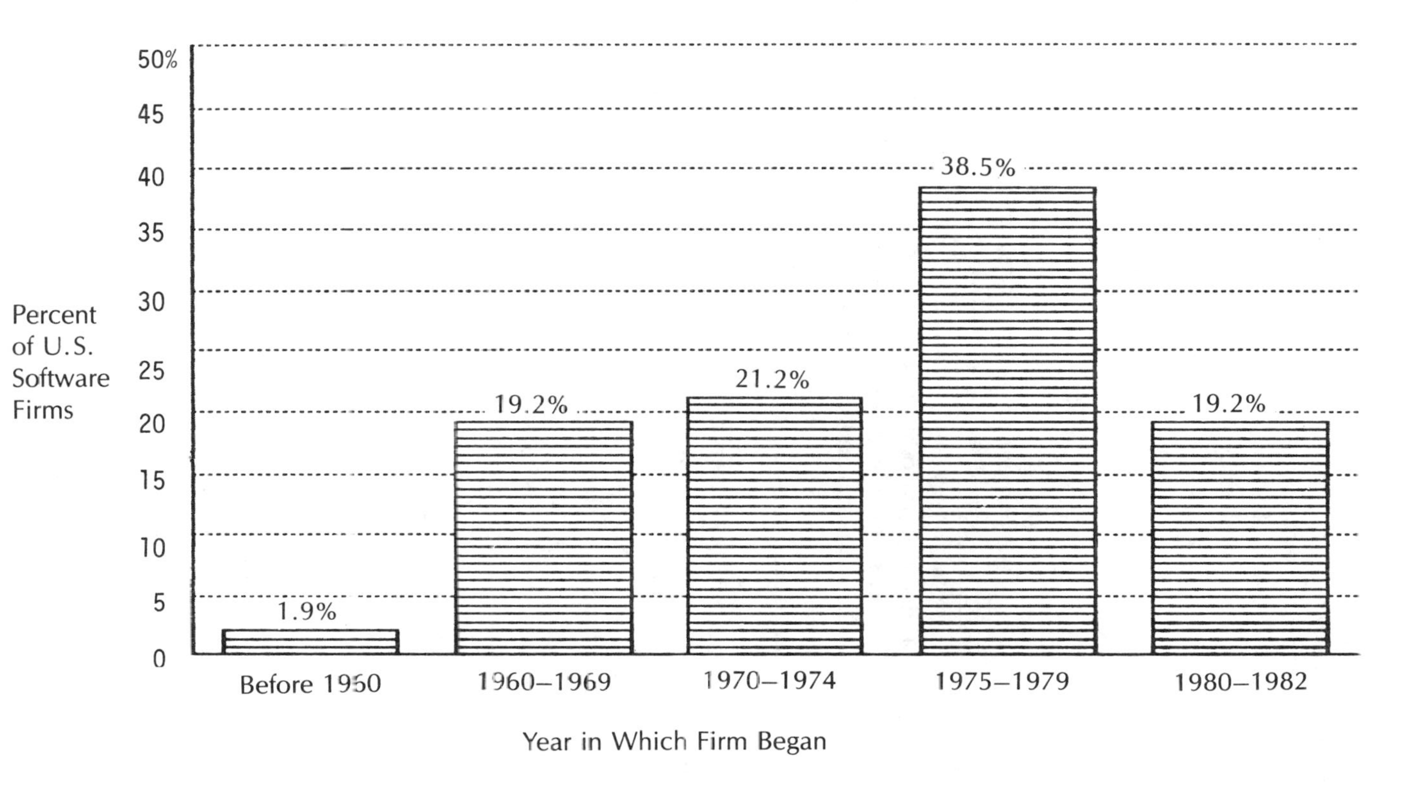

Figure 3.1. *Business formation birth cohorts among U.S. computer software firms.*

tem software and applications software. The majority (62.8 percent) of the firms specialize in applications software, such as productivity-oriented programs for business and home entertainment programs. A specialization in systems software, whose function is to control applications programs, was reported by only 9.8 percent of the responding firms. Reflecting the small scale and specialization of these businesses, 11.6 percent report production of only a single product; over half (53.5 percent) offer five or fewer products, and only a quarter (23.3 percent) offer more than ten separate products. The importance of identifying and servicing relatively narrow market niches is seen in the fact that the products of 44.2 percent of the firms are customized for the specific end-user, while only 19.2 percent of the firms rely exclusively on standard-packaged products aimed at a wider more general market. The significance of this industry to the productive activity across a wide range of goods and services producers is revealed by the fact that 23.1 percent of the software firms consider the primary market for their major products to be industrial and/or factory settings and another 36.5 percent target their products to commercial and office settings (Figure 3.2). Not one firm targets the consumer/household market principally.

Location Patterns and the External Environment: The "location" of an economic activity like a software firm can be understood at a number of levels. In this section, I examine the location patterns within this industry at three intrametropolitan spatial scales. The first scale is the largest—that is, location within the structure of the metropolitan area. The second focuses on the immediate vicinity defined as the surrounding area within a quarter-mile radius. The third is the smallest scale—that is, the specific physical setting or structure that houses the firm.

As is the case with the bulk of this industry, each of the firms in this national subsample is located somewhere in a metropolitan area. The subsequent distribution of software firms throughout the metropolitan economy reflects the decentralization process that has been restructuring our settlements and their respective economies throughout the twentieth century. Clearly, geographical centrality no longer coincides with economic centrality. Fully 51.9 percent of the firms are located in suburban jurisdictions; 32.7 percent are located just outside the central business district; and only 15.4 percent are physically located within a central business district (Figure 3.3). At the next smaller scale, we see that software firms appear able to accommodate themselves to widely varied land-use and neighborhood settings. A third (34.6 percent) are located in high-density commercial/retail settings, 26.9 percent in low-density commercial/retail settings, 9.6 percent in industrial/manufacturing settings, 5.8 percent in residential settings, and 23.1 percent in mixed land use settings (Figure 3.4).

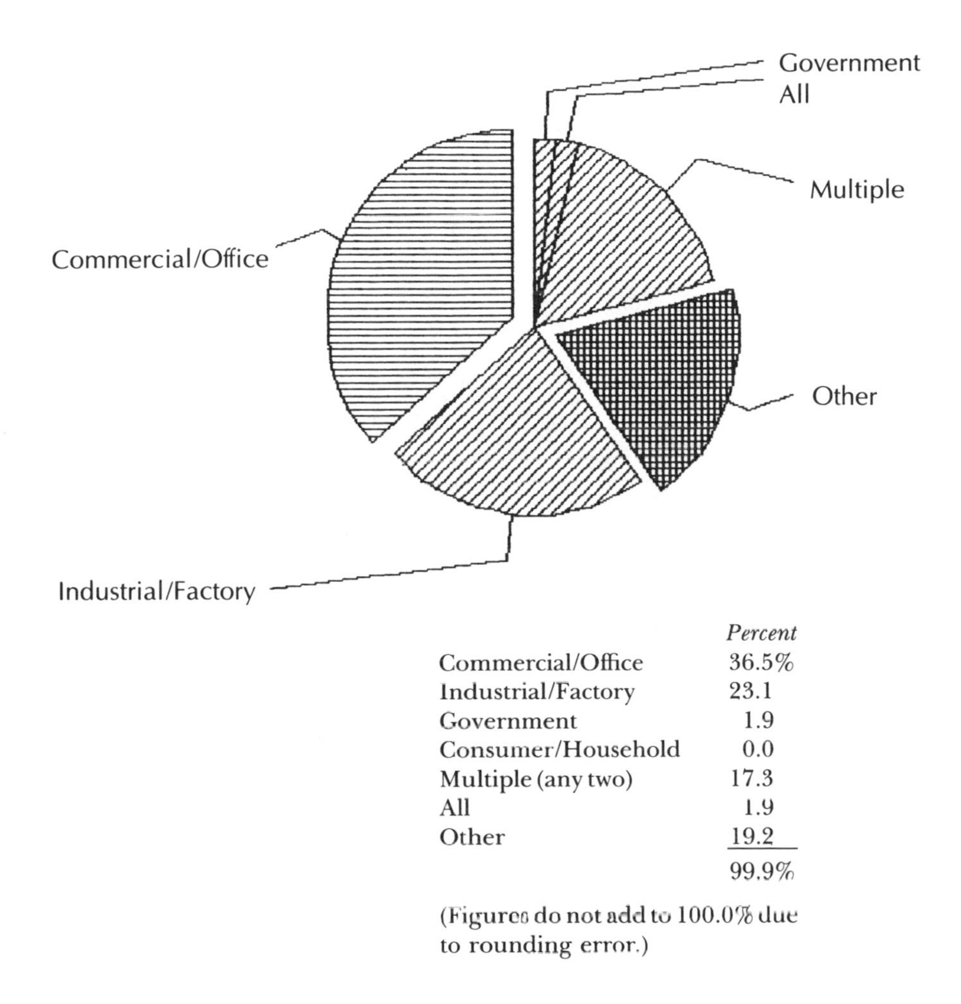

	Percent
Commercial/Office	36.5%
Industrial/Factory	23.1
Government	1.9
Consumer/Household	0.0
Multiple (any two)	17.3
All	1.9
Other	19.2
	99.9%

(Figures do not add to 100.0% due to rounding error.)

Figure 3.2. *Primary markets for U.S. software firms.*

Software establishments do not appear to require or necessarily prefer location at or even near a physical crossroads such as a central city, as was true for earlier generations of new industries. The "marketplace" to which these firms are oriented may be understood as less a physical place than a temporary abstraction to which they have primary access on demand through communication (especially telecommunication) as well as transportation. The importance of this apparent indifference to central physical location within a metropolitan economy reveals much about the weakening relationship between the "city," which reflects the productive arrangements of an older era, and those newer urban arrangements that characterize an advanced industrial economy. I will

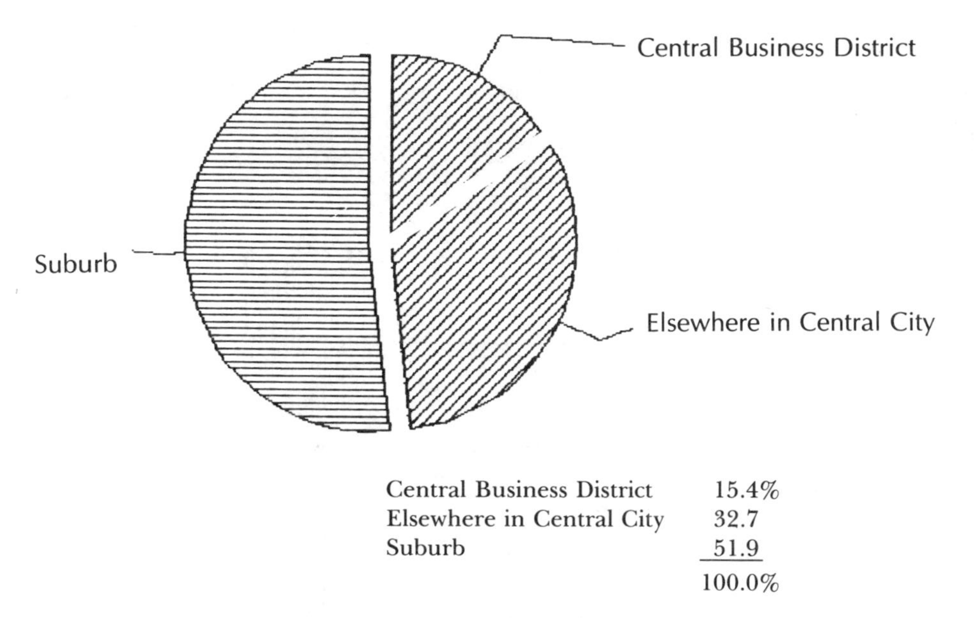

Central Business District	15.4%
Elsewhere in Central City	32.7
Suburb	51.9
	100.0%

Figure 3.3. *Metropolitan location of U.S. software firms.*

turn to such matters in greater detail in the next two chapters.

It is notable that there is little in the data on this rapidly growing high-technology service industry that appears to favor or in any way promises to restore the importance of physical high density or centrality—especially as it has been associated with the traditional central business district. The concentrated production and consumption arrangements that defined the older industrial city were reflected in the clusters of multistory factories, the commercial "downtown," and the compact residential areas; these in turn reflected the technologies undergirding the prevailing industrial, commercial, and residential activities; the transportation and distribution systems; and utility infrastructures that linked them all together.

At the smallest scale, the majority of the firms (57.7 percent) occupy leased space in a multistory office complex; 23.1 percent occupy space in a single-story office complex; and 11.5 percent occupy a building they own. That this stratum underrepresents the legions of "mom and pop" software producers is revealed by the fact that only 1.9 percent of the software firms are housed in a converted portion of a residence (Figure 3.5). A more comprehensive census of such software production settings undoubtedly would have uncovered a substantial proportion of enterprises in such home locations. The significance of such a proportion is tied to the fact that this "cottage" stratum of the industry is

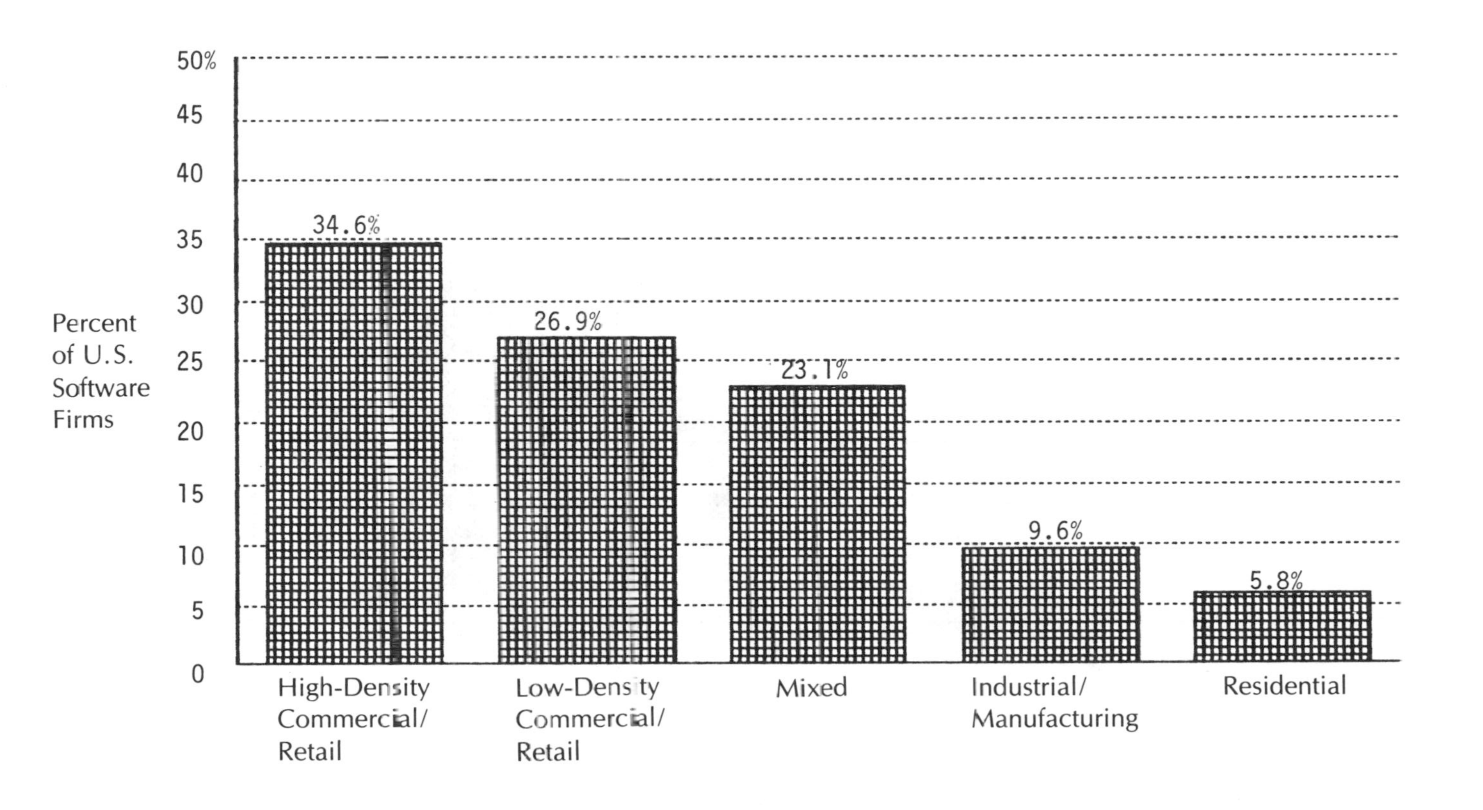

Figure 3.4. *Location of U.S. computer software firms.*

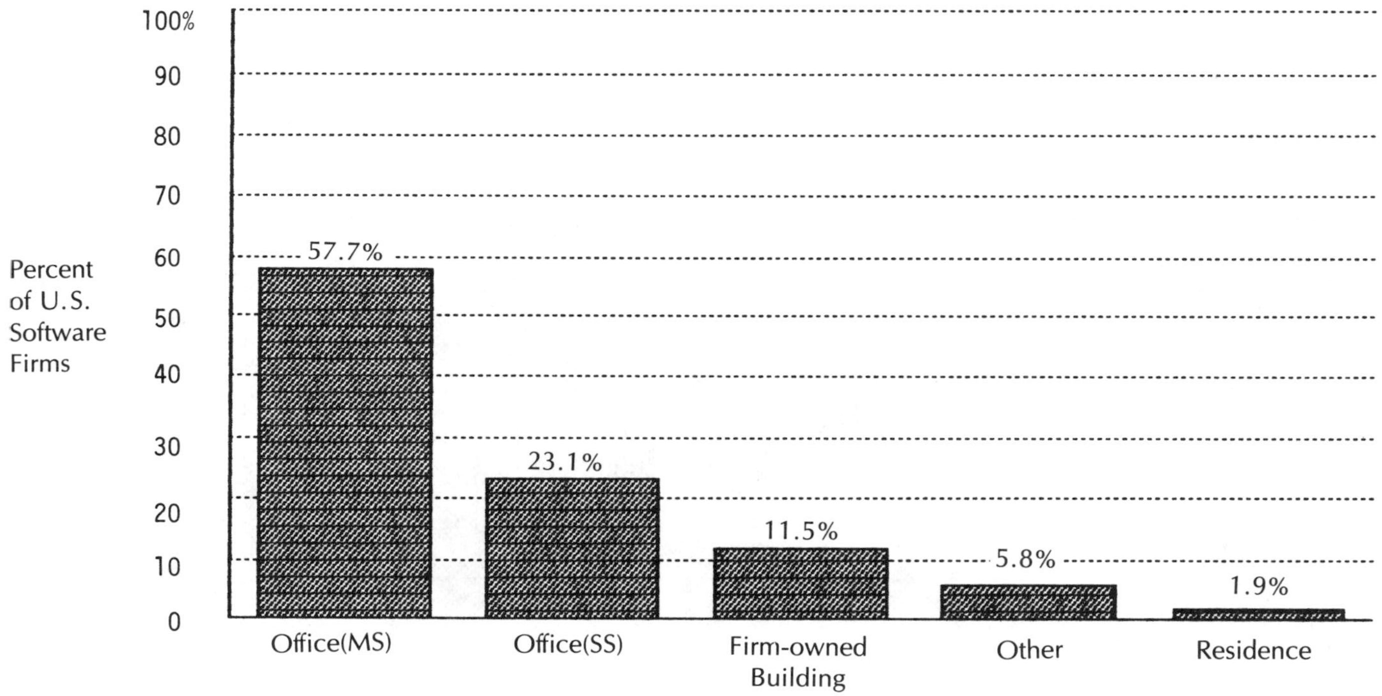

Percent of U.S. Software Firms

100%
90
80
70
60
50
40
30
20
10
0

57.7%
23.1%
11.5%
5.8%
1.9%

Office(MS)
Office(SS)
Firm-owned Building
Other
Residence

Office(MS) = Leased space in a multistory complex
Office(SS) = Leased space in a single-story office complex
Residence = A converted portion of a residence

Figure 3.5. *Physical structure in which firm is located.*

rendered nearly invisible not only because it is composed of the smallest establishments but also because the separation of workplace and home-place which has long characterized industrial societies has been rendered unnecessary as much by the technology of the production process and eventual product as by personal preference.

To the extent that increasingly large portions of the economic activity of an advanced industrial economy involve the manipulation and distribution of information rather than physical materials and can thereby accommodate themselves to locations that eliminate the necessity of job-related commuting, we can expect a substantial restructuring of our settlement patterns. The "plastic" nature of settlement patterns and their ability to reflect prevailing technologies of production, transportation, communication, and consumption expand the influence of non-economic residential preferences. While an industrial economy cannot be considered to sculpt entirely the settlement patterns that accompany it, it can be seen to establish the parameters within which noneconomic influences operate—influences such as cultural values and social preferences (like single-family dwellings, racially or class-segregated housing patterns, and so on). This relationship between place and production has major implications for land use patterns that characterize an advanced industrial economy. These and other "place"-related patterns will also be the subject of a later chapter.

Physical and Human Capital Requirements: The software industry has registered its influence on the larger economy not only through its rapid growth rate and increasing relevance to productive activities throughout the goods and services economies, but also through its characteristic capital requirements. The spatial organization of the industry is characterized by the potential for relatively high value-added derived from very little actual physical space (Figure 3.6). Given the special nature of the industry's output, the production setting does not have extensive space requirements. There is little essential physical capital to accommodate, with the exception of the hardware necessary for software creation. Bulky inventories of either raw materials or finished products are likewise unnecessary. Consequently, the space utilization of the industry appears to be modest indeed.[58]

A principal influence on space needs is the employment size of the company. One in twelve (8.0 percent) companies report having only one full-time employee working on the premises. Half (48.0 percent) report having a half-dozen or fewer employees, and three-fourths (76.0 percent) have twenty or fewer employees working on the premises. The service economy is generally characterized by significant part-time employment.[59] However, the majority (58.8 percent) of the software com-

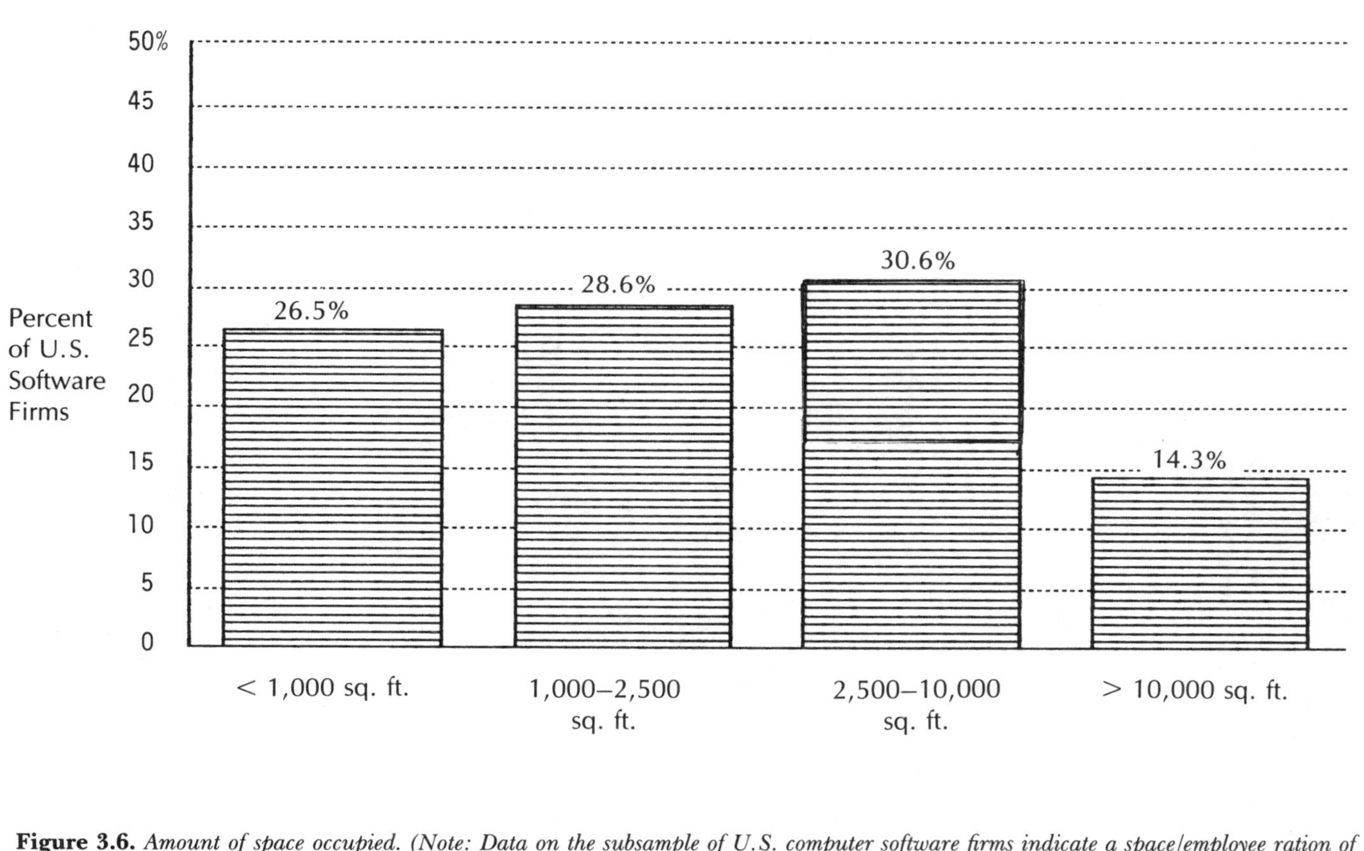

Figure 3.6. *Amount of space occupied. (Note: Data on the subsample of U.S. computer software firms indicate a space/employee ration of 362.4 sq. ft. per employee.)*

panies in this data set have no part time employees; among those who do, 61.9 percent have only one or two such workers.

In addition to the weak locational constraints on the development of this industry per se, there are also weak locational constraints on the work activity itself. Despite their small size, the majority (61.5 percent) of the software firms employ people such as independent consultants, outside contractors, free-lance programmers, and software authors who do not work on the premises. As a result, the total production of even these essentially small companies can be supported by still less physical space.

Business Climate and Locational Factors: Software production in the United States is highly dependent on its human-capital resources. It has been said that the principal assets of software firms walk out the door and go home every night! Consequently, land (even understood in the fairly narrow sense of mere location) and other physical capital can be expected to place relatively minor constraints on the overall production process. Rather, the essential human capital embodied in a range of technical skills and proficiencies looms as the most important production input.

The importance of the workforce's access to the workplace and of the company's access to appropriately skilled workers emerge as major locational factors governing a firm's preference for a particular location. Three-quarters (75.0 percent) of the firms rated as significant or very significant ease of employee access to the worksite, while 66.7 percent and 63.5 percent, respectively, placed great importance on access to a good highway transportation system and on the availability of a potential pool of labor skills (Figure 3.7). These contextual features are important largely because they can be influential in attracting and retaining the specialized workforce required by these firms as they grow. The other locational factors that at least a majority of the firms rated as significant included access to a major airport (60.0 percent), proximity to customers and markets (59.6 percent), room for expansion at the present site (53.9 percent) and in the general location (53.0 percent), and an entrepreneurial climate (51.9 percent). In contrast, the relative indifference of these companies to location understood as specific parcels of land is partly revealed in the fact that the cost of land was rated as significant by only 29.4 percent of the firms.

Viewed as of lesser importance are other locational factors generally considered to be subject to the influence and manipulation of local governments trying to attract and retain business start-ups and secondary expansions in-place. Such factors as business-tax climate (48.0 percent), cost of living (47.0 percent), the cost and availability of energy

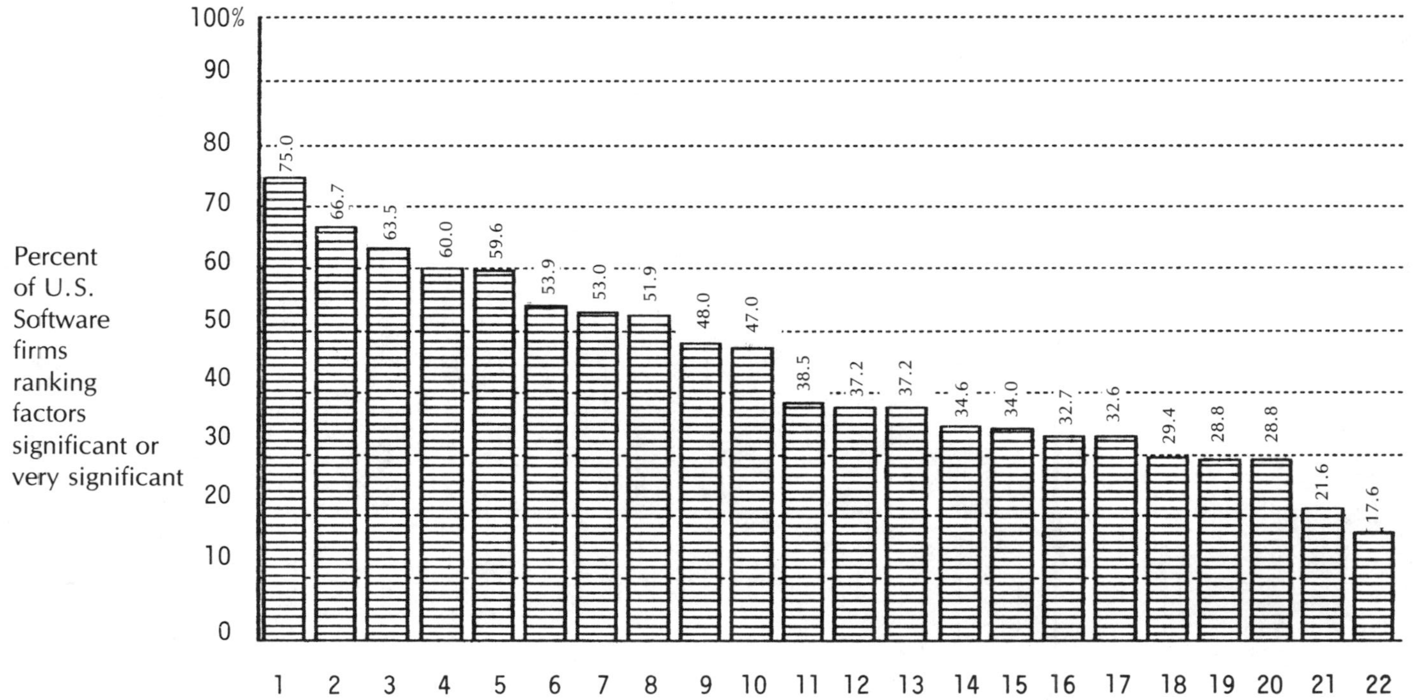
Percent of U.S. Software firms ranking factors significant or very significant
100%
90
80
70
60
50
40
30
20
10
0
75.0
66.7
63.5
60.0
59.6
53.9
53.0
51.9
48.0
47.0
38.5
37.2
37.2
34.6
34.0
32.7
32.6
29.4
28.8
28.8
21.6
17.6
1
2
3
4
5
6
7
8
9
10
11
12
13
14
15
16
17
18
19
20
21
22

1. Easy access for employees (75.0%)
2. Access to good highway system (66.7%)
3. Labor skills/availability (63.5%)
4. Access to major airport (60.0%)
5. Being close to customers/markets (59.6%)
6. Room for expansion at present site (53.9%)
7. Room for expansion in locality (53.0%)
8. Entrepreneurial climate (51.9%)
9. Business tax climate (48.0%)
10. Cost of living (47.0%)
11. Good mailing address (38.5%)
12. Energy availability/cost (37.2%)
13. Local regulation on business (37.2%)
14. Proximity to good universities (34.6%)
15. Quality of municipal/services (34.0%)
16. Climate (32.7%)
17. Proximity to good public schools (32.6%)
18. Cost of land (29.4%)
19. Proximity to other business (e.g., suppliers) (28.8%)
20. Recreational/leisure facilities (28.8%)
    (e.g., restaurants, sports, etc.)
21. Cultural amenities (21.6%)
    (e.g., concerts, museums, etc.)
22. Local public transportation (17.6%)

Figure 3.7. *Factors influencing company location.*

(37.2 percent), and local regulations on business (37.2 percent) emerged as only of moderate significance. Moreover, the proximity to good public schools (32.6 percent), cultural amenities (21.6 percent), the availability of local public transportation (17.6 percent), and the quality of municipal services (34.0 percent) are judged to be of even less significance. In part, these perceptions may well reflect the increasing willingness of owners, managers, and employees in such an industry to retreat from traditional arrangements and to use privatization strategies in education, transportation, and other services that through this century have gradually come to be viewed as public responsibilities.

The Significance of Quality Institutions of Higher Education: Recently, much importance has been assigned to the presence of high-quality institutions of higher education for nurturing and sustaining technology/knowledge-intensive business development. It has been presumed that the range of potential benefits includes access to a continually replenished pool of graduates, libraries and information systems, laboratories and computer centers, faculty consultants and their research activity, and university-based cultural activities. Other benefits, including degree programs and part-time teaching opportunities, may be of greater importance to employees in their nonworking hours. A separate question was asked of respondents concerning the importance of having high-quality universities nearby. Fully 78.9 percent of software firms in this nationwide study responded that proximity to such a university was *not* a factor in locating their business. While the presence of good universities may be an important feature of the larger-scale (for example, the local or regional) environment, it does not appear to have the capacity to influence software firms to huddle particularly close by.

The Influence of Land Use Regulations: We saw above that some of the principal features of the local environment that could be influenced by public policies were ones toward which software firms were largely indifferent. One major dimension of local governance involves the placing of restrictions on land and the uses to which it is put. Excluded here are the patterns of tax incidence that can be capitalized into land values and thereby influence spatial patterns of industrial development. Here again, however, such local regulations do not appear to be perceived as very consequential. The majority (59.6 percent) of the firms regarded state and local regulations as having an insignificant impact on business location plans and decisions. Whether focusing on zoning practices, building permit procedures, building codes, filing and inspection procedures, or general environmental restrictions, in no single case did more than 15 percent of the software firms report these to be of any significance.

Conclusion

In this chapter, two industries at very different stages of maturity were selected to illustrate our nation's industrial responsiveness to changing economic conditions. Despite their differing industrial life-cycle locations, both the steel and software industries revealed the capacity for adjustment through complex patterns of differential growth across distinct segments of the respective industries. While the importance of steel to our industrial future is undeniably being eclipsed by the rise of new industries like software, the capacity for restructuring *within* the two industries is judged to be more important to understanding advanced industrial development than is any larger-scale restructuring unfolding between them. Taken together, both the old steel and the new software industries can be seen to have considerable resiliency in the face of changing background conditions in the larger economy. Ultimately, the prosperity of one does not come from the demise of the other as though some industrial version of natural selection plays itself out. The possibility exists that to the limits provided by existing demand, both industries can thrive, although in forms and arrangements that are not historically familiar.

As we have seen, technology plays an important role in industrial change either by permitting the relatively rapid growth of selected industry segments or by spawning entirely new industries. Our industrial future will be defined by a continuing shift of capital investment and consequent growth within existing industries toward higher value-added production. Advanced production technology, then, becomes the means by which science, knowledge, information, sophisticated labor skills, and new corporate organization add value to goods and services.

The intraindustry shifts from sprawling older production settings to newer and more compact mini-mills, from blast furnaces to electric furnaces, from iron ore to scrap as raw material, and from union to nonunion work environments define the steady adjustment process underway in the steel industry. As a result, the steel industry is seen to have fragmented into at least three distinct segments. As the market, trade, labor relations, domestic political, demographic, and technological conditions defining the economic climate that long sustained a massive and monolithic steel industry began to change, the new conditions seemed to favor the gradual ascendance of previously insignificant segments within the industry. The new electric and mini-mill sector, characterized by lower costs for labor, raw materials, energy, and distribution than the older and larger integrated mills, has emerged as a far more important component of the overall industry than ever before. Ultimately, it is the uneven patterns of such growth that continue to take place within the industry, rather than ways in which an entire industry contracts, that interest us here. Moreover, the cyclical recovery of the mid-1980s not only reveals what appears to be a permanent structural shift within the industry; it also illustrates the way in which recessions nurture certain kinds of industrial growth while they retard others.

Likewise, the infant software industry was selected to illustrate the ways in which a new industry is spawned and accommodates itself to existing industrial/urban circumstances. The rapid growth of the once-small software products industry, the nontraditional ways in which it is tied to physical and human capital, and the rise of new steering factors that influence its locational features reveal a similar resilience and capacity for adjustment. In the end, the birth of new industries, combined with uneven and selective adjustment in older industries, provides the driving force behind the emergence of a new industrial structure.

All in all, it is unwise to assume that an advanced industrial economy will be composed largely of infant industries that have been spawned from the commercialization of new technologies. Certainly this view follows easily from the imagery that older basic industries are being eclipsed by entirely new ones. Commonly overlooked, however, has been the possibility that older industries can restore their competitiveness with a transfusion of advanced production technology and an orientation toward new technologically sophisticated products as illustrated by the diffusion of computer-based production technologies throughout the U.S. metalworking sector. It is a mistake to overlook or discount the possibility that the gradual diffusion of new technologies through older basic industries holds the promise of industrial renewal. It is by this route that older industries can be expected to retain or recapture key roles in an advanced industrial economy.

4

Advanced Urban Development and the Rise of Post-Industrial Places

For most of the industrial era, employment and population have tended to gravitate to common locations, with increasingly large and dense populations becoming interspersed throughout growing and developing industrial regions. Until the twentieth century, open space could not be used to expand the scale or diminish the high densities inherent in early urban development. Technologies that one day would permit economic activity to spread out—especially at smaller scales—only gradually became available, and until the twentieth century, they would not be used to augment social distance with physical distance. Instead, state-of-the-art production and linkage—transportation and communication—technologies placed effective physical bounds on the scale of social and economic activity. The technologies undergirding early industrial development, therefore, set rather narrow time and scale limits on urban development. As a result, the industrial-era city has always reflected primarily these technological imperatives and only secondarily the social preferences made possible by them.[1]

This chapter is devoted to exploring the key features of the new urban economic landscape and the emergence of a post industrial urban system that has accompanied advanced industrial development.[2] I begin by examining the patterns of deconcentration and dispersal that have had the effect of unraveling industrial-urban development and the settlement patterns to which we have grown so accustomed. In the process, we are invited to develop new standards by which to assess and respond to those forces that are hastening the passing of older and outmoded urban arrangements. This chapter proceeds by focusing primarily on the redistribution of population and employment at three spatial scales—among multistate regions, in and around central cities, and between metropolitan and nonmetropolitan areas. Since conventional economic-base analysis commonly assigns greatest importance to manufacturing employment, its role in advanced industrial development receives priority here. The extent to which new services can substitute for manufacturing also is addressed.[3] Ultimately, I intend to draw attention to those spatial features of our industrial past that are being altered as well as to those that have endured as advanced industrial development has proceeded.

I then sketch the outlines of the new post-industrial city and suggest how it differs from its industrial-era predecessor. The changes are seen best in the aggregate of urban places, since the process is far from uniform across individual localities.[4] A new conception of "place" that accompanies advanced industrial development is also explored. Industrial-era cities do not so much die as evolve into new social and economic entities to enable them to assume new roles and play old ones better in an advanced urban system. To illustrate this urban transformation in more detail, I explore the ways in which one major U.S. city—Dallas, Texas—has developed in recent years in order to play a new set of roles in the regional, national, and international economies. Part of that process has involved a shift toward high-technology manufacturing and high-skill/wage service industries. The general patterns of high-technology growth and the development characteristics of a specific high-technology service sector—computer software and data processing services—in and around Dallas will be explored in the next chapter.

Scale Adjustments for a New Urban Economy

Throughout the nineteenth century and well into the twentieth, the developing industrial status of the United States was tied to an arc of factory towns and urban-oriented agricultural hinterlands stretching south from New England down through the Middle Atlantic states, and then west below the Great Lakes through the upper Midwest. This industrial heartland constituted a core region to which all else was supportive and peripheral. The South's distinctly subordinate agricultural base and rural dominance reserved for that region the role of an economic backwater, and the area west of the Mississippi River was essentially underdeveloped and therefore inferior even to the South. By 1900, all parts of the nation had been settled and were developing, albeit far from uniformly. Today, our advanced industrial society and economy are characterized by an interregional distribution of people and jobs that differs markedly from what it was even a few short decades ago.

Long-term industrial restructuring has likewise had a distinct local dimension. Unlike that of Europe, the industrial development of the United States proceeded relatively unobstructed by entrenched pre-industrial settlement patterns. This nation's industrial-era cities grew rapidly, and their forms reflected their functions—compact clusters of factory-based jobs with workers' residences huddled close by. It was into these dense industrial arrangements of people and jobs that successive waves of displaced rural, Southern, and European immigrants were drawn and distributed. Today, both the form and function of our industrial-era cities have changed so substantially that a case can be made that the dynamics of advanced industrial development seem largely capable of sculpting the settlement patterns they require.

A redistribution of people and jobs between and within regions has accompanied advanced industrial development. The nation became predominantly urban by 1920, metropolitan by 1950, and suburban by the early 1970s. In the process, the scale of social and economic activity expanded from local to metropolitan-regional; only political activity retained its smaller scale and continued to abide essentially fixed and traditional local jurisdictional boundaries. This mismatch between the increasing locational flexibility of population and industry and the jurisdictional inertia of local political arrangements lies at the heart of those developments to which urban economic development policies—including national urban policies—have been a collective policy response. While the industrial era is commonly equated with a structural shift of people, jobs, capital, and opportunity from the farm to the factory, this chapter will focus on the locational aspects of advanced industrial development and the resulting spatial features of the deconcentration of people and jobs from industrial-urban settlements over the past half-century.

Deconcentration and Dispersal: Resettling Industrial-Urban America

It is interesting, at this point, to consider the degree to which industry concentrated in the United States, and how this concentration has changed. In 1929 there were 33 industrial or metropolitan areas in the United States. These areas embraced 97 counties, only three percent of all the counties in the country. Yet in those tiny parts of the nation were concentrated well over one-third of the population, and more than one-half of the industrial wage jobs. In only a few instances did sources of raw materials have an important bearing on this condition. But this concentration of wage jobs had remained unchanged for 31 years, [since 1898], in spite of the fact that these 33 industrial areas had more than doubled in population. In the same period, the nation's population had increased by sixty percent. In other words, for thirty years prior to 1929, the number of industrial wage jobs had been steadily increasing *in the smaller communities*, at a rate greatly in excess of that in the

> industrial areas. *Industrial decentralization, of which we hear so much, has been underway, now, for nearly forty years!*[5]

It was inevitable that the post–World War II period would spawn countless challenges to prevailing industrial-urban arrangements. During this period, several growing trends that reversed the concentration dynamics of the early industrial era finally began to register net impacts at three distinct spatial scales. First, at the largest scale, the interregional shift of population and jobs toward the West and later the South has continued all through the past two centuries. Shortly after World War II, however, the older industrial Northeast and Midwest began failing to compensate for those continuing losses, and slow relative and, later, absolute population declines commenced. Second, at the smallest scale, the suburbanization of the population had begun shortly after the Civil War; the slow movement of first manufacturing, then retail and other employment was underway by the turn of the century. Yet, industrial-era cities continued to grow until the 1950s, when for the first time many of the largest and oldest of them began to experience significant population and employment loss. Finally, the deconcentration of entire metropolitan areas began in the 1950s, although the net shifts leading to actual population declines in selected metropolitan areas and the faster growth of nonmetropolitan areas were not registered until the 1970s.

In the past decade, this nation's industrial-urban system has come to be characterized by a multiscale deconcentration of population that has not been evident for more than a century and a half.[6] By the mid-1970s, growth in nonmetropolitan areas exceeded that of metropolitan areas, thereby reversing the only major population dynamic that continued to reinforce population concentration in the twentieth century. With this shift, the demographic trends came to mirror those that had been governing industrial development and the location of manufacturing for many years. The dispersion of both population and manufacturing growth at all spatial scales both within and between multistate regions raised speculation that the new metropolitan scale of advanced industrial economies, like the urban scale tied to central cities before it, could not long be sustained. Was the capacity of the metropolitan area to contain industrial development, like that of the city before it, diminishing as a result of the continued deconcentration of people and jobs? Were regional-metropolitan economies destined to spread out indefinitely? The remainder of this section is devoted to tracing in greater detail the shifts in population and manufacturing employment that have unfolded to accompany advanced industrial-urban development.

Interregional Development Patterns: Growth and Convergence

Industrial dispersion of people and manufacturing jobs and the deconcentration of older industrial arrangements are clearly evident at the scale of multistate regions. As Figure 4.1 illustrates, over the half-century since 1929, the capacity to produce has been steadily and significantly redistributed away from the Northeast and Midwest and toward the South and West. By 1980, the major regions of the United States had all but converged with respect to regional shares of U.S. output. Until recently, even though all regions experienced continuous population and manufacturing growth, that growth proceeded at differential rates. In the past decade, however, relative declines became absolute ones as broad areas throughout the Northeast and North Central regions experienced actual contraction of population and manufacturing.[7] As the disparities in growth rates continued to widen—especially during the

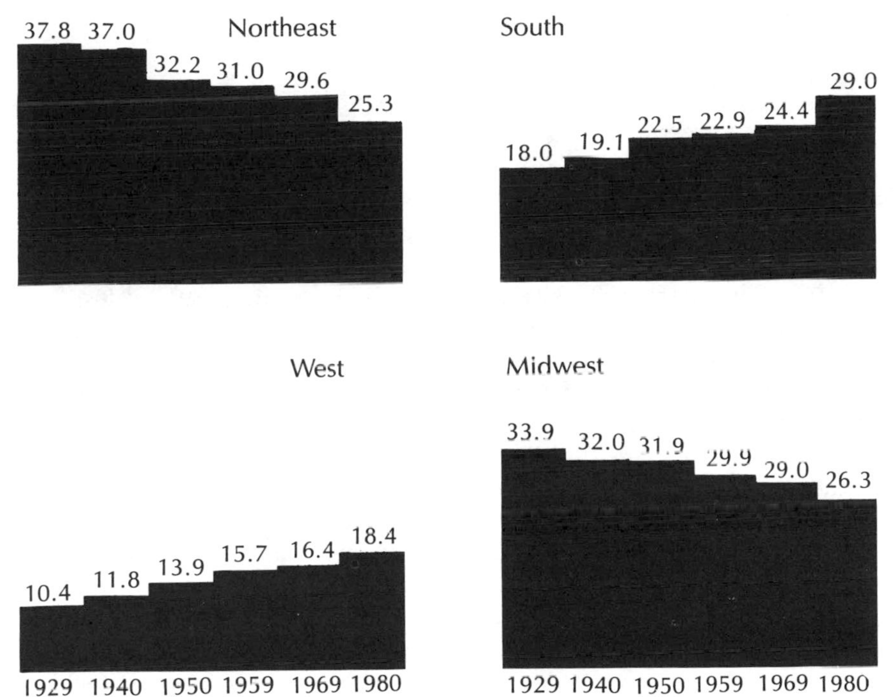

Figure 4.1. *Regional shares of U.S. output, 1929–1980 (percent). Data for Alaska and Hawaii are not included in regional totals. (Source: U.S. Department of Commerce; adapted from Office of the U.S. Trade Representative,* Annual Report of the President of the United States on the Trade Agreements Program, 1983 */Washington, D.C.: USGPO, April 1984/, p. 26.)*

recent economic recovery—the spatial features of the U.S. economy appear to have undergone major rearrangement. Let us explore the extent of these shifts by examining the interregional redistribution of population and manufacturing employment.

Interregional Population Redistribution: The dominant population dynamic throughout U.S. history has been the shift of population from east to west during the nineteenth and twentieth centuries. Against this backdrop, other major interregional shifts have risen and spent themselves. As the twentieth century began, the net migration of population from the rural South to the industrialized cities of the North was set into motion by the increasing mechanization of farming, the consequent labor surpluses accumulating throughout the rural economy, and the rising labor demands associated with the growth of a city-based factory economy. Nonetheless, the exodus from the rural South had run its course by the 1970s, by which time a reverse trend began to take shape. In recent years, the industrializing South has begun to retain its native population and even attract return migrants and newcomers from other regions.

The South and the Northeast had equal population shares between 1790 and 1820. Over the next century and a half, the South's population grew at a faster rate until, in 1980, the South had a substantially larger population (75.3 million) than the Northeast (49.1 million).[8] The North Central (Midwest) region overtook the South as the most populous region by 1870 and held that distinction well into the twentieth century before relinquishing it back to the South by 1940. Since 1940, despite net growth in all regions, the South has steadily increased its population dominance, and by 1980 fully a third (33.2 percent) of the nation's people resided in the South. Slightly more than a fifth (21.7 percent) of the nation resided in the Northeast, 26.0 percent in the North Central region, and 19.1 percent in the West. While the massive proportions of the interregional redistribution of population during the 1970s were unanticipated, the net shift of population to the South and West is expected to continue, if at a reduced rate, well into the next century.

Residential Mobility and Migration—A Fading Option?: The residential mobility of Americans is commonly thought to be on the decline; however, that conclusion is not substantiated by recent data. On balance, it is likely that long-distance migration can be expected to remain relatively robust. While one family in five has relocated annually since the 1960s, that rate did indeed drop to one in six (16.6 percent) by 1981–1982.[9] However, while relatively short-distance mobility within counties declined from 13.7 percent to 10.3 percent, long-distance intercounty migration rates actually stayed stable at 6 percent. This suggests that while short-distance relocations, generally thought to express upward mobility without a job change, may be dampened, longer-distance moves in search of new employment opportunities may remain as attractive and viable as before.

Similarly, there is growing speculation that the mobility of modern households seems to be constrained by the difficulty of renegotiating new mortgages in an unstable housing market, the rise of two-earner households, and the general aging of the population. Rises in housing prices, homeownership, and interest rates may explain part of the decline in short-distance mobility, and the gradual aging of the population has increased the proportion of Americans who are in lifestages presumed to be associated with relatively lower mobility rates.[10] However, while Current Population Survey data indicate that household mobility declined slightly between 1962 and 1976, data from the Panel Study of Income Dynamics and Annual Housing surveys report rather sharp increases in mobility since the late 1960s. There is little in the data on declining birth rates, delayed marriage and childrearing, and the rising proportion of single-person households that suggests that the mobility of U.S. households must necessarily decline in the future simply because of a gradual aging of the population. After all, lifestage events and social correlates of age, not age per se, have been most closely associated with changing mobility patterns.

It appears, then, that the exercise of a mobility option by many U.S. households can be expected to continue to play an important role in the reconfiguration of local, metropolitan, and regional economies. This is true both for households seeking jobs in new locations and for those retiring to new locations. While business-cycle downturns can dampen moves to new residences within the same location, under certain circumstances they may also encourage longer-distance moves to locations that are believed to offer greater opportunity. Historically, this mobility has taken place both within and between generations—that is, by the independent relocation of workers and their children, both in search of their fortunes in locations distant from where they were raised. Often, however, a mobility option is not equally available to all households for reasons relating to household composition, financial constraints, and lack of information about residential and employment possibilities in other locations.[11]

Recently, arguments have been advanced to broaden the traditional focus of national urban policy in ways that would strengthen and extend the mobility option. Conventional local economic development (jobs-to-places) efforts would be more conscientiously augmented by a relocation policy focus.[12] Policy shifts of this kind, however, have met with considerable resistance and have been frequently criticized as veiled attempts to "write off" distressed local or regional economies while overestimating the presumed benefits of a mobility/market-based strategy of economic adjustment. Clark, for example, offers an interpretation of recent interregional migration processes that views any such urban policy shift as both inefficient and unfair.[13] It remains, however, that residential mobility will most likely continue to be an important element in advanced industrial development as the economies of localities and entire regions continue to evolve. Any future urban policy formulations will continue to face the need to devise a new conception of local and regional economic development that protects and extends the ability of households to be mobile at the same time that it seeks to stabilize localized economies.

Interregional Shifts in Manufacturing Employment: The slowing growth of U.S. manufacturing employment tends to amplify the interregional shifts that have accompanied it. Indeed, the manufacturing share of total employment in the United States declined from 23.0 percent in 1980 to 20.9 percent in 1984, although cyclical effects account for a major part of this steep decline. Through most of this century, the bulk of the nation's industrial production has been anchored to the Northeast. In 1940, nearly three-quarters of the nation's manufacturing jobs were located in the industrial heartland with 39.6 percent in the Northeast and 32.8 percent in the North Central region. The South and West were still peripheral to the nation's industrial status in both a structural and geographic sense, among others. Since then, the spatial characteristics of U.S. manufacturing at the regional level have been altered considerably. By 1984, and after a quarter-century of lagging growth rates in manufacturing employment throughout the industrial heartland, the South led the nation in share of manufacturing employment with 30.9 percent. The North Central region had 28.6 percent, while the Northeast had only 24.3 percent. The West trailed with 16.2 percent.[14]

Differential growth rates for manufacturing have led to distinct patterns of industrial development across regions. Gradually, the gaps in the industrial structure among regions have narrowed. Between 1940 and 1960, the manufacturing share of total employment grew slowly and steadily in all regions (Figure 4.2). The regional economy of the Northeast remained dominated by manufacturing—roughly a third of all nonagricultural employment—all through the 1960s. Manufacturing employment constituted far smaller proportions of total employment in the South and West. However, after 1960, a relatively rapid process of convergence began to alter this regional pattern. A quarter-century

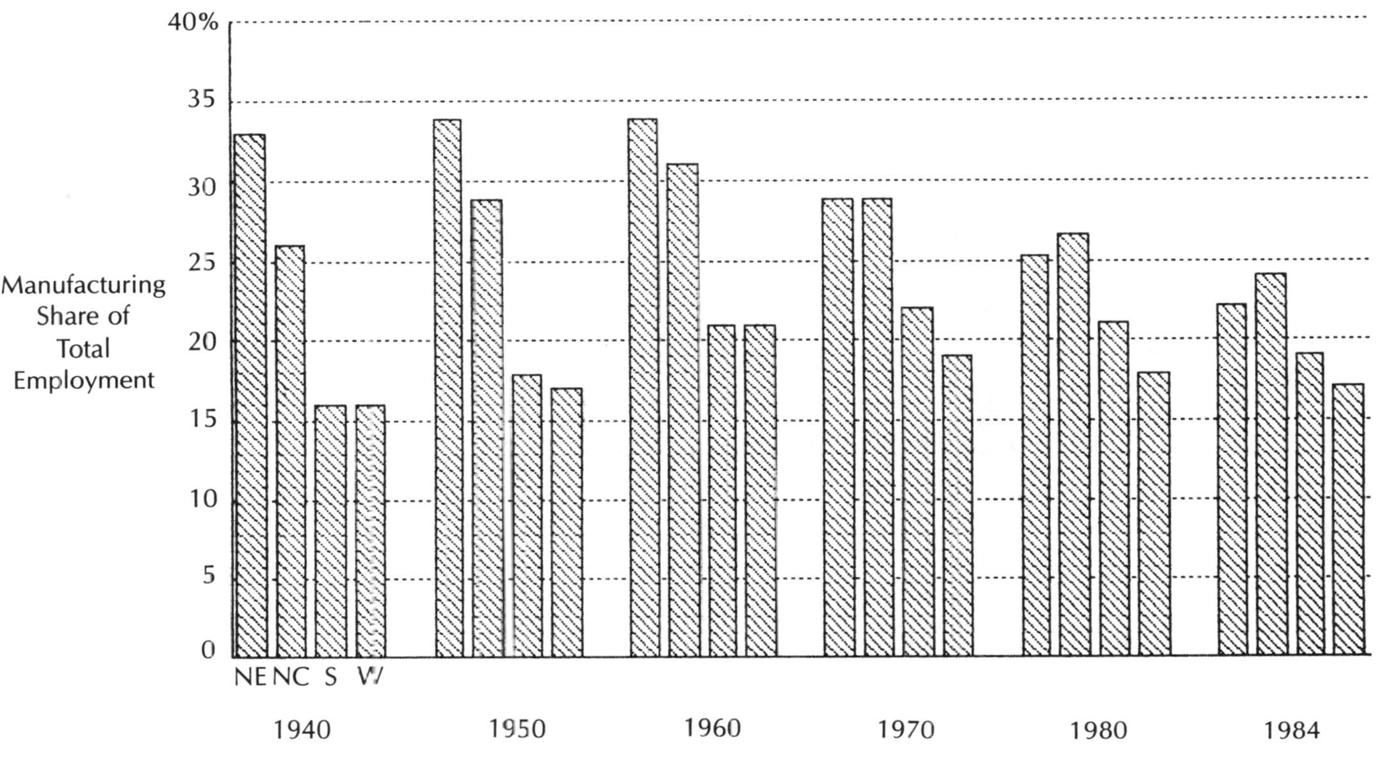

Figure 4.2. *Shifts in manufacturing share of total U.S. employment by region, 1940–1984. (Source: Bureau of Labor Statistics.)*

later, by 1984, manufacturing as a share of regional employment had declined dramatically to 22.2 percent in the Northeast and to 24.1 percent in the North Central states, while it constituted 19.2 percent of total employment in the South and 17.1 percent in the West.

As a result, the industrial structures of all regions have come to resemble each other to a greater extent than we have seen previously. Nevertheless, the Northeast and the North Central regions continue to be more dependent on manufacturing than either the South or the West. Figure 4.3 displays this convergence in industrial structure through the use of location quotients that compare each region's industrial structure to that of the nation's. It is apparent that in recent decades the impacts of industrial restructuring have been the greatest for the Northeast and the South. On the surface, the restructuring of the Northeast's economy since 1940 looks to be a reverse-image of what unfolded in the South over the same period. The industrialized North Central region alone does not appear to have undergone any major aggregate industrial restructuring since 1950. However, while major aggregate shifts at the regional level are not apparent, beneath the surface there has been considerable spatial repositioning of manufacturing, as we shall see below.

Corporate Restructuring and Capital Shifts: As I discussed in the previous chapter, a number of debates have been triggered by these patterns of regional change. A major one focuses on new forms of corporate organization that have accompanied the shift of manufacturing to once peripheral areas. High-velocity capital mobility and the managerial strategies that promote it have drawn attention to the expanded scale of industrial production. Bluestone and Harrison cite evidence indicating that the spatial deconcentration of manufacturing activities masks a more consequential set of dynamics moving in the opposite direction.[15] They draw attention to the actual and potential threats posed by increased centralization of control over capital, industrial concentration within major industries, and greater resort to acquisition and merger by corporate conglomerates. While such corporate adjustments may be viewed by some as necessary to compete in a new global economy, Bluestone and Harrison contend that the domestic repercussions on communities and households may be too steep a price to pay. In any event, there remains the view that the capital mobility which has made possible a portion of the shift of manufacturing growth to peripheral areas is accompanied by domestic costs borne within regions that are not fully compensated for by an improved national industrial capacity.

On a related matter, others have suggested that manufacturing

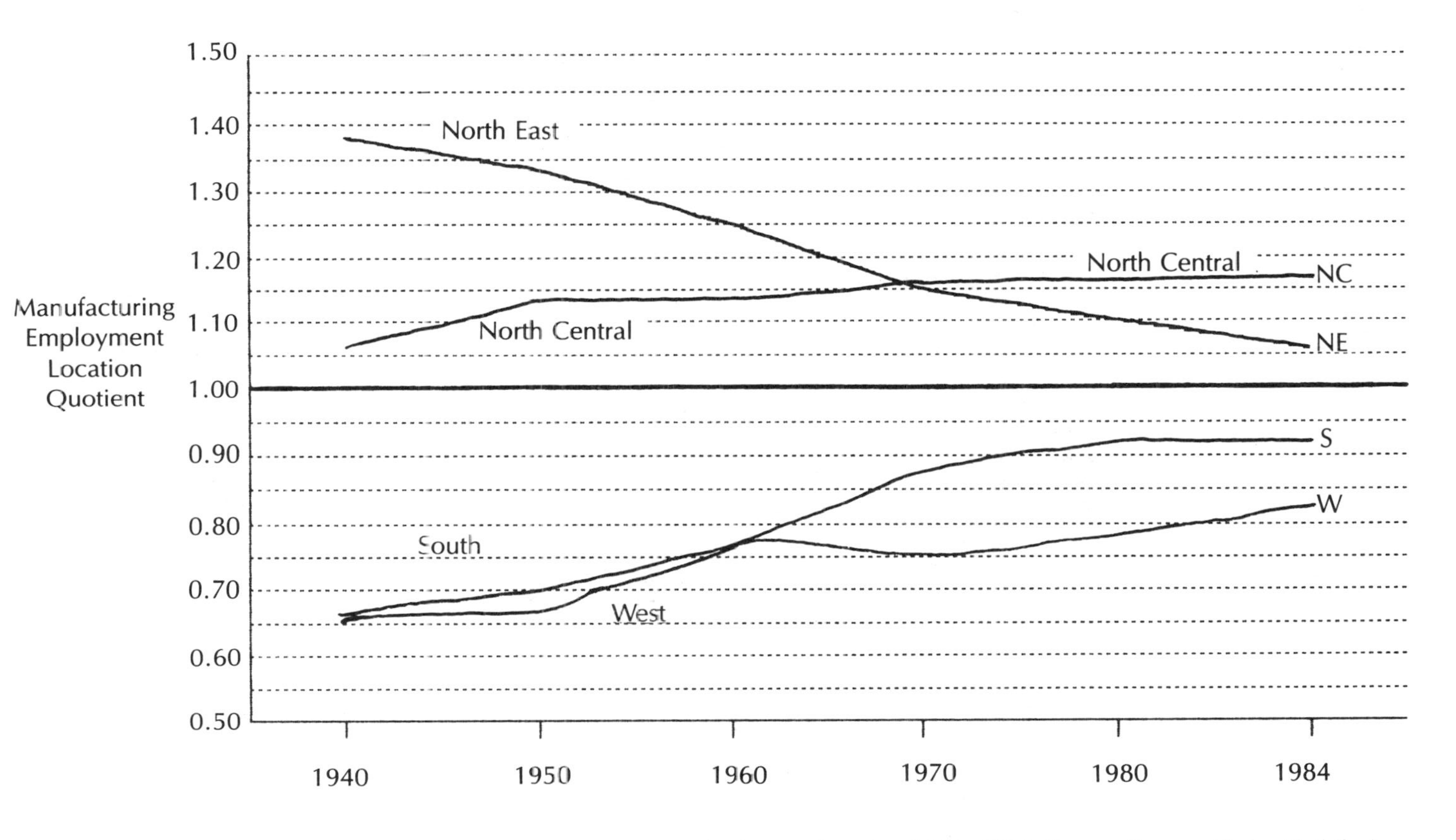

Figure 4.3. *Regional convergence and the spatial restructuring of U.S. manufacturing, 1940–1984. (Source: Bureau of Labor Statistics.)*

growth in peripheral regions like the South, West, or nonmetropolitan areas is tied to the branching or acquisition activity of larger companies in the industrial heartland, as opposed to new business formations, and to secondary expansions in-place in the South or West. In this view, the subordination of the periphery, rather than being diminished, has simply entered a new phase.[16] Thwaites has noted that plant mobility to peripheral locations in Britain cannot ensure plants' industrial salvation since the innovation and implementation of new technologies needed to sustain and improve industrial capacity face higher barriers in peripheral locations.[17] Hansen likewise does not view the shift of manufacturing growth to the South as sufficient to ensure sound industrial development within the region.[18] Headquarters plants would still exercise external control, albeit from a greater distance. Rees, however, offers evidence of significant industrial development triggered internally, rather than being orchestrated from outside by large firms seeking to execute a "Southern strategy."[19] Yet concern lingers over the regional and local economies into which corporate branching has taken place. With an economic base more closely tied to manufacturing, cyclical sensitivity increases, and branch plants may be vulnerable to layoffs and even plant closings. While this can be true, Beale notes that locally owned plants may be even less able to bear the brunt of cyclical downturns and, moreover, that branch plants—wherever they may be located—may act as a shock absorber for newly industrialized areas.[20] New and often small business, although it may be the result of indigenous business growth, typically expresses its vulnerability to cyclical shifts by high rates of death, formation, and turnover.[21] Therefore, whether or not the spatial patterns of industrial change in recent decades constitute a net economic gain or loss to the South and West remains the subject of continuing debate.

Intraregional Shifts in Population and Employment

Despite the dramatic industrial restructuring that has taken place among regions in recent decades, the advanced industrial development of the nation may be better understood by examining population and employment redistribution and resulting settlement patterns *within* multistate regions rather than among them.

Deconcentration dynamics are evident in the increasing tendency throughout the twentieth century for our largest industrial cities first to grow more slowly than other places and then eventually to contract. Yet simultaneously, concentration trends were still evident on a larger scale as population and employment continued to cluster into sprawling metropolitan economies. As a result, the scale of advanced industrial development can be said to have shifted from urban-local to metropolitan-regional. Between 1920 and 1950, the typical industrial-era city was gradually displaced by the metropolitan area as the principal organizing unit of advanced industrial society.[22]

A Century of Residential Suburbanization Continues: The full sweep of the twentieth century has witnessed efforts by Americans to leave the country's largest urban settlements—particularly older industrial central cities. This restless population had been either displaced to the cities during the nineteenth century by the rapid industrialization of the predominantly agrarian economy, especially in the decades between the Civil War and World War I, or deposited there by successive waves of immigration from Europe. During the present century, upward social mobility has commonly been accompanied by residential mobility to locations increasingly distant from city centers. The dynamics of population dispersal and deconcentration—suburbanization—in the network of urban settlements throughout the nation were clearly evident by the 1880s.[23] The 1920 census was the first to report that the majority of Americans lived in urban areas, up from only 5 percent in 1790. Yet since 1950, settlements larger than 100,000 have steadily been declining in population share; this same trend has been evident since 1930 for settlements of over one million.[24] In the nation's fifty-six metropolitan areas with populations larger than 250,000 in 1980, the central city share of metropolitan population had declined in all but the largest (New York City) and the smallest (Virginia Beach, Va.) since 1970.

Despite the building and broadening of population deconcentration, the nation's industrial cities continued to grow ever larger until midcentury. It was not until the 1950s that suburbanization began to exact at first a relative and then an absolute toll on central cities. During the post–World War II period, our oldest and largest central cities actually began to contract dramatically.[25]

The separate contributions of cultural preferences and enabling technology in triggering and sustaining suburbanization are difficult to disentangle. The technology that permitted the evolution of local transportation arrangements, from the horse-drawn trolley after the Civil War to the automobile in the 1920s, had the effect of offering to the relatively affluent an expanded range of residential options that soon included residing outside a city at night while working in it during the day. Consequently, while industrial jobs continued to concentrate within urban areas between the Civil War and World War II, residential preferences that led to the decoupling of workplace and homeplace could increasingly be expressed.[26]

The Suburbanization of Manufacturing Employment: At the turn of the century, the vast majority of manufacturing employment was located in central cities. However, Fremon has noted that the suburbanization of manufacturing has been underway since at least 1900, and Pred has reported that manufacturing ceased expanding in large U.S. cities by 1919.[27] In 1905, 94 percent of the manufacturing establishments in the Cleveland area, 99 percent in Minneapolis–St. Paul, 96 percent in Baltimore, 95 percent in St. Louis, and 91 percent in Philadelphia were located in the central cities.[28] Over time, central cities gradually relinquished their hold on both population and manufacturing employment growth. By 1977, for example, Cleveland's central city share of metropolitan manufacturing employment had shrunk to 45.5 percent, Minneapolis–St. Paul's to 24.0 percent, Baltimore's to 43.9 percent, St. Louis's to 37.2 percent, and Philadelphia's to 34.9 percent. As suburbanization proceeded, population growth in suburbs generally preceded employment growth, especially that in manufacturing.[29] The rapid expansion of localized markets outside central cities and the reduced efficiencies tied to urban agglomeration are deemed to have played major roles in attracting manufacturing growth to the suburbs.[30]

After World War II, the redistribution of employment growth away from central cities became even more widespread. Among the nation's fifty-six metropolitan areas with populations larger than 250,000 in 1980, the central city share of manufacturing employment declined in forty-four of them between 1972 and 1977. The deconcentration of nonmanufacturing has accompanied that of population and manufacturing employment. Between 1972 and 1977, the central city share of metropolitan retail employment declined in forty-seven of the largest fifty-six metropolitan areas, while wholesale employment declined in fifty-six and selected services employment declined in forty-one.[31]

During the 1940–1970 period, the growth rates for manufacturing employment slowed dramatically for metropolitan places of all sizes, while increasing in the smallest rural places (Figure 4.4). By 1970, a clear inverse relationship had emerged between manufacturing employment growth and population size of place. Carlino has reported the same inverse relationship between employment growth and place size in both metropolitan and nonmetropolitan areas between 1969 and 1979, although metropolitan areas in the South show the opposite relationship.[32] Nevertheless, even though manufacturing *growth* generally has been shifting from the largest to the smallest places, *the overall structure of manufacturing employment by place size has not been appreciably altered.* As recently as 1970, just as in 1940, the vast majority of manufacturing employment was located in large metropolitan places of at least a quarter-million population (See Figure 4.5).

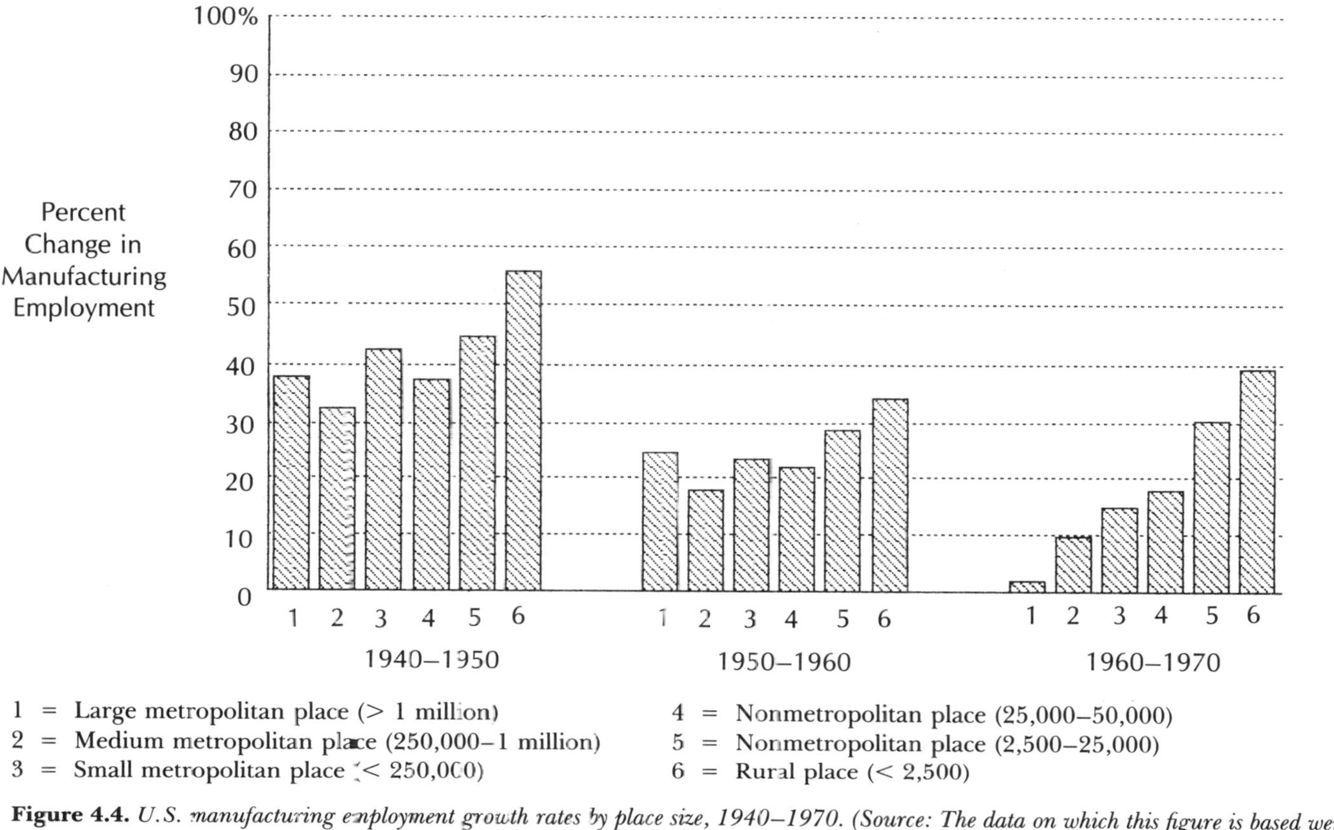

Figure 4.4. *U.S. manufacturing employment growth rates by place size, 1940–1970. (Source: The data on which this figure is based were originally compiled by Bahar Norris using Bureau of Economic Analysis files.)*

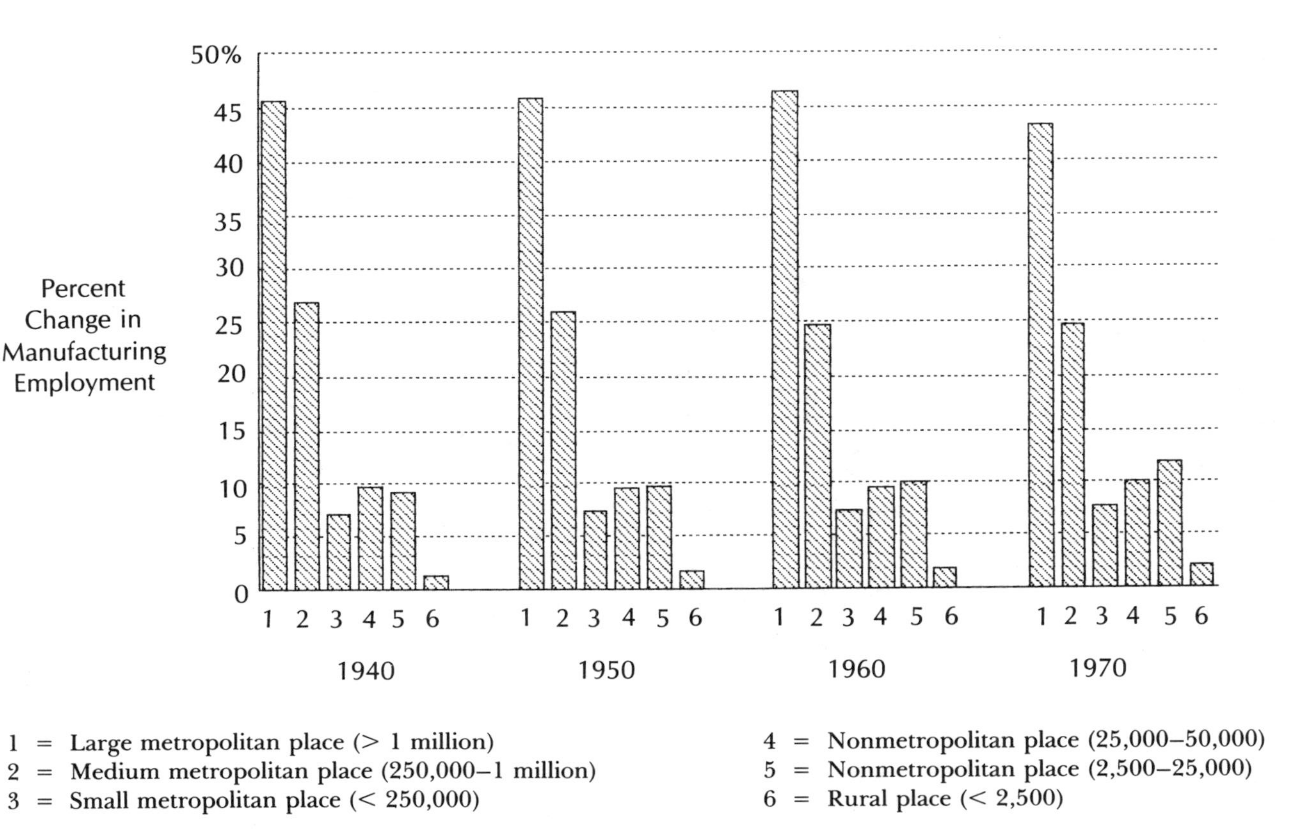

Figure 4.5. *Distribution of U.S. manufacturing employment by place size, 1940–1970. (Source: The data on which this figure is based were originally compiled by Bahar Norris using Bureau of Economic Analysis files.)*

Despite the fact that the central city no longer is the exclusive locus of the nation's manufacturing capacity and suburbs have eclipsed central cities as the principal residence of the nation's population, not all central cities have experienced continued population and employment loss. Population loss has often been accompanied by a resurgence of non-manufacturing employment growth that has hastened both the restructuring of central city industrial economies (and of suburbs too) and the redefinition of the socioeconomic and political roles played by them in the larger society.[33] Since the early 1960s, many large cities have ceased registering job losses. Vernez et al. discovered that of the 388 cities larger than 50,000 in 1975, absolute employment decline was evident in only 11 percent, those being older industrial cities. While the proportion of cities losing population had risen from 28.8 percent during the 1960s to 46.4 percent between 1970 and 1975, the proportion of cities losing jobs had declined from 21.1 percent to 10.1 percent over the same period.[34]

Post-Industrial Development and the New Metropolis

It was inevitable that the continued shifting outward of urban resources—capital, people, business growth, jobs, income, tax base, and political allegiences—would fundamentally transform selected older industrial central cities. Do larger-scale metropolitan economies face the same prospect? To what extent is the new metropolitan scale of industrial-urban organization itself vulnerable to being unraveled by continuing deconcentration and dispersion?

Population Dispersal from Metropolitan Areas: The expansion of urban areas by suburbanization set the stage for increasing metropolitanization. By the time the United States had become a predominantly urban nation in 1920, a system of metropolitan organization was already well defined. In that same year, roughly a third (34.0 percent) of the population lived in metropolitan areas, and a quarter of that (25.5 percent) lived outside central cities. Thirty years later, by 1950, the majority of the nation (56.1 percent) lived in metropolitan areas, with 41.5 percent of that population living outside central cities. After only another thirty years, by 1980, fully 72.8 percent of the population lived in metropolitan areas. Moreover, the bulk (59.6 percent) of that population lived outside central cities.[35] Indeed, America today is predominately a suburban nation, although many suburbs have by this time evolved away from their historical subordinate relationship to central cities. After 1970, more Americans lived in suburbs than in either central cities or nonmetropolitan/rural areas. Central city population actually declined in the United States after 1970, after having lagged behind overall U.S. population growth since the 1930s.

During the 1970s, population deconcentration continued to the point that it simultaneously enlarged the metropolitan system, as smaller cities earned a metropolitan designation, even as it eroded it. Gradually, entire metropolitan areas began to experience net population loss, rather than simply low or no growth. By 1974, growth in nonmetropolitan areas exceeded that in metropolitan areas, and a nonmetropolitan population growth rate of 15.8 percent eclipsed the 9.8-percent rate for metropolitan areas between 1970 and 1980.[36] Fifteen of seventy-eight metropolitan areas with populations over half a million lost population between 1970 and 1980. All were in the Northeast or North Central regions. Nonetheless, this "turnaround" was widely distributed throughout all regions; while 45.4 percent of the 2,469 1970 nonmetropolitan counties grew between 1960 and 1980, an additional 35.8 percent of them grew between 1970 and 1980.[37]

The shift of urban populations out of densely concentrated central cities as well as away from urban core regions may now be regarded as a feature of advanced industrial development.[38] Furthermore, Vining et al. have discovered similar patterns of population dispersion away from core regions in a number of advanced industrial nations.[39] Nevertheless, just as early suburbanization was more than compensated for by rapid rates of natural increase well into the 1950s, the birth rate has been declining and the rate of household formation rising, which in turn has led to a decline in the average household size; these factors appear to be far more influential than changing migration patterns between metropolitan and nonmetropolitan areas, the rise of a mobile retired population, or the changing spatial structure of U.S. industry.[40]

The Shift of Manufacturing Growth to Nonmetropolitan Areas: The eclipse of metropolitan population growth rates by those in nonmetropolitan areas did not emerge until the 1970s, even though the trend had been building for some time. By comparison, nonmetropolitan manufacturing growth has paced that in metropolitan areas since the 1950s. A "leapfrog" relationship between employment and population growth rates is evident as we move between scales. While jobs followed people to the suburbs, people appear to have followed jobs from metropolitan to nonmetropolitan areas. Small and often remote locations were not only attracting migrants from elsewhere but often reversing the historical loss of residents to larger places. Carlino notes that while manufacturing jobs in nonmetropolitan areas expanded by 21 percent in the 1970s, metropolitan areas recorded only a 2.6-percent gain, with 23 percent of the new jobs captured by nonmetropolitan areas.[41]

Metropolitan Dominance Endures

The post–World War II era has indeed witnessed a revitalization of nonmetropolitan areas and a dampening of growth in the largest metropolitan areas outside the South. The revival of nonmetropolitan growth has been accompanied by a slight decline in the proportion of Americans living in metropolitan areas from 76.0 percent in 1970 to 75.3 percent in 1980, with projections for a continued decline to 74.8 percent by the year 2000.[42] Yet trends such as renewed population growth beyond metropolitan areas in no way herald the decline of those areas in any broader sense. Indeed, that surge may well have dissipated already. Recent data indicate that while metropolitan areas grew at an annual average rate of 1.5 percent over both the 1970–1980 decade and the 1980–1982 period, nonmetropolitan growth has dropped from a 1.3-percent average annual increase to a 0.8-percent increase over the same periods.

Might the apparent energy and time efficiencies inherent in deconcentration further erode metropolitan organization? Indications are that they will not. There are those who had predicted that this surge in nonmetropolitan growth would be short-lived because it would render the lengthened commutes both inefficient and costly in the changed energy environment of the late 1970s. Yet Morrison and Abrahamse have discovered that on average, migrants from metropolitan to nonmetropolitan areas shorten, rather than lengthen, their commutes between work and home.[43] Bowles and Beale likewise note that the commuting time required, as well as the distance covered, is less for nonmetropolitan workers than for metropolitan workers.[44] Consequently, from a demographic perspective, the metropolitan scale of advanced industrial society appears likely to endure for the forseeable future.

Nonmetropolitan Industrial Development Trends: Recent decades have also witnessed the dramatic industrialization of nonmetropolitan areas, at least in the aggregate. In a process similar to that discovered among regions, the economies of both metropolitan and nonmetropolitan areas have slowly developed in ways that have caused their industrial structures to converge. As Figure 4.6 illustrates, by 1970 manufacturing came to constitute nearly equal proportions of the economies in and beyond metropolitan areas. This was achieved in large part by the relatively faster growth of manufacturing in nonmetropolitan areas as the economies of small and rural places began to diversify:

The metropolitan scale of industrial-urban organization does not appear to be greatly threatened by this surge in nonmetropolitan development, however. While first manufacturing employment and then population growth in nonmetropolitan areas have come to outpace the

Figure 4.6. *Manufacturing share of U.S. metropolitan and nonmetropolitan total employment. (Source: Bureau of Labor Statistics data.)*

growth in metropolitan areas, employment continues to be even more concentrated in metropolitan areas than is population (Table 4.1). Moreover, this shift of growth to nonmetropolitan areas has not altered the overall distribution of manufacturing, as we have seen. As Figure 4.5 illustrated, the bulk of manufacturing remains tied to the nation's largest settlements. During the 1970s, employment in nonmetropolitan areas grew at a 2.3-percent average annual rate, while metropolitan employment expanded by only 1.9 percent annually.[45] The manufacturing share of metropolitan employment declined steadily from 30.3 percent in 1960, to 26.2 percent in 1970, to 22.8 percent in 1979, and these compositional shifts were mirrored in nonmetropolitan economies. The manufacturing share of nonmetropolitan areas grew from 22.7 percent in 1960 to 26.5 percent in 1970, only to decline again to 23.1 percent by 1979. Therefore, manufacturing growth outside metropolitan areas has triggered considerable nonmanufacturing (especially services) employment growth as well.[46]

The share of employment located in metropolitan areas has always fluctuated over time. Nevertheless, metropolitan areas appear to have largely retained their commanding hold on both total employment and manufacturing employment. The proportion of jobs located in metropolitan areas declined slightly over the 1951–1979 period in all sectors—except agriculture![47] Beneath the surface, however, while metropolitan employment shares declined in seven of nine sectors between 1951 and 1959, between 1959 and 1969 employment shares actually increased in six of the nine sectors. Therefore, in these data there does

Table 4.1
Shifting U.S. Metropolitan Employment Shares by Sector, 1951–1979

	1951	*1959*	*1969*	*1979*	*Directions of shift, 1951–1979*	*Directions of shift, 1969–1979*
Total	80.9%	81.8%	81.5%	80.5%	–	–
Agriculture	71.0	70.5	71.9	81.1	+	+
Mining	44.9	43.5	46.8	46.5	+	–
Construction	83.9	81.7	82.4	81.0	–	–
Manufacturing	82.0	81.6	79.7	77.1	–	–
Transportation	84.8	84.2	85.2	84.5	–	–
Wholesale trade	86.8	86.4	87.3	83.5	–	–
Retail trade	77.7	78.2	79.5	79.0	+	–
Finance	89.5	88.9	88.4	86.6	–	–
Services	83.7	85.0	84.8	84.8	+	NC

Source: Adapted from G. Carlino, "New Employment Growth Trends: The U.S. and the Third District," *Business Review,* Federal Reserve Bank of Philadelphia, Sept. –Oct. 1983, p. 13.

not appear to be justification for expecting a substantial spatial reordering of employment in the United States that signals a jettisoning of metropolitan-scale industrial organization.

Looking ahead, manufacturing employment is expected to increase by approximately 10 percent by the year 2000, a relatively low rate of growth compared to that in other sectors, especially services. In regional economic forecasts for the National Planning Association, Holdrich has projected that over the thirty-year period from 1970 to 2000, the hold by metropolitan areas on manufacturing will loosen only very slightly, from 80 percent to 77 percent. Moreover, no major spatial shifts in manufacturing are expected. Northern cities will still be the primary location of manufacturing employment in the year 2000. However, much shifting can be expected at smaller spatial scales as some older metropolitan areas fail to retain manufacturing jobs that they have had for a long time, and as new manufacturing employment in high-technology sectors takes root in new and rapidly growing metropolitan areas. Nonetheless, Holdrich notes that "by 2000, more people will be employed in manufacturing in Chicago, New York, Boston and Detroit than in any other city except Los Angeles. After California and Texas, New York, Pennsylvania, Ohio, Illinois, and Michigan will have the most manufacturing jobs."[48]

Consequently, not only does the population "turnaround" of the 1970s appear to have been short-lived, but for the forseeable future the metropolitan dominance in manufacturing does not appear threatened either. Trends such as we have observed may simply obscure an important reality about the dominance of metropolitan organization in advanced industrial development. Renewed population growth and a spread of selected structural characteristics of an advanced industrial economy to nonmetropolitan areas may merely redirect attention to other significant factors on which metropolitan and nonmetropolitan industrial environments continue to differ. While advances in production and linkage technologies no longer constrain manufacturing to concentrated metropolitan areas, an array of new location factors that apply more to workers than to businesses may be in ascendance.[49] These new factors tend to favor existing metropolitan-scale settlement patterns.

There may well be a new set of location factors governing traditional manufacturing—and other older industrial development—that reduces locational constraints and allows selected manufacturing activities to filter more easily than ever before into nonmetropolitan areas. Norris has noted that all through the post–World War II period, it has been primarily advanced-stage, mature, low-growth durable goods manufacturing industries, including nonelectrical machinery (SIC 35), electrical and electronic machinery (SIC 36), transportation equipment (SIC 37), and fabricated metal products (SIC 34), that have led the shift of manufacturing growth to nonmetropolitan areas. This situation has occurred largely as the result of a quest for greater profit margins achievable through lower production costs to compensate for the reduced productivity of standardized production technologies.[50] And it has occurred despite the fact that those growth rates have steadily declined over the same period. Yet new manufacturing and allied industrial development may still find metropolitan areas more attractive seedbeds for early growth, if not necessarily for the same reasons that they once did. As we will see in the next chapter, metropolitan areas appear to retain their attractiveness for successive waves of new high-technology industries and the employment growth tied to them. While dispersion dynamics are evident in high-technology development, the vast bulk of this development is still easily captured within metropolitan areas.[51] One reason for this is that advanced industrial development is critically dependent on selected forms of human capital and the special skills of workers to innovate and implement new product and process technologies. New location factors may be biased in favor of where pools of such people prefer to live, and these residential preferences are often more easily accommodated within existing metropolitan areas than outside them. People may be leaving very large cities, but they appear to prefer to retain easy access to them.

Consequently, the spatial restructuring of U.S. manufacturing has been clearly evident at the interregional scale. The filtering of new industrial development into the once peripheral West and South has dramatically altered economic, social, and political relationships among regions. In contrast, intraregional trends have yielded a more varied set of outcomes. While the significance of metropolitan areas may be diluted somewhat in the very long term, advanced industrial development promises to remain principally a metropolitan-scale activity well into the next century. Manufacturing may be growing rapidly outside metropolitan areas, but this has not yet led to a redistribution of total manufacturing employment down a place-size hierarchy from large places to small ones or from metropolitan areas to places outside, as happened between the Northeast and North Central regions and the South in particular. At the smallest spatial scale, central cities do appear to have lost their grip on total manufacturing to places elsewhere within metropolitan areas. In the next section, I will illustrate in greater detail how those small-scale patterns have unfolded.

Recycling the Industrial-Era City[52]

> Enforced perpetuation of an originally profitable combination of interests, and the perfect preservation of old conditions in particular, creates a museum, which like all such institutions, requires large sums for its maintenance. As soon as the breaking up of old combinations is economically justified, every attempt to obstruct it means a sacrifice, however, that may be vindicated now and then by the fact that it helps to preserve the political and cultural existence of an economic landscape for a while even though it has passed its economic prime. Those who have to bear the burden in this case are the inhabitants, who, prevented from migrating, are forced to put up with a lower standard of living. In the long run they are also politically endangered thereby. If it is a case merely of a depressed area within a country, the prosperous areas generally bear the cost. This is not always a wise policy, even when the importance of extra-economic causes is freely admitted. It would often be much better to *facilitate the breaking of an old combination of land, people, and economic activities and seek systematically for a new and vital one;* that is to promote adaptation rather than to obstruct it.[53]

We have seen that in certain respects, advanced industrial development organizes the resources and the capacity for production in ways that differ significantly from those of an earlier era. Restructured local economies require and prepare central cities to play new socioeconomic roles; these roles, in turn, call forth new architectural arrangements and land use patterns as advanced industrial development begins to register its impacts visibly in specific locations. The key departures from the past include an altered physical form and socioeconomic function. These transformed central cities look different, then, because they serve us differently than before. A network of post-industrial places now administers an advanced industrial national economy. The essential outlines of these new urban arrangements only began to be detectable in the late 1950s and early 1960s. Today, following a surge in commercial real estate activity that began in the late 1970s, the central business districts of many of our older industrial-era cities have become the best examples of post-industrial places that advanced industrial development has yet produced. What are some of these changes and how do they further illustrate the larger dynamics of advanced industrial development?

The New Services-Dominated Urban Economy: Despite a preoccupation during the past quarter-century with what has been perceived and interpreted as urban "decline" and even the "death" of our cities, many of our preeminent industrial-era cities have exhibited considerable economic resilience in the face of population and manufacturing contraction. Many transformed older and larger central cities retain an economic dominance in many respects, but in these cases it is likely to be based on a new mix of strengths and advantages in relation to locations elsewhere in the metropolitan area. Sectoral shifts—especially the shift to a services-dominated economic base—have proceeded more rapidly within central cities than they have at larger spatial scales. The mainstay of many of these recycled central city economies has become services rather than goods production, especially high-value-added producer services that typically retain their affinity for central locations and high densities.[54]

These shifts have not proceeded without considerable misgivings concerning their effects on those whose lives are influenced by them. The extended high-rise office construction boom of the past decade offers a graphic illustration of the reality that the social function of central cities is less closely tied than ever to the provision of entry-level factory jobs for legions of unskilled workers. There is little doubt that the changing demography of disadvantage, as is illustrated by an urban underclass of the unskilled, poorly educated, minority young and by the growing ranks of female-headed households, interrelates in important ways with the restructuring of central city economies. However, as post-industrial cities continue to evolve, there is as yet little justification for assuming that these sectoral changes necessarily constitute a net welfare loss either for those who live in them or for the larger society.[55]

The role of recessions in economic adjustment is revealed at this local level as well. The employment and income effects of the recessions of the early 1980s on our older industrial cities have been buffered considerably by the shift-to-services that has been rapidly transforming their local economies. A study of cyclical impacts on cities revealed that central cities, counties, and localities with slow population growth were in general relatively insulated from business-cycle downturns. In contrast, as manufacturing and construction have dispersed, the cyclical sensitivity of the receiving locations, such as rapidly growing suburbs, has increased.[56] To the extent that recessions have accelerated the redistributions of capital, population and employment to new locations, they have functioned to stabilize the economies of older industrial cities.

The Retreat of the "Central" City: This shift to services-dominated local economies within older industrial cities may hold the promise of rejuvenating these settings of a century or more of monumental, if increasingly obsolete, public investment and historical significance. Traditionally, metropolitan systems have been conceptualized as being oriented toward a central city core economy. In contrast, the role of the central city in an advanced industrial economy has often receded in importance relative to locations elsewhere in the larger metropolitan system. The central city becomes less "central" as geographical centrality no longer constitutes economic centrality, and it likely retains smaller shares of metropolitan population and employment than ever before. Often a central business district is joined by one or more other growth poles within and around the central city, even as the central city itself enters into direct competition with increasingly industrialized and independent suburbs. As a result, the new urban settlement system is increasingly polycentric or multinodal and linear rather than concentric.[57]

The Central Business District as an Electronic Crossroads: Frequently, these recycled central cities are being retrofitted with a new kind of infrastructure tied to information, telecommunications, and the administration of an expanded urban field. Once primarily a physical crossroads, the new metropolis is likely to become more important to us in the future as an electronic crossroads for information processing and exchange. Unlike before, however, central cities no longer possess the unique locational advantages that favor them in competition with suburban and other locations seeking to assume a "switchboard" role.[58] Consequently, while new activities may cluster in these retrofitted central cities, especially in older central business districts, there is no reason to expect population to pour back in as well.

The new technologies do not so much eliminate the need for space in restructured central cities as they influence cities to organize and use space differently. While, in general, automation technologies have the capacity to disperse factory and back-office activity into smaller and more compact plants and more remote physical settings, they also unleash complex and conflicting influences in newly reconstructed central cities. The continued thinning out of central cities can be easily obscured by the physical thickening of older central business districts. Feverish high-rise construction has filled many older industrial cities with taller buildings and ever-higher-density land uses. However, the high densities of commercial real estate construction commonly mask the ever-lower densities of what is going on inside these buildings. A century-old process of horizontal urban sprawl has been joined by a new process of "vertical sprawl" in which high-density land use is accompanied by lower-density space use inside. It is estimated that since 1977 the amount of space allotted to the average office worker has increased 14 percent from 195 to 223 square feet.[59] In part this reflects the proliferation of office machinery of all kinds as the essential production tools in a new office environment. It may also reflect that as overconstruction results in high vacancy rates, new tenants find it increasingly attractive to "bank" space in preparation for future expansion. In any event, the high-rise architectural reconstruction of many older central business districts is at best likely to slow, but not reverse, the larger dispersal of jobs and people away from central cities.

The Industrial Recycling of Dallas, Texas[60]

The major aim of this section is to trace in greater detail the patterns of capital investment and disinvestment to their points of impact on particular places. This place orientation can be important because the costs and benefits of industrial change are ultimately tallied up in specific places. The goal is to illustrate that places, like the industries, products, and production arrangements that define them, are continuously in transit. Advanced industrial development inescapably involves patterns of settlement and mobility that are eventually reflected in population and employment trends for particular places.

To illustrate the kind of small-scale industrial shifts that restructure localities, I have chosen the city of Dallas, Texas. Although it was incorporated in 1857, the older industrial origins of the city of Dallas are frequently obscured by the fact that its most rapid growth came after World War II. Today it is recognized as the anchor for one of the nation's premier "sunbelt" regional economies, which has been responsible for the rise of the Southwest since the 1970s. The Dallas-Fort Worth (D-FW) metropolitan area has been the locus of significant growth in recent years with population growing 6 percent during 1980–1982 alone. While the proportions and rate of this growth have recently captured the attention of the nation, there is little reason to expect that the city of Dallas, as the central city of a large metropolitan economy, should be exempt from the deconcentration and dispersal trends that have been unfolding in central cities of less rapidly growing metropolitan areas across the nation.[61] However, to spot these incipient trends, aggregate growth must be spatially disaggregated. As I intend to illustrate below, the city of Dallas as a whole—and, even more so, many of its geographic subareas—is for the most part failing to grow as rapidly as the larger region. Let us turn to the trends for population, housing, business establishments, and employment.

Dallas is the seventh largest city in the United States. During the 1970s, the D-FW region's population mushroomed from 2.3 million to nearly 3 million, for an increase of 24.7 percent. This growth has been distributed far from uniformly throughout the region. The proportion of the metropolitan population residing within the city of Dallas declined during the 1970s from 54.3 percent to 30.5 percent, thus illustrating the relative accessibility and growing attractiveness to the waves of new population growth and business expansion of locations in the region beyond Dallas. The city of Dallas, which anchors the eastern end of this metropolitan region, grew from 844,000 to 904,000 (7.1 percent) over the same period. Set amid a rapidly growing region, the more modest growth of the city was accompanied by a 15-percent decline in population density between 1970 and 1980.[62] Although population growth lagged significantly behind that for the larger metropolitan area, Dallas was one of only two dozen U.S. cities of over a quarter-million population to experience a net increase during the 1970s.

This aggregate growth rate masks the spatial features of population shifts that occurred throughout the city. As is clear in Map 4.1, population growth was largely confined to the outer edges of the city, especially in the northern sector. Close-in neighborhoods throughout the center of the city experienced widespread and often dramatic population losses. The race-ethnic composition of Dallas was likewise changing. The distinct tri-ethnic character of the central city became much more evident during the 1970s. The Anglo share of the city population shrank from 74.2 percent to 61.4 percent, while the black share increased from 24.9 percent to 29.4 percent. The remainder of the shift was dominated by the increasing share of Hispanics. Somewhat unexpectedly, the spatial trends that, tied to residential mobility, defined population growth led to a dramatic decline in patterns of residential segregation in Dallas during the 1970s.[63]

Housing and a Declining Residential Function: The majority of neighborhoods that lost population lost housing as well (Map 4.2). The architectural reconstruction of the central business district and selected inner-city neighborhoods in recent years is simply the most visible aspect of the changing residential function of the city of Dallas, and evidence is beginning to hint that the physical transformation of the CBD has probably run its course.[64] Many neighborhoods throughout the center of the city experienced dramatic shrinkage in housing stock during the 1970s. What the implications of this localized population and housing contraction might be for business growth and expansion within the city are unclear, however, since the new high-rise office buildings and the expanded services employment they may promise are accompanied by a far less visible eclipse and elimination of the older businesses and jobs that made way for them. Nonetheless, major parts of Dallas appear to be slowly surrendering their residential function to neighboring suburbs through a process most visibly underway in the inner-city neighborhoods.

Shifting Business "Microclimates": Shifting patterns of growth and decline have gradually transformed Dallas's economic geography. Between 1972 and 1977, the city of Dallas's share of metropolitan manufacturing employment declined from 46.6 percent to 41.8 percent. The same pattern was evident in several services categories. Over the 1967–1977 decade, the share of retail trade employment declined from 62.1 percent to 40.1 percent, for wholesale trade from 90.4 percent to 56.1 percent, and for selected services from 85.8 percent to 59.3 percent. While economic diversification was radiating outward from the central city, it was principally the northern suburbs that captured the bulk of the new expansion within these sectors.

The uneven distribution of new growth throughout the city underscores that even in a growing city like Dallas, there is no such thing as

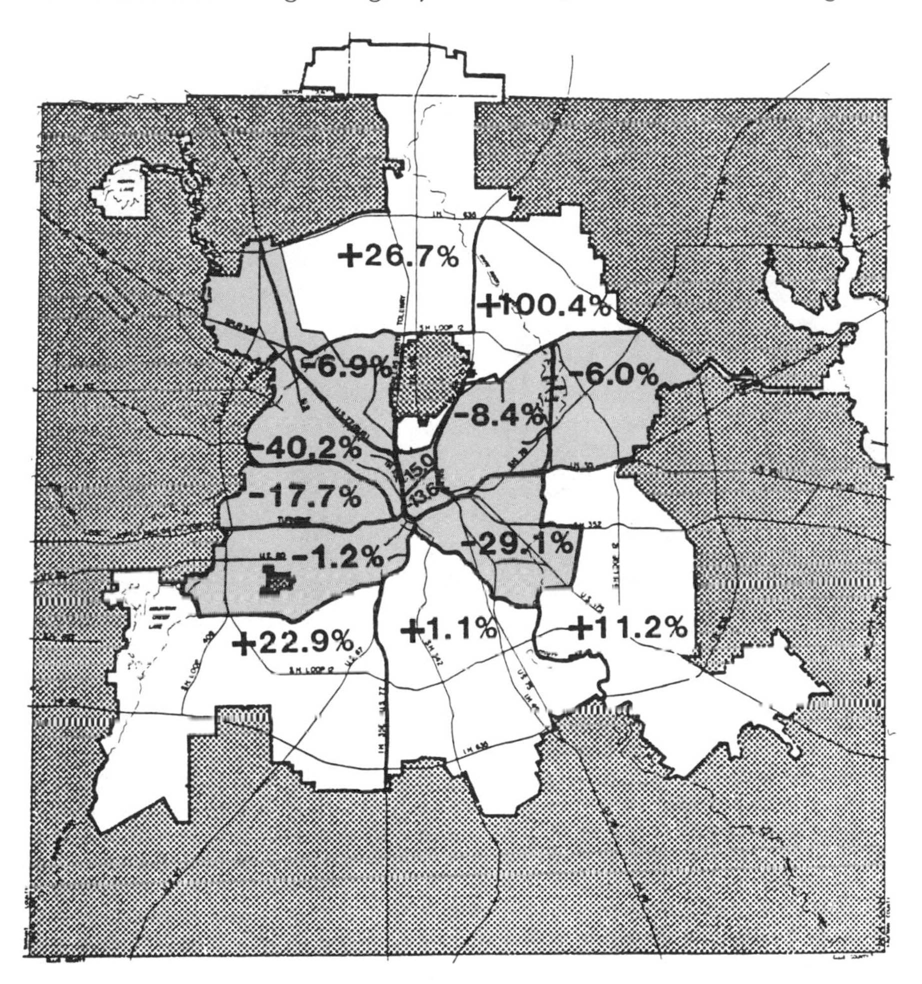

Map 4.1. *City of Dallas population change: 1970–1980.* (*Source:* Dallas Data Book, *Department of Planning and Development, City of Dallas, August 1983, p. 3b.*

a uniform business climate. Rather, a mosaic of "microclimates" exists in which decline and contraction are integral to overall growth and development as growth is drawn and propelled to new locations. Dallas's "textured" business climate provides structure to the incubation, expansion, and retention of industrial development. Consequently, as these subareas experience differential rates and directions of change, both the city and the larger region as a whole are transformed in major ways. The spatial features of uneven growth rates may even result in the new economic center of an urban area being dislodged from its traditional geopolitical center.

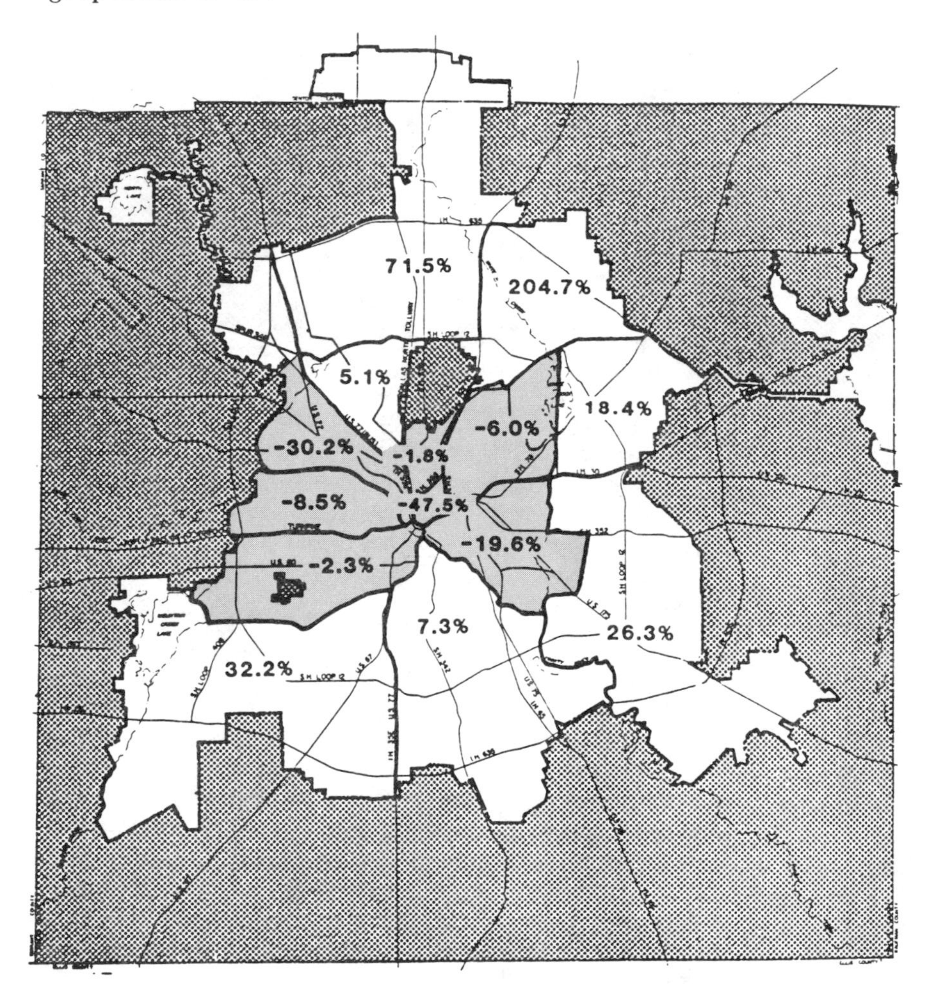

Map 4.2. *City of Dallas housing unit change: 1970–1980. (Source:* Dallas Data Book, *Department of Planning and Development, City of Dallas, August 1983, p. 33b.)*

In a recent study, Dallas was divided into fourteen geocoded economic subareas.[65] The results illustrated in Map 4.3 indicate that between 1977 and 1982 there was a contraction in the number of business establishments in ten of the fourteen subareas. These declines, which reflect net shifts only, resulted in an estimated loss of 5.8 percent of the Dallas's business establishments over the five-year period. Establishment growth was generally confined to the northern sector of the city, while slow net growth and contraction characterized the innermost parts of town. Over the same period, Dallas's employment expanded approximately 3.7 percent. In Map 4.4 it is evident that the distribution of net employment shifts was likewise far from uniform across the economic landscape of the city. Of the fourteen subareas, eight were estimated to have experienced employment contraction between 1977 and 1982.

While the large-versus-small business mix and component expansion and contraction rates are concealed by these net figures, other evidence indicated that the losses were greatest in the small business stratum as new business starts and expansion failed to keep pace with business closings in many parts of town. The desirability of this aggregate trend can be viewed in alternative ways. Net contraction in business establishments is typically equated with urban decline and deterioration. Yet it is also consistent with a situation in which a few businesses expand so dramatically that their growth displaces and dislodges several smaller ones. While the transformed economy of the subarea may provide more employment and be more productive than before such recycling, the new jobs may be poorly matched to the skills and backgrounds of those displaced from failed businesses or left unemployed either by businesses that did not expand or by new start-ups that never materialized.[66] In the end, the desirability of an outcome must be determined from more than data alone. Nevertheless, such a scenario suggests that the business microclimate of the subarea may no longer be an attractive setting for the incubation and future expansion of new small businesses that can provide flexibility to a locality's economic base over time. This possibility can translate into a growing disadvantage for a city over the long run and is examined in greater detail in the following chapter.

Reinterpreting Urban "Decline" and the Eclipse of Urban and Regional Policy

> [W]ith the "end of urbanization" practically the entire population of many countries is urban in character. Therefore, . . . all government policies have become urban policies. This can become an unnecessary source of confusion in decision-making and at the limit could lead to government paralysis. In

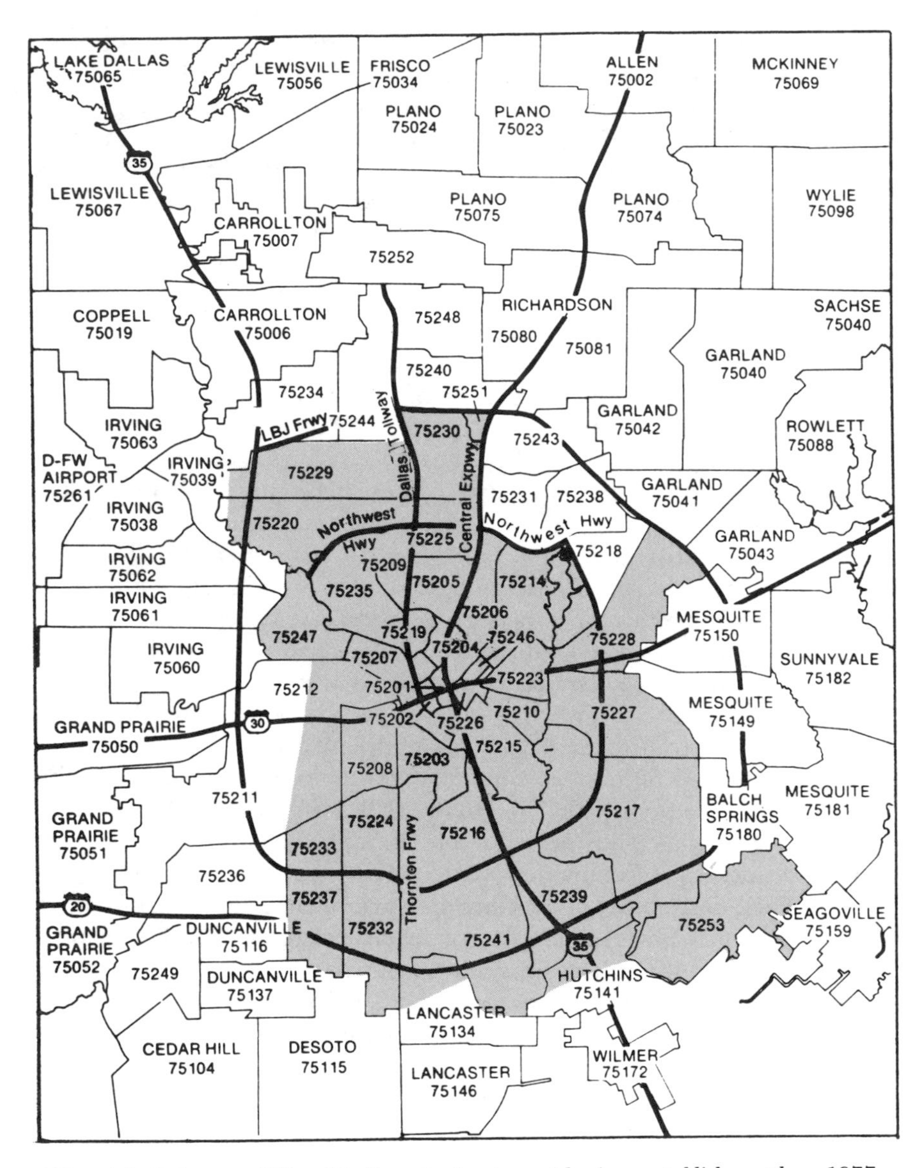

Map 4.3. *Subareas of City of Dallas experiencing net business establishment loss: 1977–1982. (Source: Donald A. Hicks and Joel H. James, "The City of Dallas as a 'Textured' Business Climate: Establishment and Employment Contraction and Expansion in the City of Dallas Between 1977 and 1982," Center for Policy Studies, University of Texas at Dallas, November 1983.)*

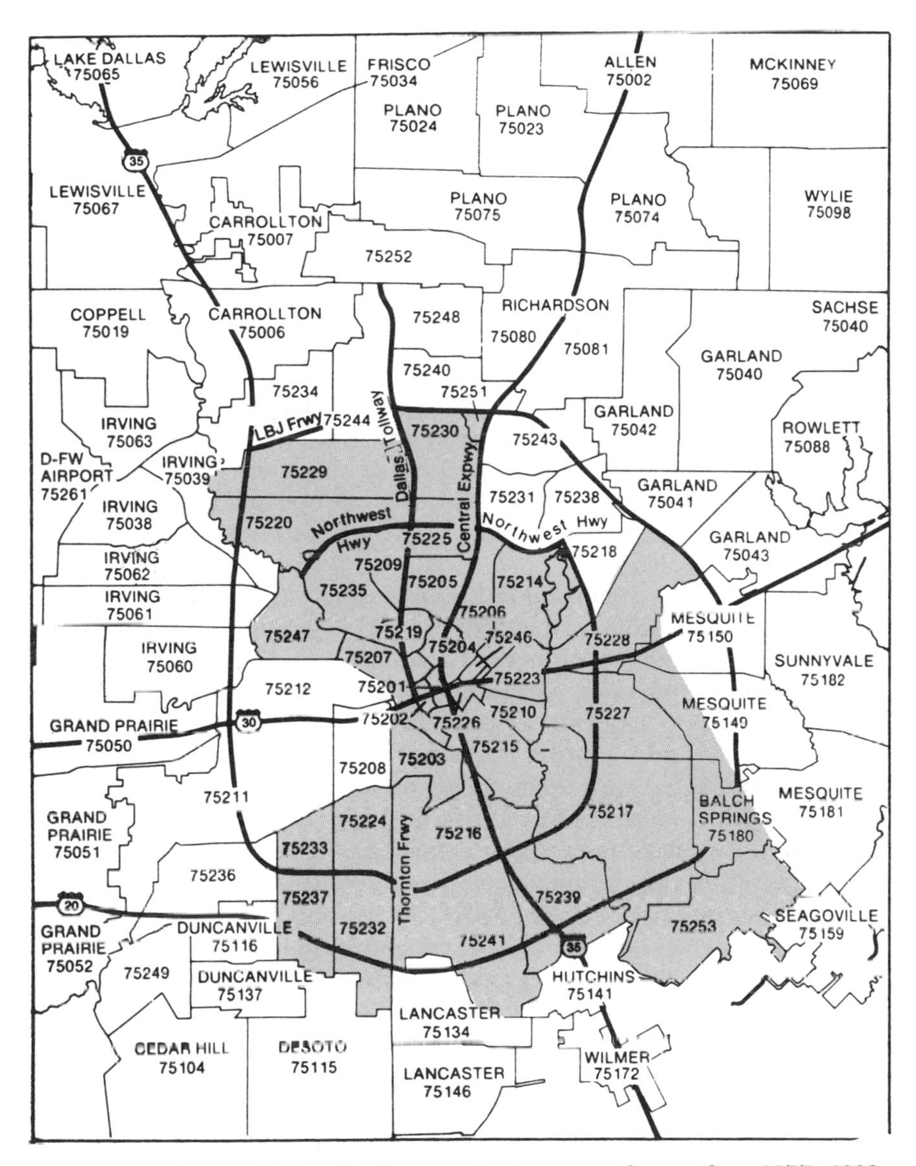

Map 4.4. *Subareas of City of Dallas experiencing net employment loss: 1977–1982. (Source: Donald A. Hicks and Joel H. James, "The City of Dallas as a 'Textured' Business Climate: Establishment and Employment Contraction and Expansion in the City of Dallas Between 1977 and 1982," Center for Policy Studies, University of Texas at Dallas, November 1983.)*

fact, it is most unlikely that urban policy objectives will take precedence over those of... industrial policies or trade policies in the 1980s.[67]

Arguably, the dominant perception Americans have of the changes being experienced by U.S. cities has been shaped by the notion of urban "decline." Factors responsible for the population and employment losses that principally lend meaning to the term have been identified, categorized, and tested in a burgeoning urban economics literature.[68] There have been numerous systematic efforts to devise strategies whereby the public sector might reignite traditional forms of investment in central cities through direct intervention[69] or by creating incentives for the private sector to steer investment into mixed use, compact, and infill development.[70] Still, market forces reflected in the increased access to, and attractiveness of suburban and exurban locations for firms and households remain the principal shapers of U.S. cities. While nonresidential development in many central cities has boomed in recent years, residential revitalization—urban gentrification—has been more heralded than realized.[71]

There were in the 1960s and 1970s increasingly ambitious and sophisticated urban policy efforts to "save" our cities—as distinct from efforts to assist needy residents of them in more direct ways. Had the failure of urban policy initiatives not been enough to dissuade us from even more urban policy initiatives in the future, the slow appreciation of the strengths of new post-industrial cities might also have led to the demise of urban planning. In any event, these trends, of which the new political agenda of the 1980s is both a cause and consequence, have led to a dramatic lessening of interest in explicit, spatially sensitive urban-regional policies. In this section, I shall explore further this gradual redefinition of our cities.

The conceptual building blocks of conventional urban economic analysis include, as we can see, trends applying to multistate regions, settled places that either constitute or fall outside metropolitan areas, and central cities and their suburbs. Analyses that involve movement between scales are useful to the extent that they permit us to isolate and identify processes, such as population and job redistribution, that are at work at one scale but might be missed or absent at another scale. However, this same strategy is necessarily flawed by its unavoidably mechanical nature. In short, it must make the prior assumption that cities or regions cannot exist apart from the political or administrative borders that demarcate them. As messy as the prospect of such misplaced concreteness might be, this assumption can be challenged. The implications of failing to do so include being blinded to alternative interpretations of what is happening to our industrial settlements, why, and with what effects. We are also led to embrace policy responses that may be either ineffectual or counterproductive or both. Finally, unless we are willing to reexamine the concepts and assumptions of conventional urban economic analysis, the substance of advanced industrial development may come to be mistakenly associated with abiding—if not actually encouraging—the "death" or decline of older industrial-era urban arrangements, without at the same time seeing the coherence of what is emerging to take their places.

As we examine trends at successively smaller scales, it is easy to lose sight of the fact that the "stepping stones" for that journey—regions, states, counties, and cities—are inevitably economically artificial. They are the historical resultants of principles of state sovereignty, city chartering arrangements, annexation impulses, and record-keeping conventions of government bodies. As such, what has political or administrative reality does not necessarily have economic reality. Cities and regions are ultimately political and administrative overlays that cause us to treat certain places as jurisdictions and others only as abstract units of analysis. Yet both have only the meanings that we give them. While shifts in population and manufacturing employment—among many trends that we routinely monitor within and among these levels and units of analysis—are invaluable in helping us separate certain urban "facts" from fiction, final interpretations do not flow easily from such data. Patterns of investment and settlement are ever shifting at all levels. From the perspective of advanced industrial development, we may be nearing the end of the time when we can uphold the fiction that cities and regions should have any meaning beyond that associated with political custom and tradition.

Consider the phenomenon of urban "decline." We have seen that advanced industrial development is characterized by the redistribution of existing and new population and employment at several spatial scales. In the process, certain cities, metropolitan areas and even multistate regions have experienced stagnation and contraction on key indicators in recent years. Must such contraction mean decline in a more profound sense than simply population loss? Not at all. Even the extent to which decline exists depends in large part on how we draw our boundaries and to what level of analysis we assign importance. Bradbury et al. have demonstrated that urban decline is widespread, especially at the city level. While 38.6 percent of the cities in their analysis lost population during the 1960s, the proportion jumped to 62.1 percent during 1970–1975 and declined slightly to 56.2 percent during 1975–1980[72] Recognizing the importance of the metropolitan context for that decline, a companion analysis of population trends during 1970–1975 involving central city–metropolitan area clusters discovered that fully 78.5 percent of these cities were experiencing either population growth, stagnation, or decline, but doing so within *growing* metropolitan areas. Moreover, of all those cities whose population either stagnated or shrank during 1970–1975, the majority (66.2 percent) were located in growing metropolitan areas.[73]

How might we interpret *these* facts? Indeed, the loss of population by cities became increasingly pervasive through the 1960s and early 1970s, only to moderate by 1980. Yet in the majority of cases, cities were embedded in larger-scale metropolitan areas that continued to grow. Are such trends not equally consistent with the interpretation that cities were involved less in a process of decline than of equilibration or adjustment to the growth dynamics of the larger urban areas? (Recall that only fifteen—19.2 percent—of all metropolitan areas with populations exceeding half a million experienced population loss during the 1970s.) At the level of cities, urban decline takes on proportions and a reality that do not emerge from metropolitan-scale analyses. Therefore, population "loss", urban "decline," and similar terms have meaning only after administrative or jurisdictional boundaries have been imposed. The redistribution of population and jobs is not without consequence, but the metaphor of "decline" can distract the observer from entirely valid alternative interpretations.

Does it make sense, then, to conclude from the patterns of slow growth and contraction that the city of Dallas is in "decline" in some fundamental way? I do not think so. Rather, an older industrial Dallas appears to be receding, while a new post-industrial Dallas is taking its place. The process involves redistributions both within and around the central city. The city of Dallas is growing more slowly than the larger metropolitan economy of which it is a part. Yet, as we saw, the trends for the smaller geocoded areas within Dallas yielded a net result that did not represent very well what the city itself is experiencing. In the Dallas case, there is little doubt that problems relating to perceptions of traffic congestion, poverty, crime, quality of public schools, and a shortage of close-in affordable housing are being reflected in these trends. It may be time to reconsider local economic development strategies that stress industrial recruitment and promotion at the expense of business retention and the expansion of existing business. However, the transformation of the city of Dallas is at least as much an illustration of the way in which advanced industrial development is recasting Dallas's role in a larger regional economy as it is the result of a growing list of Dallas's urban problems.

Urban Economic Development Policy Reconsidered: Does this mean that the loss of population by central cities, counties, states, or even regions is of no consequence? By no means. However, the consequences are neither more nor less than what we decide they will be. The political implications of population decline often mean that cities and states as administrative units face increasing difficulty in delivering services to their residents. Yet the consequences of population decline for a city, state, or even region are determined by the functional assignments in our federal system and the federal-fund flows that are triggered by population levels and rates of change. At a time when the internationalization of business is instructing us that even our national economy is no longer a self-contained entity, it should become easier to see that treating Dayton, Detroit, or Duluth—or even Dallas—as freestanding economic entities as well as political ones only preserves a view of a relationship between people and the economy of an industrial-era settlement that no longer exists.

There is no conceivable data set that is able to aid us in distinguishing between urban "decline" and a more benign process of urban transformation. Ultimately, the same trends as we have reviewed for the city of Dallas—which can be seen even more dramatically in dozens of major central cities around the United States—will be judged against some perceived notion of what a "city" should be and what trends should constitute its "health." However, there is a social and economic role for "decline" via the deconcentration of people and jobs that continues to go unappreciated, and consequently hinders our understanding of what is happening to our major industrial cities. It only becomes apparent when we see that metropolitan areas have emerged to replace central cities as the principal units at the scale at which new urban economies are organized. As Noyelle and Stanback have demonstrated, a "new urban system" has emerged in recent years that draws together older and newer industrial-era urban economies into new sets of functional relations. The dominant national, regional, and specialized urban economies are increasingly characterized by new industry and occupational mixes with high proportions of producer services and well-paying white-collar jobs. To be sure, cities and metropolitan areas with mixes of slower growth industries and occupational skills in less demand will face an often difficult transition in the years to come. Nonetheless, it is important to underscore that *population and employment contraction have been integral parts of the urban economic development process that has resulted in this new urban system.* No longer is there reason to equate urban health with city or even regional population growth.

A Need to Defend Community Against Capital?: Attempts to account for urban "decline"—as with "deindustrialization" in earlier chapters—commonly view capital and community as antagonists in an increasingly consequential struggle. From an annotation on a work by Carnoy and Shearer: "Cities should not be regarded simply as productive spaces to be disregarded when capital decides them to be obsolete, but as places to be stabilized and prized—places where community can be strengthened."[74] Ultimately, that is what urban and regional policies have always been about: attempts to identify and defend a negotiated trade-off between the ever less fettered mobility of capital and the presumed stability of community. The past quarter-century has witnessed a lock-step succession of federal policy proposals, from urban renewal in the 1950s to urban enterprise zones in the 1980s, all designed to use public leverage to achieve specific results in specific places.[75] As a result, urban and regional policies have been conceived of as primarily *place oriented* and spatially sensitive, with both cities and regions as the key spatial referents.

Traditionally, the outcomes of these policies have been so broad as to blur the distinction between those targeted to places and those targeted to the people living in those places. The prevailing logic has been that local economic development strategies that assist places inevitably would have their benefits redound to the benefit of those who lived in them. Yet, increasingly, amid the changing demographics of disadvantage in central cities, the eventual trickle-down of benefits can no longer be assumed. Where a nation's urban policy is expected to function as a de facto welfare or employment policy, the problems of people become subtly subordinated to those of place and jurisdiction.[76] The intention of aiding people indirectly by aiding places directly no longer makes a great deal of sense now that the scale of new urban economies has become so thoroughly decoupled from that of older industrial-era cities.

Not surprisingly, urban-regional policy responses to the redistribution of people and jobs have been largely reactive, protectionist, and in no small measure defensive. In part, this has been because new post-industrial urban economies now spill over the jurisdictional borders of industrial-era cities and have in a certain sense neutralized them. While older urban constituencies have declined, no new political coalitions and cultural constituencies have emerged yet to fight for the newly organized urban system, either locally or nationally.

As advanced industrial development both requires and permits new forms and scales of urban settlement, the logic of the full range of past urban-regional policies bears reexamination. Invariably, at the heart of all varieties of urban policy is the attempt to achieve a specific balance between capital and community, notions that are presumed to be inherently in conflict. Competing images of industrial change are the result. Tension flows from the belief that mobility naturally enhances the efficiency of capital, while it is perceived as being deleterious to human communities. Patterns of capital investment, disinvestment and reinvestment are commonly presumed to set a pace that humans in communities cannot and should not be expected to keep up with.[77]

What is the notion of "community" that varieties of urban and regional policy serve? No longer is there justification for assuming that a single such notion exists. Perhaps what is being challenged by the spatial dynamics of advanced industrial development is not community per se, but a monolithic, outdated, and homogenous mental image of community that is assumed to be coterminous with territorial units like cities and regions. Such a notion is at considerable odds with the imperfectly overlapping "communities," defined by ethnicity, lifestage, lifestyle, generation, occupation, and indicators of social class, that coexist in most urban areas.[78] And, as such, this kind of complexity is impossible to defend and preserve using spatially oriented public policies. Community is no longer—or even primarily—a territorial phenomenon, but a social, psychological, and cultural one. It is the ability of people primarily, and places and jurisdictions only secondarily, to make the transition that should be the goal of public policies.

Ultimately, the activites of people and how they are organized and changed determine the substance of places and the boundaries of community at any given time. Uneven patterns of business development, settlement, and mobility invariably accompany the recycling of an urban economy at any scale. The resulting spatial patterns of their adjustments are wisely not imposed or assumed in advance. Initial industrial development gave us our central business districts, suburbs, and metropolitan-scale urban economies in the first place. It should neither surprise nor alarm us that advanced industrial development under the regimen of new technologies and cultural preferences can be expected to be accompanied by new urban-industrial arrangements in and around our human settlements.[79]

Today, the notion of urban policy, much as Renaud has suggested, appears to languish as attention has shifted to the possibility that the larger national economy has reached an industrial crossroads. The past assertions that urban outcomes were largely a matter of political will have been considerably muted. Where in the nation economic growth takes root has become less important than that it takes root at all. The very real distributional concerns that remain are probably better addressed through spatially neutral social welfare policies than implemented through the complex mechanics of place-oriented urban and regional policies seeking to influence adjustment patterns integral to advanced industrial and urban development.

5

Local-Regional Features of High-Technology Industrial Development

> [I]n the last twenty years there have developed four new major industries in America—the automobile, the motion picture, the electrical and chemical industry—that, unknown twenty years ago, today furnish a livelihood for 30,000,000 people. There is but one way to gauge the future and that is by what has happened in the past; hence, it is certainly safe to believe that with the inventive genius of the American people, with their resourcefulness, with the vast wealth now owned by this Nation, that the next twenty years will bring about the development of other vast industrial enterprises that will absorb, take up and provide a livelihood for the millions of people who will make up our increased population.[1]

At any one time, a national economy has a phalanx of industries whose technological underpinnings place them at the leading edge of industrial change. Throughout the industrial era of the past century and more, there has been a continuous succession of such technologies.[2] In their own times, the steel, railroad, telegraph, automobile, telephone, aviation, and television industries have been among our "high" technologies. Today, that status is temporarily reserved for semiconductors, lasers, fiber optics, biotechnology, photovoltaics, and computer hardware and software, among others, with their endless more tailored industrial and commercial applications. Ten years from now, as these current frontier technologies continue to diffuse through production processes and final products, new clusters of technological innovations will likely be led toward commercialization by small and large firms in a variety of industries. Inevitably, *high-technology status resides not in any inherent characteristics of particular innovations but rather in the patterns of investments and impacts that come to be associated with their development within and diffusion throughout (and beyond) an industry.*

The anticipation of economic growth and rising levels of prosperity in the wake of technological innovations and their commercialization is likewise not a new phenomenon. New technologies, the new industries they create, and the older ones they rejuvenate have long been linked to the larger process of industrial development, as the opening quotation, from a past president of the Dallas Manufacturing Association in 1925, clearly illustrates. Today, as was the case more than a half-century ago, the potential employment and productivity effects of technological change deserve careful consideration. Uncertainty concerning the labor-displacement effects of automation technologies inevitably accompanies discussion of whether a particular technology will induce employment expansion along with industrial growth, and if so, what the new job mix will look like.[3] Similarly, the spatial and locational features of technology-induced industrial growth have long been closely linked. Curiously, contemporary discussions of the implications of today's high-technology growth commonly reinforce the illusion that "high-tech" industries have somehow sprung *de novo* from an otherwise mature economy. Frequently, such discussions overlook the longer-term evolutionary features that culminate in a sector whose industry members circulate and are reconstituted continuously.

Development Versus Discovery: The single feature of advanced industrial development that has probably received more attention during the 1980s than any other has been its association with so-called high-technology industries. While the growth of selected high-technology industries has often been rapid, their discovery and subsequent visibility, more so than their development, have been recent. Inevitably, the determinants of, and the prospects for growth in these industries have been seized upon by countless economic development efforts across the nation seeking to reverse years of slow or no growth in local and regional economies. Ironically, it was the sharp and extended cyclical downturn that first sensitized many states and localities to the longer-term structural features of their economic bases, which in many cases ensured slow growth and relative decline even during periods of economic recovery. Yet the special esteem that has been accorded this newly demarcated sector has often been granted uncritically and without careful scrutiny.

Only a small portion of the nation's labor force—estimates range between 3 and 13 percent—is employed in this sector, depending on which industries are defined into it.[4] Yet employment and productivity growth rates for this sector have long exceeded those for the rest of the economy. Approximately three-fourths of the employment growth in U.S. manufacturing over the 1955–1979 period was accounted for by this sector.[5] Not only has the apparent vitality of the sector been enhanced by the harsh cyclical impacts experienced by many "low-tech" industries, but a reasonable case can be made for including some of our oldest basic industries—most conspicuously the automobile industry—in the sector. Nonetheless, several high-technology industries, including plastic materials and synthetics, paints and allied products, hydraulic cement, ordnance and accessories, electrical transmission and distribution equipment, household appliances, radio and television receivers, and motor vehicles and equipment, have experienced steady employment contraction since the early 1970s. Still other industries in the sector have relatively tiny employment bases obscured by high rates of employment growth.[6] By whatever definition, high-technology industrial development is not synonymous with significant employment growth.

Industrial Cycles and Regional High-Technology Development

The premier feature of the high-technology sector is that the firms and employment within it are distributed very unequally throughout the nation. Unequal distribution, as we saw in the previous chapter, has long been recognized as a feature of industrial development. Nonetheless, the spatial inequality associated with this sector is expressed differently than that of the past. The unequal intraregional distribution of frontier industries during the early industrial era was accompanied by a similar inequality among regions. The result was the ascendancy of the northeastern quadrant of the nation as the early industrial heartland. Today intraregional inequality is no longer so clearly reinforced by a similar pattern among regions at the larger scale. The seedbeds for high-technology industrial development are geographically scattered throughout the country. A 40-mile-long corridor known as the Silicon Valley in northern California, the Route 128 complex outside Boston, the Research Triangle in the North Carolina Piedmont, and the Dallas-Fort Worth Metroplex in North Central Texas are simply the largest and most visible of these locations, which also include the northern suburbs of New York City, the southern suburbs (northern Virginia) of Washington, D.C., and concentrations in and around Atlanta, Detroit, Austin, Minneapolis-St. Paul, Denver-Boulder, Salt Lake City, San Diego, and Portland.

In an advanced industrial era, these scattered locations constitute a multinodal "new industrial heartland" which, unlike the relatively compact and unified one of an earlier era, is defined by locally clustered development that is no longer simultaneously concentrated regionally.[7] New and renewed industries whose growth and development depend on frontier technologies often appear to respond no longer to older mixes of location factors, even though what constitutes the new mix of factors that steer high-technology development is not yet clearly discerned and is still the focus of important research. Whatever that new mix is discovered to be, it is likely to be more industry-specific than we have known before. The high-technology sector will undoubtedly be revealed to be far less homogeneous than commonly thought.[8] And increasingly, multiple scattered locations around the nation will be discovered to be capable of meeting these new industrial location requirements.

Spatial Patterns of High-Technology Industrial Development

What are the dominant spatial features of new waves of high-technology development? At the largest scale, once-peripheral regions such as the South and the West increasingly appear able to nurture new technologies and the business development tied to them, a capacity once largely restricted to older industrial regions. Rees and Norton and Rees see in this a spatial manifestation of the product-cycle model, whereby the capacity not simply to absorb and *implement* new technologies, but to *innovate* as well, has in recent decades descended a regional hierarchy from the Northeast and Midwest to the South and West.[9] Therefore, advanced industrial development implies the ability not only to host new industrial development but to incubate it as well. Consequently, what are the major spatial patterns that have emerged *within* regions? Is there any evidence that the locational features of high-technology development differ substantially from those of manufacturing in general?

In the previous chapter, we saw that the spatial structure of industrial production—specifically manufacturing—has been altered substantially during the post–World War II era. Yet the greatest changes were registered among, rather than within, multistate regions. The bulk of manufacturing employment has migrated from the older industrial North to the South. Nonetheless, manufacturing still represents a greater share of the employment of the North Central and Northeast regions than it does in either the South or the West, even though those gaps have narrowed considerably, especially since 1960 (Figure 4.2). In contrast, other spatial features of industrial structure have largely resisted significant change. For instance, despite the fact that manufacturing growth has been greatest in the smallest places in the United States, data to 1970 on the distribution of manufacturing by place size indicate that the largest metropolitan areas have retained their hold on the bulk of manufacturing employment (Figure 4.5). This is true even though manufacturing employment growth has typically shifted out of central cities to other locations within metropolitan economies (Figure 4.4). Moreover, this pattern has unfolded even as nonmetroplitan areas have converged with metropolitan areas with respect to the manufacturing share of total employment (Figure 4.6). Consequently, despite continuing intraregional deconcentration, *manufacturing is still substantially dominated by metropolitan areas, usually very large ones.*

How do high-technology development patterns compare? Several years ago Berry raised the possibility of "the hierarchical diffusion or filtering down of innovations from urban areas to peripheral areas."[10] As nonmetropolitan economies have diversified and converged in structure with metropolitan economies, can we likewise expect the capacity to innovate and the industrial development tied to it to descend a hierarchy from large metropolitan places to smaller nonmetropolitan ones? Or, rather, are the size, scale, and ancillary features of metropolitan organization still vital to the incubation of new high-technology industries, such that they are likely to remain an essentially metropolitan phenomenon? Similarly, within metropolitan areas, what important relationships between high-technology growth and the renewal of central city economies might be gleaned from observing how high-technology industrial growth distributes itself within a metropolitan area?

In order to explore these and related questions, I will present the results of a series of research efforts that describe patterns of high-technology growth within and among the metropolitan areas of Texas, one of the nation's more prominent locations of high-technology seedbeds. In the first study, I will explore the development features of a relatively new advanced service industry—computer software and data processing services (hereafter CSDPS)—as it has taken root within and among major Texas metropolitan economies. Since the bulk of this industry has emerged only in the past decade and a half, its development may hold some clues regarding how a new high-technology industry evolves and positions itself within an existing economy. In a second study, I will both broaden and restrict the focus to explore the longer-term development features of the entire high technology sector as it has emerged over the past quarter-century in the Dallas-Fort Worth regional economy. No claim is made that what characterizes the development of the entire high-technology sector—or any specific industry within it—can be casually generalized to other regions or to the nation at large; but these development patterns offer important and useful insights as we consider high-technology development elsewhere. This possibility exists in part because the computer services industry is the fastest growing high-technology industry in the United States; furthermore, as noted earlier, the overall Dallas-Fort Worth high-technology sector is the third largest and the fastest growing in the nation.

High-Technology Industrial Development in Texas

The emergence of high-technology growth and development in Texas must be viewed against the backdrop of a longer-term restructuring of the entire state economy.[11] High-technology growth is only one part of that larger picture. Historically, Texas was dominated by an agricultural economy tied to farms and ranches. A fledgling oil and gas industry was born at Spindletop and Santa Rita No. 1 at the turn of the century and quickly developed in both East and West Texas in response to the building energy demands of an industrializing nation. Texas, too, was slowly industrializing, although until recently it has lagged behind the rest of the nation in that process.

In recent decades the Texas economy has restructured significantly. Increasingly, the state became heavily dependent on its energy sector, a complex of industries that accounted for nearly a quarter of the gross state product by the 1980s. Simultaneously, manufacturing expanded rapidly. During the 1970s alone, Texas added more manufacturing jobs than did any other state in the nation, and the growth of its services economy proceeded even more rapidly.[12] Today, those development trends continue in Texas. The energy sector is beginning what constitutes a long-term decline in both output and employment. Just before the end of this century, Texas is expected to become a net energy importer, following nearly a century of having served the nation as a net energy supplier. Moreover, the advanced industrial development of the state will likely be accompanied by a slow rate of employment growth, a declining employment share, and a rising output share attributed to manufacturing. As a result, considerable interest is building once again around the question of which new industries will emerge to compensate the Texas economy for the contraction of its energy mainstay and the slow expansion of its manufacturing base.

Texas reacted with relative nonchalance when federal government studies began reporting in the early 1980s that it was one of several states nurturing a significant high-technology sector. Depending on the scope of definition, Texas ranked either second, third, or fourth in levels of high-technology employment, although it ranked considerably lower than other states in high-technology share of total employment.[13] Given the concentrated nature of the seedbeds for this sector, however, Texas was found to host several regional economies whose structures clearly registered the growing impact of this expanding sector. Recently, reeling from the softening of the energy sector amidst a larger economic recovery during the early 1980s, Texas, like state and local governments throughout the nation, has become intrigued by the prospects for new employment growth and new competitive strengths from a variety of frontier technologies. That new technologies have the capacity to spawn new industries and thereby create new employment presently dominates regional economic development discussion and planning. Interest continues unabated in devising economic development strategies that have some chance of recreating in new locations the conditions that have come to be associated with the rapid growth of high-technology industries in a few selected places.

Metropolitan Dominance and High-Technology Services

For a half-century, the centripetal pull of economic activity into increasingly large metropolitan economies has tended to underscore the importance of large urban concentrations in patterns of industrial development.[14] However, the past decade has revealed that manufacturing in the nation's smallest nonmetropolitan locations has been growing faster than in the larger metropolitan locations through much of the post–World War II period, and that population growth followed the same course by the mid-1970s; this discovery has prompted a reconsideration of metropolitan dominance and industrial development.[15] For manufacturing, in particular, the importance of agglomeration economies in advanced industrial development is widely thought to be much diminished from what it once was.[16]

The structural convergence of metropolitan and nonmetropolitan economies and the diffusion of new production and linking technologies are both the cause and the consequence of the relaxation of traditional locational constraints on much economic activity. Yet, as we have seen, the metropolitan dominance of total employment and employment in most sectors (Table 4.1) has not diminished appreciably despite several decades of relatively faster growth in nonmetropolitan areas. Moreover, the population turnaround has apparently turned around yet again; in the 1980s, metropolitan areas are once again growing faster than nonmetropolitan ones, even though selected older and larger metropolitan areas continue their population shrinkage first registered in the 1970s.

What do these trends suggest regarding the continued metropolitan dominance of advanced industrial development in general and high-technology growth in particular? From the data presented below, it appears that this dominance has not been appreciably diminished, and metropolitan areas and even central cities continue to be important to the incubation of high-technology goods and services industrial development. Metropolitan areas, if not always the largest among them, are still settings for the bulk of advanced industrial development in which new and sophisticated technologies—and the skilled workers to develop, test, apply, and commercialize them—play a central role. In time, as a technology standardizes and the industries dependent on it mature, a dispersion away from these metropolitan areas may take place. But the metropolitan dominance of technological innovation and industry tied to it appears to continue as it has throughout the previous industrial era.

In this section, I explore the development characteristics of the fastest growing high-technology service industry in the United States—computer software and data processing services. The goal is to examine the locational features of the development of this industry at several spatial scales within Texas. The data from Texas indicate that the development of the computer services industry has been highly concentrated in a few locations. Furthermore, while a new set of location factors appears to be responsible for steering this high-technology service development among and within locations, metropolitan dominance of this growth remains. For the CSDPS industry, these new location factors may well operate directly to attract and pool appropriately skilled workers and only indirectly to attract computer firms to locations where these workers prefer to live.

Computer Services Industry Growth and an Accelerated Industry Cycle

Whereas the computer manufacturing industry experienced its infancy during the 1950s, the industrial origins of the computer services industry are nearly two decades closer. The explosive development of the computer services industry has been compressed into a remarkably short period of time.[17] Today, the computer services industry (SIC 737) is one of the fastest growing industries in the U.S. economy. Together with telecommunications and computer equipment manufacturing, this industry plays a key role in what Porat has called the "information economy," a broad sector which itself consists of those industries that produce, process, and distribute information goods and services; as recently as 1976, this sector accounted for over half of U.S. wages and nearly half of the U.S. Gross National Product.[18] Factors affecting supply and demand within the computer services industry are having a synergistic effect resulting in the rapid evolution of the industry. A recent Standard & Poors analysis credits sweeping office automation and greater and more sophisticated user need as the main impetuses behind its explosive growth.[19] Also, the perennial shortage of skilled programmers, which often has been identified as a barrier to the expansion of this industry, has probably functioned equally as an inducement for accelerated technological development. Increasingly, the prospects are bright that shortages of critical labor skills will be compensated for by encoding greater amounts of language assembly directly onto chips. This technology will make the software configuration more accessible to the end-user as well as progressively less dependent on a stratum of skilled programmers and language assembly architects and technicians. In this way, technology advances figure prominently in the labor-adjustment processes of advanced industrial development.

Computer Services: An Industry Overview

High-technology services have not yet received much attention in recent policy research supporting economic development discussions. This neglect is due in large part to the traditional acceptance of manufacturing as the key element in the economic base of both regional and national economies. High-technology services are frequently ignored even in those instances where the output and delivery characteristics of a service industry reflect high and increasing dependence on advanced technology and specialized labor skills. The computer services industry experienced a rate of employment growth of 235.1 percent over the 1972–1982 period—a rate higher than that of any other high-technology industry.[20] The industry is composed of three principal segments. Computer programming and software (SIC 7372) includes firms that provide computer programming, systems design and analysis, and other computer software; this segment is currently the smallest, yet it is the most rapidly growing. Data processing services (SIC 7374) firms provide data-base processing services and comprise the largest and slowest growing segment of the industry. And computer-related services (SIC 7379) is a residual category including firms that provide customized programming systems and data-base management, consulting, and the rental, leasing, and repair of computers and peripherals.

Spatial Features of Computer Services Industry Development: Commerce Department figures indicate that employment growth in computer services more than doubled (142.4 percent) during the 1974–1983 period. Currently, the industry is represented in the South more so than in any other region. A third (32.0 percent) of U.S. total employment in computer services is in the South, with nearly equal portions in the Northeast (24.8 percent) and the West (24.5 percent); the North Central region trails with 18.7 percent. While a slow regional reranking is underway, the South's domination of this industry's employment does not appear to be threatened. Employment in computer services expanded 158.9 percent in the South during the 1974–1983 period, while expanding 195.6 percent in the West. These growth rates far exceed those for the Northeast (125.5 percent) and the North Central region (102.9 percent).[21]

At every spatial scale within Texas, development within the computer services industry has proceeded in a very uneven manner geographically. Approximately one-fourth (25.4 percent) of Texas's high-technology employment is in the computer manufacturing and the computer services sectors alone.[22] While employment in Texas's high-technology sector expanded from 1979–1982 by 5.1 percent, employment in these computer-related industries expanded by 20.0 percent and 13.3 percent, respectively. Figure 5.1 provides evidence of the growth trajectories of computer-related employment in Texas between 1972 and 1981. In the computing equipment manufacturing industry (SIC 357), employment had been steadily growing until 1978; after that date, the rate of employment growth increased dramatically, with a 43.9 percent increase in 1979 alone. Prior to 1978, the computer services industry was larger than its manufacturing counterpart. Hardware production employment surpassed computer services employment in 1978, although by 1981 it appeared that the software industry was on a trajectory to take the lead back once again. Among the CSDPS segments, data processing services has been the leader throughout, with its lead increasing over time.

Testifying to the continuing relevance of urban concentrations and accompanying agglomeration economies in attracting and nurturing growth and development in frontier service industries of advanced industrial economies, the CSDPS industry in Texas appears to be largely a creature of very large metropolitan organization. Table 5.1 indicates that only the very largest metropolitan areas—Dallas-Ft. Worth, Houston, San Antonio, and Austin—have served as the seedbeds for the development of this industry, with the Dallas area as the premier location in the state. More than 90 percent of establishments and employment in all three CSDPS industry segments are metropolitan based. As of September 1982, 62.3 percent of the firms and 68.2 percent of the employment in the CSDPS industry were located in the Houston and Dallas Primary Metropolitan Statistical Areas (PMSAs). The thinness of the distribution of the industry beyond the very largest Texas metropolitan economies indicates that size, growth rate, and their correlates are likely to be the general features of metropolitan organization that influence the development of the CSDPS industry. Moreover, this distribution underscores that computer services makes up one of several rapidly growing producer services that are particularly essential for providing specialized business services as intermediate inputs to a wide variety of final goods and services producers that themselves are highly clustered in large metropolitan areas.

Since the 1970s, a distinct division of labor has developed among the

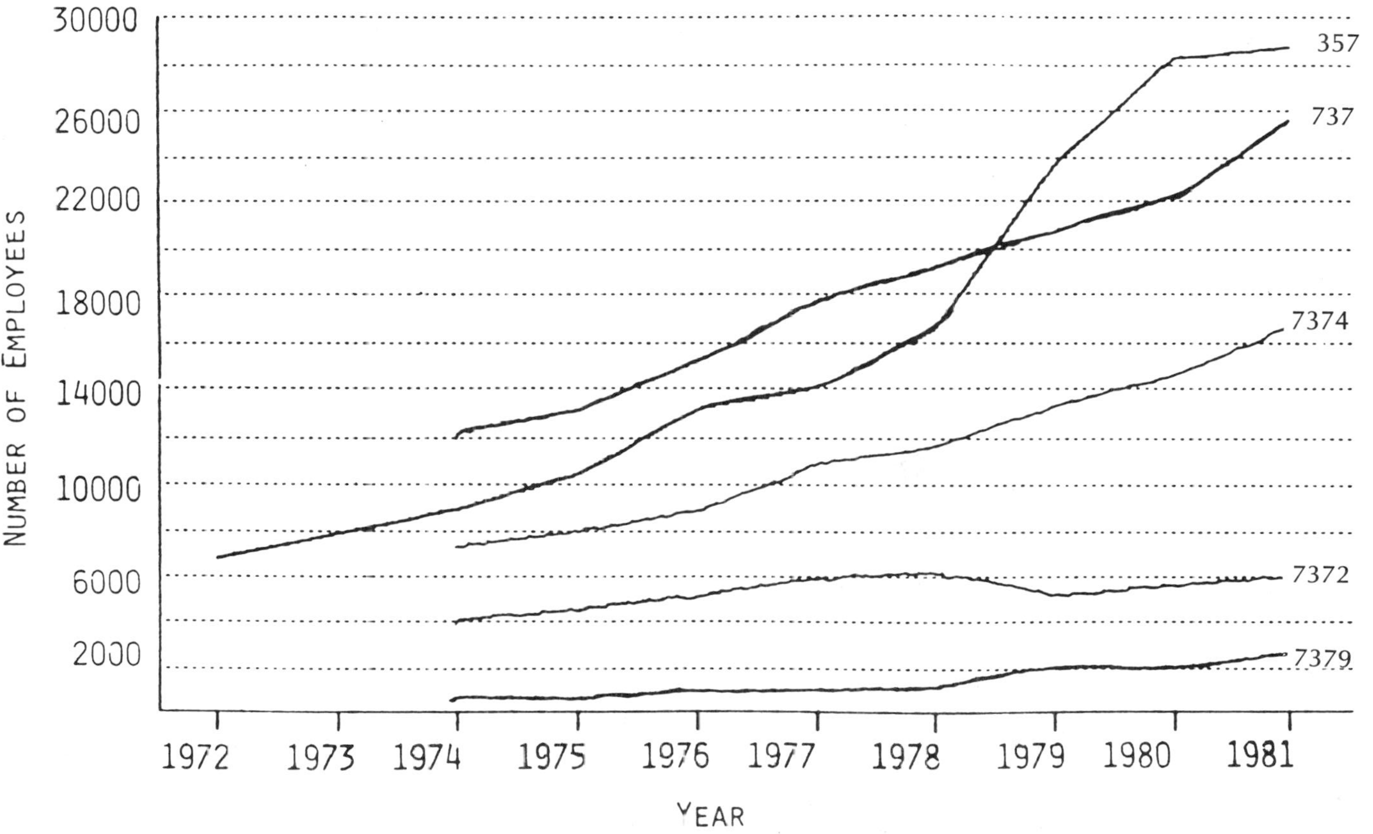

Figure 5.1. *State of Texas: computer-related industry employment, 1972–1981.*

Table 5.1
Metropolitan Distribution of Software and Computer Services Industry in Texas

	Software products		*Data processing services*		*Professional computer services*	
Metropolitan Area	*Number of firms*	*Employees at location*	*Number of firms*	*Employees at location*	*Number of firms*	*Employees at location*
Consolidated Metropolitan Statistical Areas (CMSAs)						
Houston-Galveston Brazoria CMSA	207	4,345	174	5,655	109	2,489
Dallas-Ft. Worth CMSA	251	6,300	185	11,981	133	1,999
Metropolitan Statistical Areas (MSA) and Primary Metropolitan Statistical Areas (PMSA)						
Population 1,000,000 or more						
Houston PMSA	203	4,281	168	5,616	102	2,444
Dallas PMSA	200	5,823	156	10,904	103	1,528
San Antonio MSA	42	394	39	1,225	19	142
Population 250,000–1,000,000						
Ft. Worth Arlington PMSA	51	477	29	1,077	30	471
Austin MSA	51	1,749	22	4,081	16	243
El Paso MSA	11	1,252	6	253	4	34
Beaumont-Port Arthur MSA	6	125	5	63	2	7
Corpus Christi MSA	4	16	8	61	2	10
McAllen-Edinburg-Mission MSA	6	38	6	92	2	23
Lubbock MSA	10	75	6	193	1	10
Brownsville-Harlingen MSA	1	25	2	27	1	4
Galveston-Texas City PMSA	2	9	5	33	5	25
Amarillo MSA	4	22	6	125	6	35
Waco MSA	4	58	3	36	2	44
Brazoria PMSA	2	55	1	6	2	20
Killeen-Temple MSA	1	15	—	—	—	—
Tyler MSA	6	31	6	51	1	35
Wichita Falls MSA	3	48	4	48	—	—
Odessa MSA	2	6	1	9	—	—
Abilene MSA	5	39	5	68	2	13
Texarkana MSA	2	9	—	—	1	3
Population under 250,000						
Longview MSA	3	15	1	12	1	7
Laredo MSA	3	19	—	—	1	3
Bryan-College Station MSA	9	61	2	10	—	—
Sherman-Denison MSA	2	16	2	13	—	—
San Angelo MSA	1	12	3	37	2	22

Table 5.1 (*cont.*)
Metropolitan Distribution of Software and Computer Services Industry in Texas

	Software products		*Data processing services*		*Professional computer services*	
Metropolitan Area	*Number of firms*	*Employees at location*	*Number of firms*	*Employees at location*	*Number of firms*	*Employees at location*
Midland MSA	6	58	9	551	2	9
Victoria MSA	1	10	2	91	—	—
Metropolitan	641	14,738	497	24,682	307	5,132
Total	97.9%	99.5%	93.6%	95.3%	97.8%	99.0%
Nonmetropolitan	14	78	23	174	7	54
Total	2.1	0.5	9.4	0.7	2.2	1.0
State	655	14,816	520	24,856	314	5,186
Total	100.0%	100.0%	100.0%	100.0%	100.0%	100.0%

This table reports the number of establishments and employment levels for each of the three CSDPS industry subsectors as of September 1982.

four largest metropolitan areas. Computer services—especially data processing services—have gravitated to the Dallas-Fort Worth and Houston areas, while the computer-related industrial sector in San Antonio and Austin has come to be dominated by computer equipment manufacturing, as Figures 5.2 a–d illustrate.

Computer Services Industry Development In and Around Dallas

At the smaller scale of the Dallas PMSA, once again the computer services industry is far from evenly distributed. As Table 5.2 indicates, an estimated 71.2 percent of the establishments and 89.2 percent of the employment in the industry are located in the city of Dallas proper. Despite the tendency for Dallas to grow more slowly on traditional indicators than other locations in the metropolitan area, this high-technology service industry appears to be finding the central city a particularly accommodating setting for its early establishment and growth. Even so, it is likely that in the future, suburban locations will become increasingly attractive seedbeds for business formation and employment growth in this industry. As suburban economies continue to diversify and industrialize, as congestion in commuting corridors discourages long commutes, and as skilled workers have greater access to affordable housing in suburbs and beyond, both push and pull factors may conspire to move the spatial center of the industry slowly out of the central city of Dallas.

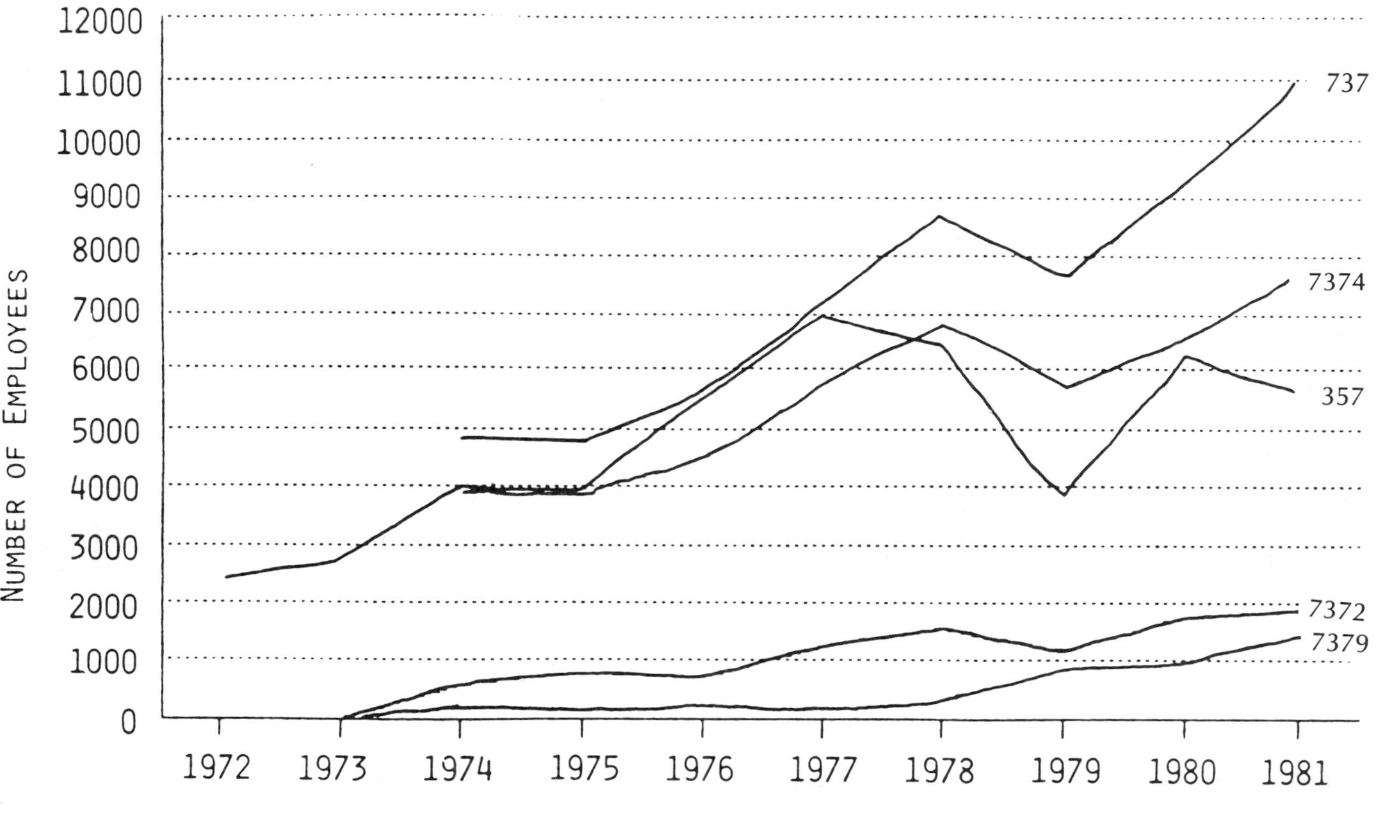

Figure 5.2a. *Dallas-Fort Worth: computer-related industry employment, 1972–1981.*

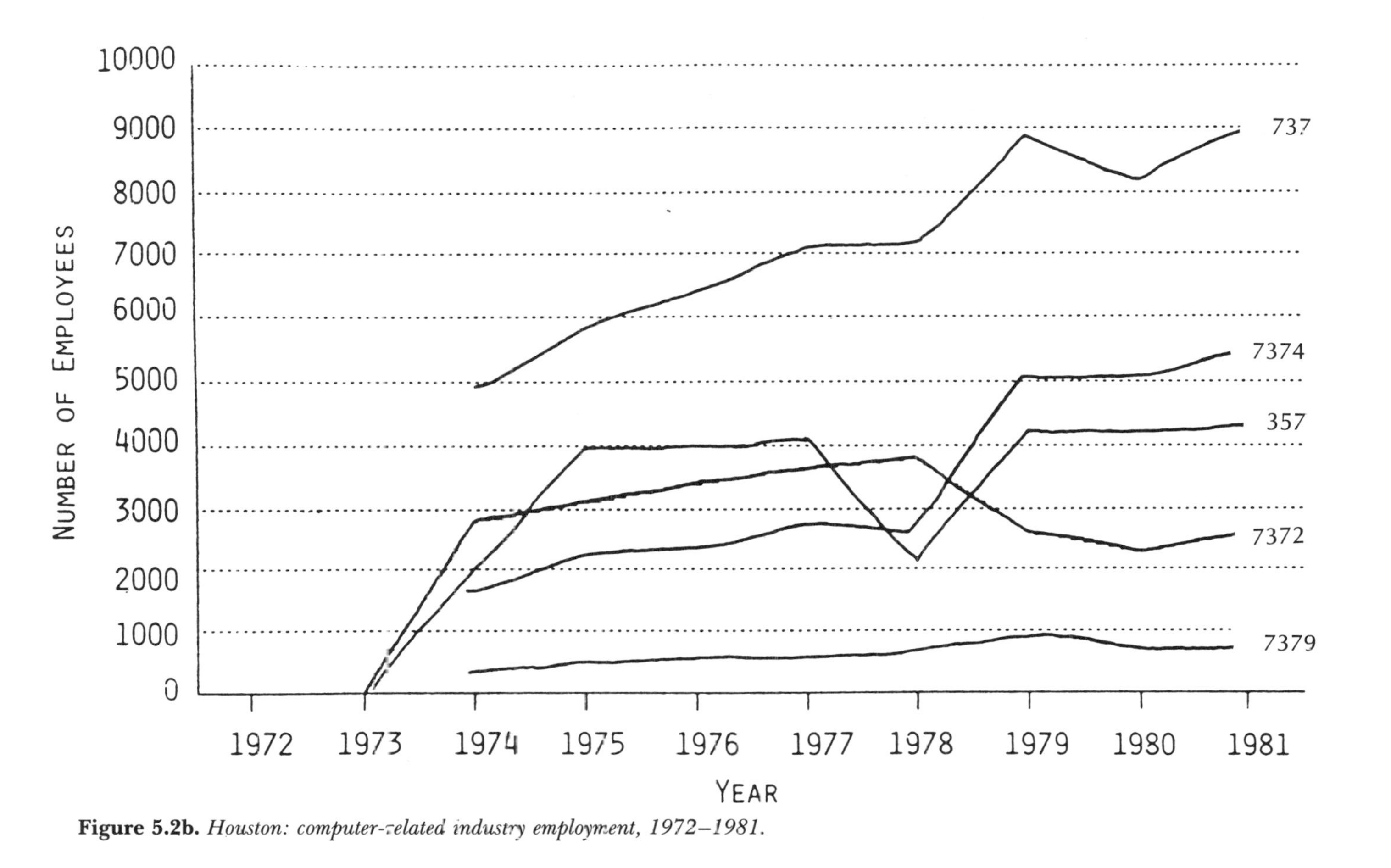

Figure 5.2b. *Houston: computer-related industry employment, 1972–1981.*

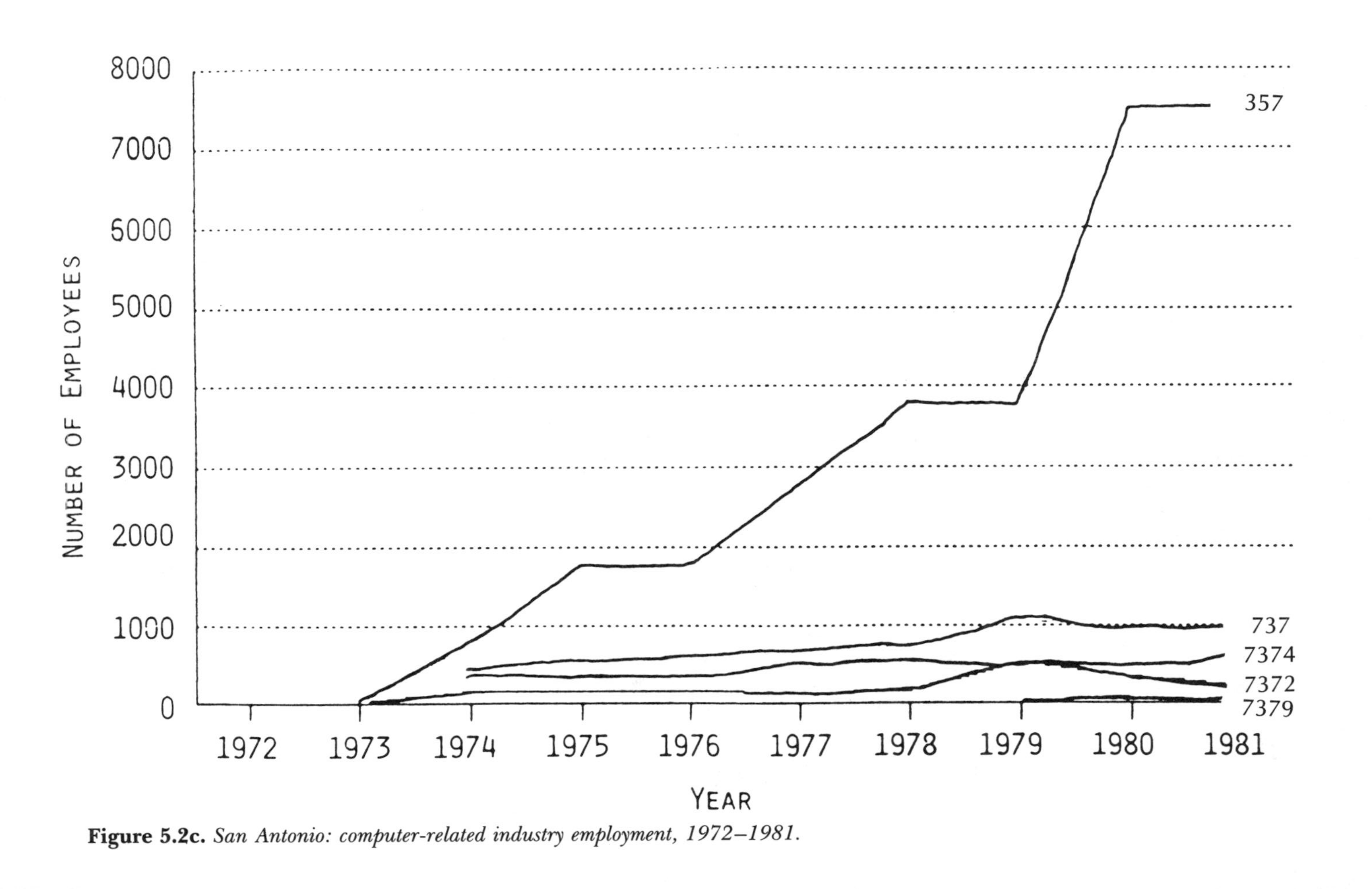

Figure 5.2c. *San Antonio: computer-related industry employment, 1972–1981.*

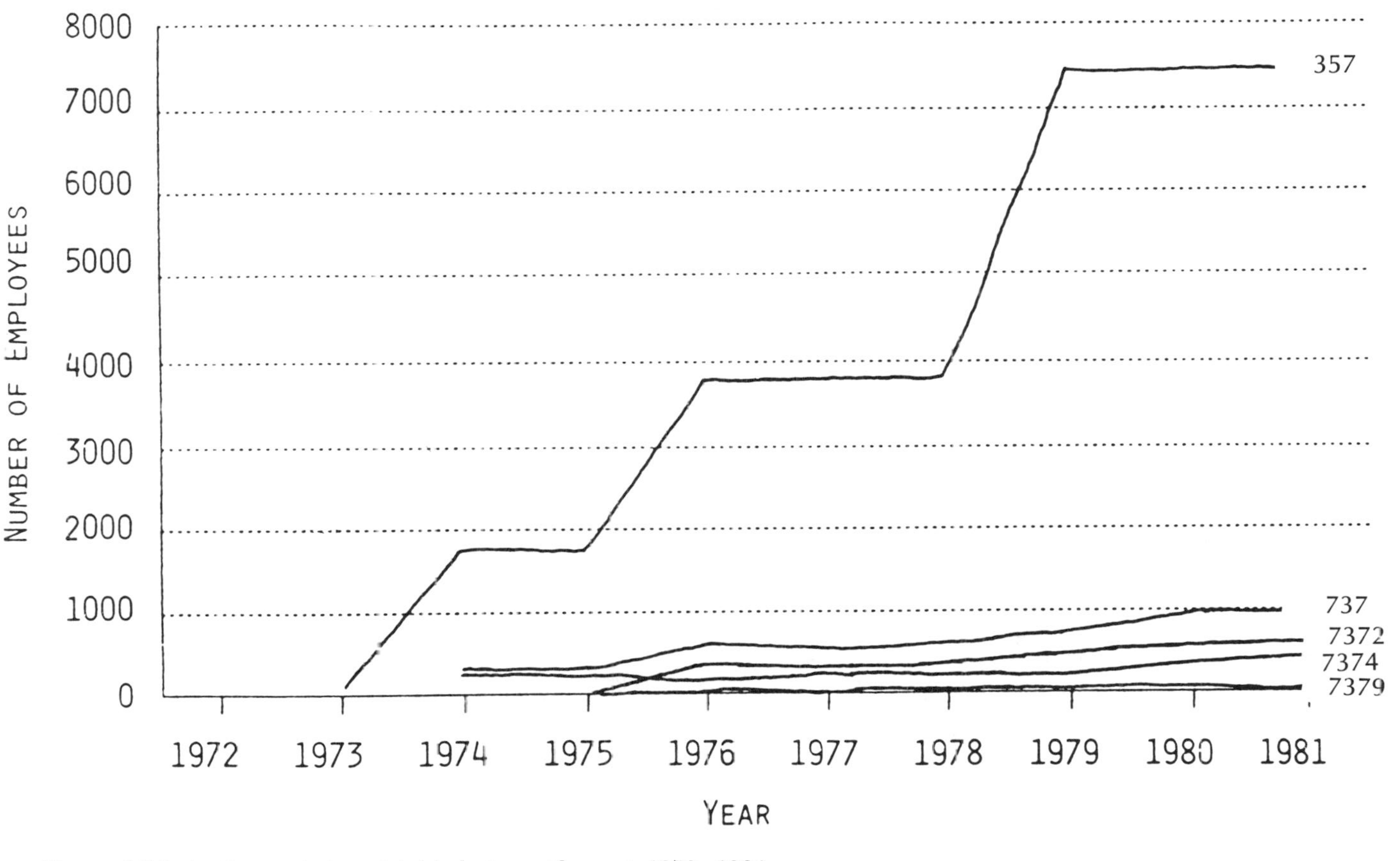

Figure 5.2d. *Austin: computer-related industry employment, 1972–1981.*

Table 5.2
Industry Establishments and Employment by City in the Dallas PMSA

City	*Software products*		*Data processing services*		*Professional computer services*		*Industry totals*			
	Number of firms	*Local employment*	*Number of firms*	*Local employment*	*Number of firms*	*Local employment*	*Firms total*	*Total employment*	*Percent of PMSA firms*	*Percent of PMSA employment*
Addison	1	4	1	6	1	3	3	13	0.7%	0.1%
Carrollton	3	15	3	24	2	6	8	45	1.7	0.2
DALLAS	136	5,061	117	9,986	74	1,232	327	16,279	71.2	89.2
Denton	2	7	—	—	—	—	2	7	0.4	0.0
Duncanville	1	8	—	—	1	4	2	12	0.4	0.1
Ennis	—	—	2	35	—	—	2	35	0.4	0.2
Garland	8	187	5	30	2	28	15	245	3.3	1.3
Grand Prairie	—	—	2	62	1	3	3	65	0.7	0.4
Irving	9	66	9	241	5	44	23	351	5.0	1.9
Kaufman	—	—	1	3	—	—	1	3	0.2	0.0
Lancaster	—	—	1	3	—	—	1	3	0.2	0.0
Lewisville	1	4	1	5	2	13	4	22	0.9	0.1
McKinney	1	4	—	—	—	—	1	4	0.2	0.0
Mesquite	3	12	—	—	1	3	4	15	0.9	0.1
Plano	5	48	1	4	6	29	12	81	2.6	0.4
Richardson	27	378	13	505	8	163	48	1,046	10.5	5.7
Rowlett	1	5	—	—	—	—	1	5	0.2	0.0
Southlake	1	20	—	—	—	—	1	20	0.2	0.1
Waxahachie	1	4	—	—	—	—	1	4	0.2	0.0
Total	200	5,823	156	10,904	103	1,528	459	18,255	99.9%[a]	99.8%[a]

[a]Totals do not add to 100.0% due to rounding error.

Finally, at an even smaller spatial scale, we are able to observe the locational patterns of the CSDPS industry *within* the city of Dallas. Table 5.3 displays the distribution of establishments across fourteen geocoded subareas of the city. Only six of the subareas (Downtown/Oaklawn, Love Field/Industrial, White Rock/Greenville, North Dallas II, North Dallas III, and Farmer's Branch) appear to offer business microclimates attractive enough to claim approximately a 10-percent share of the city's CSDPS establishment total. The area in and around the Dallas central business district (CBD) does offer a business environment attractive to a sizable portion of the establishments in the city. The Downtown/Oaklawn area is the location of a larger proportion of establishments (13.7 percent) than employment (8.8 percent), which indicates that the CBD to this point has continued to retain the capacity to attract new and small businesses in this rapidly growing industry.

Nearly half (47.6 percent) of the establishments are located within the three tiers of North Dallas and the northernmost Farmer's Branch area. Computer services development, therefore, appears to have followed the same general trajectory of residential, commercial, and industrial development within the city in the past decade. Given the presence of a handful of major computer services firms in the city of Dallas, employment is distributed even more unevenly than business establishments. A third (32.6 percent) of CSDPS employment is located in North Dallas II, a fourth (26.0 percent) in the Love Field/Industrial corridor, and one in eight employees (12.0 percent) works in the Farmer's Branch subarea on the far northwest edge of the city. The development patterns for this specific high-technology service industry raise the possibility that what a central city may incubate it may not necessarily be able to retain in the long run. Yet there appear to be clear limits to the deconcentration within this industry and sector. Despite the increasing diversification of nonmetropolitan economies in all regions, high-technology services development—perhaps even more so than high-technology goods production—appears to require large metropolitan settings, at least initially. We turn now to an exploration into the ways in which that might be true and why.

Locational Features of Texas Computer Software Industry Development[23]

A key feature of advanced industrial development is that while certain factors may be crucial to production, distribution, or consumption, that dependence need not necessarily be expressed so rigidly as before in terms of physical proximity. Backward linkages to suppliers and forward

Table 5.3
Computer Software and Data Processing Industry Establishments and Employment in the City of Dallas, 1982

Geocoded Subsectors[a]	*Software products*		*Data processing services*		*Professional computer services*		*Industry totals*			
	Establishment	*Employment*	*Establishment*	*Employment*	*Establishment*	*Employment*	*Establishment*	*Employment*	*Percent of firm total*	*Percent of employment*
1. Oak Lawn/Downtown	16	165	19	802	11	246	46	1,213	14.3%	9.4%
2. Love Field/Industrial	19	1,794	23	1,274	12	204	54	3,272	16.8	25.4
3. Southwest Dallas	4	62	9	278	—	—	13	340	4.0	2.6
4. Redbird	1	5	1	10	1	6	3	21	1.0	0.2
5. East Oak Cliff	—	—	1	10	—	—	1	10	0.3	0.1
6. Fair Park	—	—	1	7	—	—	1	7	0.3	0.0
7. Pleasant Grove	2	66	—	—	—	—	2	66	0.6	0.5
8. East Dallas	1	10	2	35	1	10	4	55	1.2	0.4
9. White Rock/Greenville	14	186	9	142	9	166	32	494	9.9	3.8
10. North Dallas I	6	58	6	85	8	112	20	255	6.2	2.0
11. North Dallas II	21	1,655	13	2,489	9	52	43	4,196	13.4	32.6
12. North Dallas III	16	159	12	897	6	38	34	1,094	10.6	8.5
13. Vickery-Lakewood	5	208	5	46	3	57	13	311	4.0	2.4
14. Farmers Branch	30	595	11	609	15	345	56	1,549	17.4	12.0
Totals							322	12,883	100.0%	99.9%[b]

[a]The subeconomies that comprise the city of Dallas were created by clustering zip code areas. The subareas were created as follows: (1) Oak Lawn/Downtown: 75204, 75219, 75201, 75246, 75221, 75222, 75202, 75250; (2) Love Field/Industrial: 75207, 75247, 75235, 75245, 75258; (3) Southwest Dallas: 75212, 75208, 75236, 75211, 75249; (4) Redbird: 75224, 75237, 75232, 75233; (5) East Oak Cliff: 75203, 75216, 75241, 75239; (6) Fair Park: 75226, 75210, 75215, 75223; (7) Pleasant Grove: 75227, 75217, 75253; (8) East Dallas: 75228; (9) White Rock/Greenville: 75218, 75238, 75231, 75243; (10) North Dallas I: 75220, 75225, 75205, 75209, 75275; (11) North Dallas II: 75229, 75230, 75251; (12) North Dallas III: 75248, 75240, 75252; (13) Vickery-Lakewood: 75206, 75214; (14) Farmers Branch: 75234.

[b]Values do not add to 100.0% due to rounding error.

linkages to distributors and consumers can lengthen dramatically as materials substitution occurs; as the product weight, mass, and scale decline; as transport costs drop; and as intermediate and final demand markets expand geographically. As greater portions of goods production and services provision come to reflect the value added by information and specialized knowledge, telecommunications can replace transportation as a means of pooling factor inputs and distributing outputs. Warehouses and factories give way to switchboards and computers. Distance and time then lose much of their traditional capacity to dictate the spatial features of economic activity. Presumably, all this should be especially evident in certain high-technology sectors such as the software industry.

As will be shown below, Texas computer software firms do not appear to be restricted in their preferences to locations at or even near existing economic crossroads such as a high-density commercial setting or a traditional central business district. Rather, they appear willing and able to accommodate themselves to a wide variety of general and specific settings. While these data provide a cursory overview at best of the development distribution of software firms across a range of alternative locations, there does not seem to be any industry or business characteristic so compelling that it rigidly constrains the location decision-making process. While there is ample evidence of the continuing importance of face-to-face communication in certain aspects of even this business, the industry also appears to be able to filter into locations where the volume of such activity may be quite low. Yet, one does not have to leave a metropolitan area to discover such low-density locations. The importance of this apparent indifference to physical location within the metropolitan economy reveals a great deal about the changing relationship between the "city," which reflects the productive arrangements of an earlier industrial era, and newer industrial arrangements that can characterize an advanced industrial economy.

To what extent can we observe this relative indifference to small scale location in high-technology services, specifically the computer software industry? If the constraints on location in advanced industrial development have indeed been so relaxed, what then accounts for the high degree of larger-scale geographical concentration that characterizes the larger computer services industry observed above? From what we have seen, location decisions in the computer services industry are neither random nor haphazard. Yet perhaps "location" is perceived differently among firms in this new and rapidly emerging industry than has typically been the case for firms in older industries. Does advanced industrial development reassign primacy to new rankings of influences on location within metropolitan economies? For answers to these and related questions, let us search for what factors, if any, influence the location of computer software firms in Texas.

Location Patterns and the External Environment: Export Orientation Within Texas Computer Software Firms

> Contrary to conventional belief, a significant portion of the services sector is not merely "residentiary" or "nonbasic" activity, but is export-oriented and locationally sensitive to city size and other indications of urbanization.[24]

An initial insight into the locational preferences of Texas computer software firms flows from data on the market orientation that characterizes the industry. It is reasonable to expect that firms that orient themselves predominantly toward local consumption may place great significance on the decision concerning where to locate. Traditionally, service industries have been considered subordinate and of largely secondary importance to goods production. At least part of this orientation stems from the perception that goods production involves more export-oriented components of an economic base, thus providing an extra margin of employment supported by the possibility of producing to meet demand generated above and beyond that tied to local consumption. We have already seen that the vast majority of computer services firms in Texas are located somewhere within a few selected large metropolitan areas and therefore appear to be "locationally sensitive to city size and other indications of urbanization." To what extent are these same firms export oriented and thereby influenced to select locations that will facilitate their development of markets external to their large-metropolitan production platforms?

As a high-technology business service, the Texas software industry reveals a considerable export orientation.[25] Among Texas software firms, there is evidence that despite their often small size, fully 37.5 percent primarily market their products and services nationally and 18.8 percent market internationally (Figure 5.3). Another 15.0 percent target markets within the multistate region, and 28.8 percent restrict their marketing primarily to locations within the state of Texas. While patterns of actual sales may differ, a purely local marketing orientation is virtually nonexistent among Texas software firms. It is unlikely that firms in a newly emerging industry with such an expansive marketing focus will judge their narrow physical locations within large metropolitan settings to be a particularly important constraining factor, at least for business-related reasons.[26]

Location and Business Expansion Plans: A second piece of evidence even more directly discounts the importance of small-scale physical

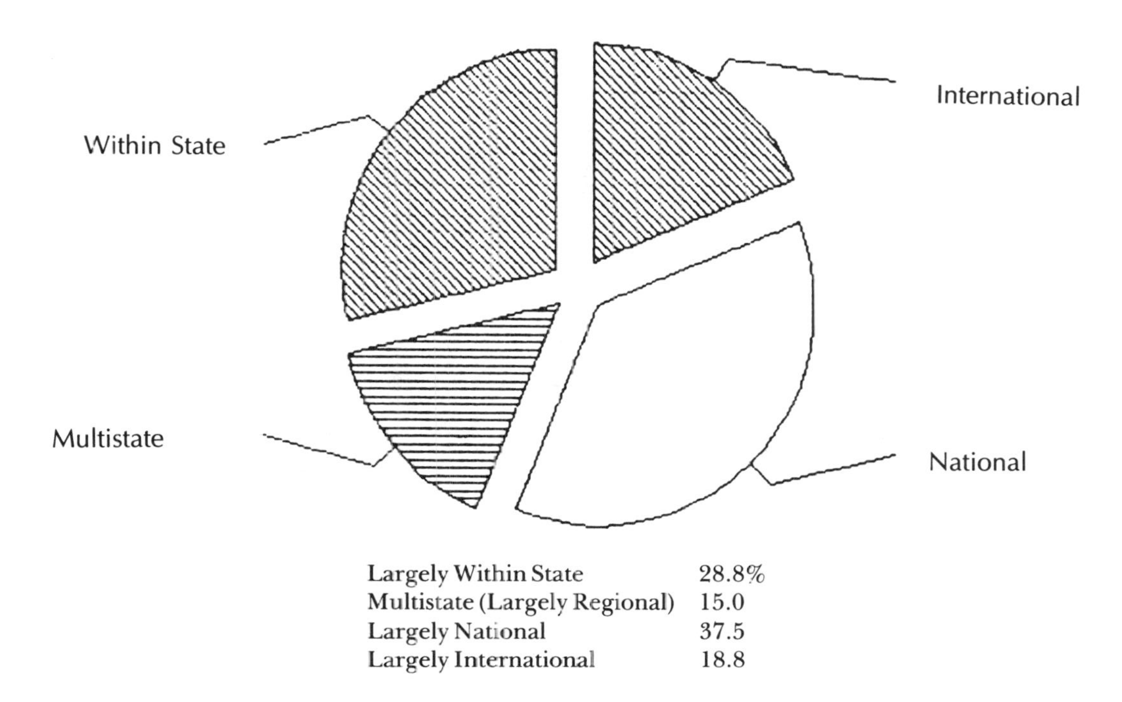

Figure 5.3. *Target market for major product.*

location factors among Texas computer software firms. Respondents from these firms were requested to rank potential barriers to their future expansion plans. Business formation and expansion decision making cannot be explained in isolation from an understanding of both the actual production processes within an industry and the important contextual features of the locations or settings in which they occur. While the decisions analyzed here all benefit from actual business experience, the outcomes can be both fragile and unstable given the speed with which evolving technology in the industry can alter the mix of physical, financial, and human-capital requirements over a short period of time.

Respondents were asked about the major barriers their companies faced in future expansion plans. Reflecting the minimal dependence on physical location associated with activity in this industry, location emerged as the least influential factor of all. A poor physical location was mentioned as a potential barrier to future growth by only 5.5 percent of all the companies (Figure 5.4). Of more immediate concern was the development of a successful marketing strategy (84.2 percent), access to long-term financial capital (61.3 percent), the cost associated with product development (60.0 percent), increasing competition within the industry (56.8 percent), problems associated with research and development (52.7 percent), and finally, problems related to growth management (50.0 percent).

Business Climate and Location Factors

Next, let us consider how firms "read" and evaluate a business environment and orient themselves toward its important features. The perceptions and assessment of "business climate" by firms in an industry or sector can provide clues regarding the features of economic and political economic landscapes that exert influence on location decisions. In light of their expansive market orientation, how have Texas computer software firms come to perceive and define *business climate?* What factors appear to exert a "pull" on individual firms' location decisions and thereby ultimately influence the spatial features of development within this industry? As we shall see in the data below, computer software firms appear to be concerned less with discovering a specific optimal location and more with locating within a broad range of alternative locations while in no way hindering their access to essential pools of labor skills.

Twenty-two factors—the same used in Figure 3.7—were used as stimuli for respondents to assess. They included circumstances that were relevant to the conduct of business in general; to the business

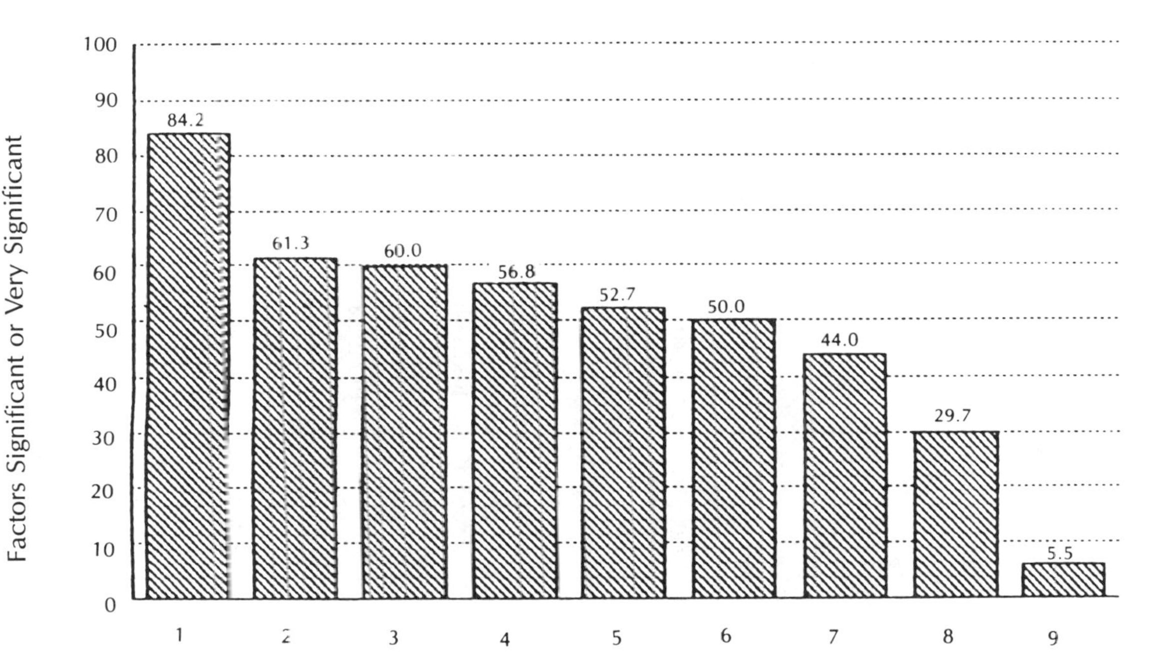

1. Marketing
2. Long-term financing
3. Cost of program development
4. Increasing competition
5. Research and development [costs]
6. Management of growth
7. Uncertainty within hardware industry
8. Lack of skilled labor
9. Poor location

Figure 5.4. *Barriers to future growth.*

establishment itself; to the lives of employees both during and after the workday; and to the relationship between production, consumption, and distribution within the sector and the larger economy.

The importance of access by employees to the workplace and access by the firm to appropriately-skilled workers emerged as the major influences that steered a firm across the economic landscape (Figure 5.5).[27] The vast majority (81.2 percent) of the firms rated as significant or very significant easy access to the worksite for employees, and 79.8 percent and 66.7 percent, respectively, placed great importance on the availability of a potential pool of labor skills and access to a good highway system. A low cost of living, which is arguably of more immediate significance to employees than to the employing firm, was cited as important by 63.3 percent of the companies.

Taken together, these first four business environmental features are important largely because they can be influential in attracting and retaining the specialized workforce required by these companies as they compete and grow. Other factors that at least a majority of the firms rate as important include room for expansion in the locality (62.3 percent) and access to a major airport (61.7 percent). In the latter case, once again, there is evidence of the importance assigned to convenience for employees regardless of whether or not business-related travel may be involved.[28]

The majority of factors 7 through 16 relate principally to specific features of the firm itself and the conduct of routine business. Several of these factors involve proximity to traditionally important resources in the production-distribution-consumption sequence. However, for the Texas software industry, these proximity-related factors carry relatively less weight than do factors that have the potential to influence whether an ample supply of appropriately skilled workers is likely to prefer to reside in the location.

Public Policy–Manipulable Factors: Those traditional business-climate factors generally considered to be manipulable by local governments in their efforts to attract and retain business start-ups and to encourage secondary expansions in-place appear to be of only moderate importance to computer software firms. Such factors as business-tax climate (57.7 percent), local regulations on business (54.4 percent), and proximity to good public schools (43.4 percent) emerged as having a relatively modest influence on the business-climate evaluation. In addition, the cost and availability of energy (34.2 percent) emerged as being of relatively minor significance. Finally, of all the factors commonly targeted for public policy intervention, the quality of local public transportation systems was rated as significant by the smallest proportion

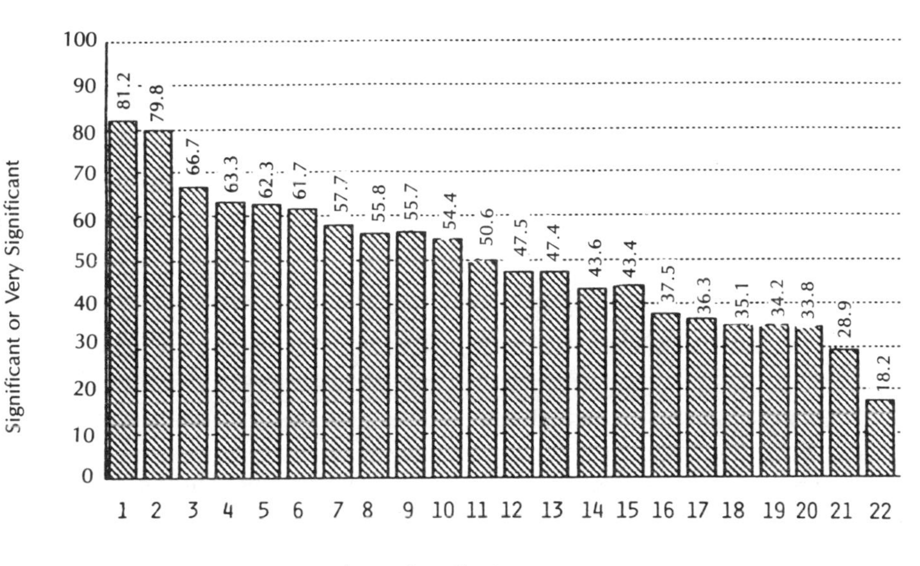

(1) 1. Easy access for employees
(3) 2. Labor skills/availability
(2) 3. Access to good highway system
(10) 4. Cost of living
(7) 5. Room for expansion in locality
(4) 6. Access to major airport
(9) 7. Business-tax climate
(6) 8. Room for expansion at present site
(15) 9. Quality of municipal services
(13) 10. Local regulation on business
(16) 11. Climate
(5) 12. Being close to customers/markets
(8) 13. Entrepreneurial climate
(20) 14. Recreational/leisure facilities (e.g., restaurants, sports, etc.)
(17) 15. Proximity to good public schools
(19) 16. Proximity to other business (e.g., suppliers)
(14) 17. Proximity to good universities
(18) 18. Cost of land
(12) 19. Energy availability/cost
(11) 20. Good mailing address
(21) 21. Cultural amenities (e.g., concerts, museums, etc.)
(22) 22. Local public transportation

Note: Numbers in parentheses are the rankings of factors from the larger data set on the U.S. computer software industry reported in Chapter 3 (Figure 3.7). A Spearman's rank-order correlation coefficient (rho) was computed between these two rankings to discover whether they were related. The value of rho was equal to +0.782, which is statistically significant at the .001 level with 20 degrees of freedom.

Figure 5.5. *Factors influencing firm location.*

(18.2 percent) of these firms. As was the case with computer software firms across the United States (Figure 3.7), this may well reflect the capability of valued employees and prospective employees to express their preferences for private automobile transportation. Such employees may well find local public transportation systems largely irrelevant to their decisions about where to reside and work while retaining satisfactory access to these and other selected locations in a relatively low-density metropolitan environment.

Work Characteristics, Labor Skills, and Human-Capital Requirements

Software production and the services allied with it are known to be very labor-intensive. That special skills required by the industry are often in short supply, even in labor surplus areas, illustrates the problem of increasingly segmented labor markets. Consequently, physical capital requirements can be expected to place relatively minor constraints on the production process and its ultimate location, whereas the essential human capital embodied in a range of technical skills and proficiencies looms as a more important production input. Worker preferences, rather than other characteristics of the production process itself, appear to have greater leverage over the location of the firm than has traditionally been the case in older industries this early in their life-cycles. Since software industry employment is largely dominated by programmers and analysts with relatively specialized skills and training, the industry may gravitate less to those locations where there is sufficient demand for its services than to those locations that offer firms and skill pools easy access to one another.[29]

Worksite Location Flexibility: In addition to the weak constraints on the physical location of software firms, there are also relatively weak locational constraints on specific work activities themselves within the industry. Despite their often small sizes, half of these firms (48.8 percent) employ such people as independent consultants, outside contractors, free-lance programmers, and software authors, all of whose work does not require them to be on the premises. Therefore, the total production of even these generally small firms can be supported by quite modest amounts of physical space, given the complex and variable working arrangements that characterize this industry.[30]

The Role of University Proximity

As noted in Chapter 3, much importance has been assigned to the presence of quality research and educational institutions for nurturing high-technology business development in regional and local economies. In a 1982 study of location factors for high-technology development, proximity to high-quality universities emerged as highly influential in discriminating among regions, although *not* among locations within regions.[31] The data reported here for Texas computer software firms appear to confirm this general finding.

The range of potential benefits presumably derived from access to and even dependence on high-quality universities includes a continuously expanding pool of graduates with state-of-the-art education and training, libraries and information systems, research facilities such as laboratories and computer networks available for industry-university cooperative research, faculty consultants for specific off-campus projects, and general university-based cultural activities. Other benefits may be of greater value to employees in their off-the-job pursuits, while still retaining their job-related value. These include advanced degree programs, nondegree continuing education programs, and part-time teaching opportunities. Despite the presumed appeal of these university-based assets, for the software industry in Texas the role of universities in the development process is neither direct nor explicit. Fully 77.3 percent of the firms responded that proximity to a university—quality-related concerns notwithstanding—was *not* a factor in their location decisions. Among those attributes considered important by the remaining firms were the view of a university as a continuing source of college graduates (52.0 percent) and as the source of continuing educational opportunities for college graduates (37.0 percent).

The implications of these findings are not clear-cut. The nature of "proximity" lends imprecision to the analysis, since access to the benefits of a university may not be reliably understood in terms of physical distance. The degree of interaction and interdependence between industries and universities may be little influenced if pools of potential and appropriately skilled labor are available for other reasons. The historical capability of Dallas and Houston, for example, to compete well for new graduates from major universities in and outside the state indicates that the computer software firms can retain access to value-added by university education in Texas and beyond despite the fact that the pool of appropriately skilled college graduates is not "home-grown" in the strictest sense. The presence of quality institutions of higher education may well be an important feature of the larger-scale environment; nevertheless, this factor does not appear to constrain software industry development by influencing firms to huddle particularly close to universities or at least to do so in a calculated way. Nonetheless, the circumstances that have caused the attraction and recruitment of skilled labor may very well be changing now as living costs rise and urban congestion steers new residential development well away from major employer locations.[32]

"Nested" Locations and the Distribution of Texas Computer Software Firms

The "location" of economic activities can be understood at several different levels. In this section, I examine the resulting "nested" locational patterns that characterize the software industry within Texas at three interrelated spatial scales. The first scale is the largest—that is, location within the structure of the larger metropolitan area. These findings will be interpreted in light of the earlier findings reported in Chapter 3 on the larger computer services industry in the United States. The second scale focuses on the immediate "neighborhood," defined as the area within a quarter-mile radius of the firm. The third is the smallest scale—that is, the actual physical setting along with the structure that houses the firm.

If Texas computer software firms have such a strong export orientation, view physical location as relatively inconsequential in future expansion plans, and value access to pools of appropriately skilled labor above all else, how do these factors influence their aggregate spatial distribution? In general, we should expect to find that software firms are relatively unconstrained by historically influential locational factors and thus can accommodate themselves to a wide variety of intrametropolitan locations. And that is indeed what we find.

As Figure 5.6 reveals, fully 42.0 percent of the software firms report suburban locations; 32.1 percent are located in a central city, although just outside the CBD; and 22.2 percent are located in the downtown area of a central city. The progressively larger proportions of firms encountered as one looks beyond the CBD reflect not only the relatively recent expansion of the industry, but also patterns of residential development and the resulting tendency for these firms to locate near to where valued employees are most likely to reside or otherwise congregate, either by preference or by necessity. At the next smaller scale, firms appear capable of accommodating themselves to widely varying neighborhood settings (Figure 5.7). Over half (57.5 percent) are located in what they perceive to be predominantly commercial or retail settings, 13.8 percent in industrial-manufacturing settings, 3.8 percent in residential settings, and 23.8 percent in mixed land use settings. Finally, at the smallest scale, nearly equal proportions occupy leased space in a multistory office complex (40.0 percent) or in a single-story office complex (38.8 percent) (Figure 5.8). Only 15.0 percent occupy a com-

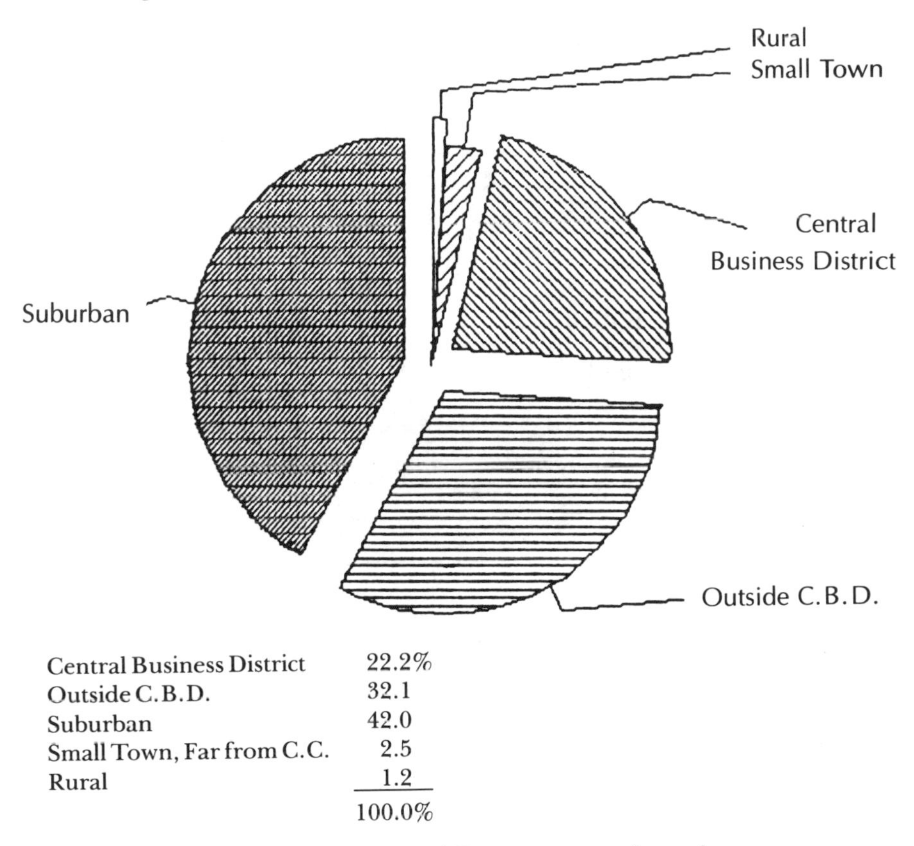

Central Business District	22.2%
Outside C.B.D.	32.1
Suburban	42.0
Small Town, Far from C.C.	2.5
Rural	1.2
	100.0%

Figure 5.6. *Intrametropolitan location of Texas computer software firms.*

pany-owned building.[33] And only 3.8 percent are housed in a converted portion of a residence.

High-Technology Industrial Development and Central City Change[34]

> There is no reason to believe that aging industrial cities will be able to revitalize unless they are able to develop a post-industrial high-technology or service activity base.[35]

The crucial role of large urban areas in the incubation and diffusion of innovations of all kinds is well known.[36] From factories for traditional manufacturing to teleports for telecommunications, successive waves of new technologies long have found in large cities a highly nurturative environment. Ironically, while the size, scale, and high densities of urban centers have traditionally provided the spark for technological innova-

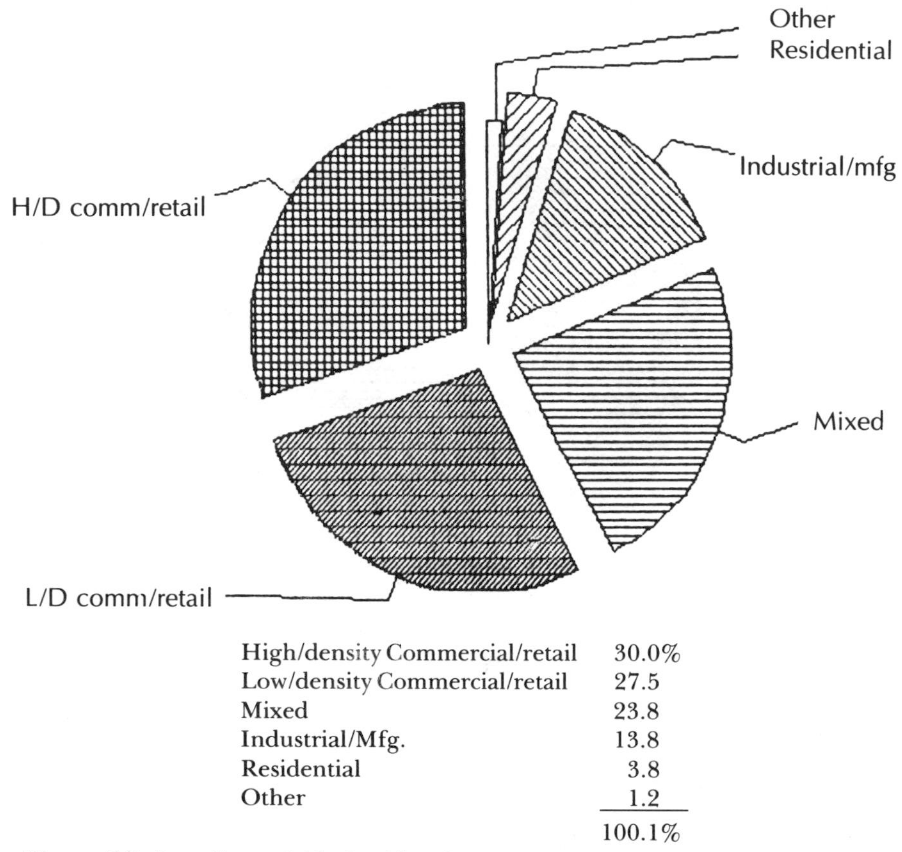

High/density Commercial/retail	30.0%
Low/density Commercial/retail	27.5
Mixed	23.8
Industrial/Mfg.	13.8
Residential	3.8
Other	1.2
	100.1%

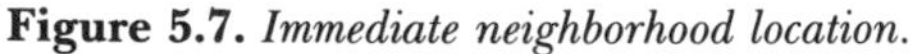

Figure 5.7. *Immediate neighborhood location.*

tion and diffusion, the succession of technologies in transportation, communication, and industrial production have gradually reinforced cultural preferences by expanding the scale of advanced industrial economies and thereby promoting the deconcentration of economic activities into lower-density arrangements. Does this dispersal of urban activities and the recasting of urban economies at larger spatial scales necessarily imply the erosion of precisely those features of older industrial settlements that are commonly believed responsible for enabling them to serve in the first place as a continuing source of new innovations and the industrial development tied to them? Have our large industrial-era cities relinquished their role as the seedbeds of new technological innovations so vital to the ongoing industrial development of the nation? Has the very capacity to innovate been yielded by industrial-era central cities to new and different locations? Or has the innovation process itself that underlies newly emerging and renewed older industries perhaps become less dependent on the features of particular places

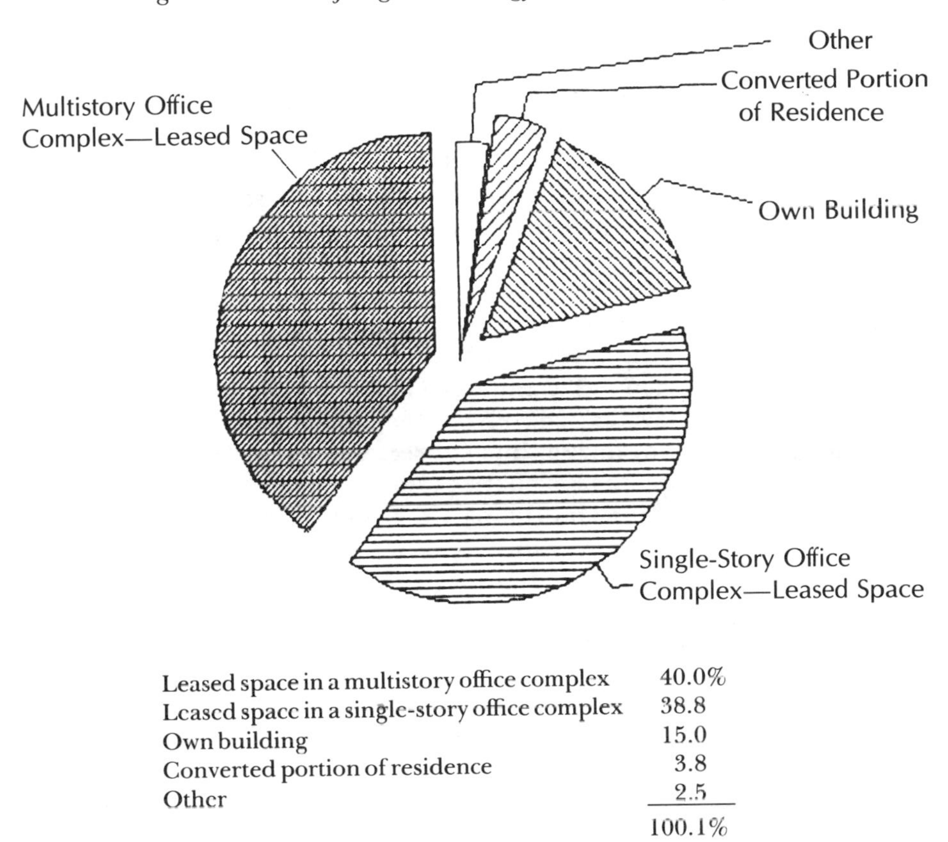

Leased space in a multistory office complex	40.0%
Leased space in a single-story office complex	38.8
Own building	15.0
Converted portion of residence	3.8
Other	2.5
	100.1%

Figure 5.8. *Physical setting.*

than ever before? In short, does innovation remain a place-sensitive process? It is to these and related questions that the following section is devoted.

As we saw in Chapter 4, advanced industrial development has commonly used population and employment contraction to reshape the form and function of large older cities in a manner similar to the way in which an earlier industrial era used the opposite processes of population and employment growth to create and shape the local urban arrangements it required. The result has been a virtual redefinition of the roles played by these cities in larger urban economies. Part of this process is thought to involve both the diffusion of the capacity for technology innovation and its implementation away from older industrial locations and toward new industrial settings. At least part of the explanation for the failure of older industrial cities and regions to compensate for the widespread losses of population, employment, and business growth is thought to involve their diminishing capacity to incubate new waves of technological development necessary to provide the basis for future industrial growth.[37] Any alteration in that historic role would erode significantly yet another basis for the centrality of these cities in a new industrial order. As we shall see, however, industrial-era central cities are not categorically doomed to be eclipsed in this respect. Many may well retain the capacity to continue to be the sources, and not simply the destinations, of new technologies, even though other indicators of urban activity are declining.

In this section I explore the question of whether or not there are important locational features of high-technology development that will allow us to anticipate the ways in which advanced industrial development and urban development will continue to influence each other. Below I trace in both employment and establishment terms the origins and defining features of the high-technology sector in the Dallas-Fort Worth regional economy over the last quarter-century up to 1984. A basic goal is to illustrate how deep and complex the roots of such a sector are. While widespread interest in this sector is new, its development has been underway for many decades. Key features of that development, viewed against the backdrop of the rapid expansion of the larger regional economy, are outlined here. In the end, what I hope to offer is a deeper appreciation of the long-term developmental dynamics of a localized high-technology sector. The emergence of a cluster of industries that meet today's criteria for being defined as high-technology follows essentially the pattern stated aptly and succinctly in 1938 by E.A. Wood, a former Director of the Texas Planning Board: "Industrial development of whatever nature is largely a matter of development from within . . ."[38]

The Origins of the Dallas-Fort Worth High-Technology Sector

Development from within—that has been the key to the development of the high-technology sector that has been steadily emerging in the D-FW regional economy over several decades. By and large, this sector's constituent industries have not been recruited from outside or grafted on through strategic planning efforts or carefully crafted industrial development schemes; rather they have developed and expanded largely unnoticed over many years. By one estimate, over five hundred firms have spun off from such local high-technology pioneers as Texas Instruments and Rockwell International alone.[39] However, how long these industries and their subsectors retain their competitive advantages for themselves and this region may now depend on selected and more conscious development strategies in the future.

Thus far, the D FW region has been fortunate. Its economic base has long been tied closely to many of those industry groups that nurtured the enterprises that constitute today's high-technology industry elite. To a considerable extent, the resulting high-technology sector overlaps with the sprawling defense industry complex defined principally by the aerospace and defense systems and the electronics, communication, and computing equipment industries. Together, this military-industrial complex received 56 percent of the FY 1984 defense procurement expenditures in Texas. The highly concentrative effect of defense spending is illustrated by the fact that 94 percent of the eleven-county region's defense expenditures were funneled to Dallas County. And at a smaller scale, while the city of Dallas has only a 24.5-percent share of the defense contractors in Dallas County, together they received 69.5 percent of the FY 1984 defense spending.[40] To appreciate better the extent to which the "high-tech" mantle has been more thrust upon the region than won by it, let us examine the development patterns in the region's overall economic base more closely.

Growth and Development of the D-FW Manufacturing Base

The origins of the D-FW region's high-technology sector are best viewed in the context of the growth and development of the larger regional economy over the past two decades. Approximately a fifth (20.4 percent) of D-FW regional employment in 1982 was in manufacturing, down sharply from 24.0 percent in 1980. These employment estimates reflect the cyclical effects of the back-to-back recessions of the early 1980s, as well as the slow growth of manufacturing employment relative to that of services. Table 5.4 reports employment trends in the D-FW manufacturing sector since 1964. Between 1964 and 1980, the growth rate of manufacturing employment accelerated in the regional economy. Manufacturing employment expanded 30.5 percent between 1964 and 1972 and 43.9 percent between 1972 and 1980 before slow growth and a declining share commenced after 1980. Indications are that manufacturing growth has accelerated once again as the present economic recovery continues.

The bulk of the region's manufacturing employment is in a handful of durable goods industries. Durable goods employment more than doubled between 1964 and 1982 as manufacturing growth shifted toward these industries. The manufacturing employment share of durable goods increased steadily from 61.0 percent in 1964, to 63.7 percent in 1972, to 66.3 percent in 1980, and 69.4 percent in 1982. All through the 1964–1982 period, three of the industries that host high-technology

Table 5.4

Total Manufacturing and Export Employment in the Dallas-Fort Worth SMSA, 1964–1982

SIC	INDUSTRY	1982 PERCENT OF TOTAL	1982 TOTAL EMPLOYMENT (000s)	1982 ESTIMATED EXPORT EMPLOYMENT	1982 PERCENT EXPORT EMPLOYMENT	1980 TOTAL EMPLOYMENT (000s)	1980 ESTIMATED EXPORT EMPLOYMENT	1980 PERCENT EXPORT EMPLOYMENT
	MANUFACTURING: DURABLE GOODS							
36	Electrical/Electronic Machinery	17.6%	55.7	13,378	24.0%	58.7*	23,559	40.1%
37	Transportation Equipment	15.4	48.8	12,172	24.9	46.3*	14,117	30.4
35	Non-Electrical Machinery	14.9	47.0			44.2*	730	1.7
34	Fabricated Metal Products	8.0	25.2			22.5*		
32	Stone, Clay, Glass & Concrete Products	3.3	10.4			8.9*		
38	Instruments & Scientific Equipment	2.6	8.3			5.3*		
24	Lumber & Wood Products	2.3	7.4			7.4		
33	Primary Metal Industries	2.1	6.5			7.8*		
25	Furniture & Fixtures	1.8	5.7			6.1*		
39	Miscellaneous Durable Goods	1.4	4.3			--		
	MANUFACTURING: NONDURABLE GOODS							
27	Printing & Publishing	7.8	24.5			22.8*	828	3.6
20	Food & Kindred Products	6.8	21.6			21.3*		
22	Textile Mill Products[1]	4.6	14.7			1.0*		
23	Apparel Products	---	---			20.9*		
30	Rubber & Plastic Products	3.5	11.1			11.4*		
28	Chemicals & Allied Products	3.5	11.1			8.5*		
26	Paper & Allied Products	2.8	8.7			8.0*		
21	Tobacco Manufactures	---	---			0.0		
29	Petroleum Refining & Related Industries	---	---			2.5		
31	Leather & Leather Products	---	---			2.7*		

39	Miscellaneous Nondurable Goods	1.6	5.2			--		
39	Miscellaneous Manufacturing Total[2]	--	--			6.1		
	TOTAL MANUFACTURING EMPLOYMENT	100.0%	316.2			312.4		
	TOTAL EXPORT EMPLOYMENT			25,550			39,234	
	EXPORT EMPLOYMENT MULTIPLIER			8.1%			12.5%	
	TOTAL SMSA EMPLOYMENT		1549.5			1299.3		

[1]The data for SIC Codes 22 and 23 have been combined for 1982.

[2]In this table the data for miscellaneous durable and nondurable manufacturing are combined for all dates except 1982.

*Adjusted for disclosure

**Data on employment are basically comparable for major SIC groups (2-digit). However, due to changes in the 1972 Standard Industrial Classification manual, data for some 3- and 4- digit SIC codes may not always be strictly comparable from 1964 to 1980. Any resulting distortions are judged to be small in most cases.

Source: 1982 figures derived from Bureau of Labor Statistics data; 1964–1980 figures from *County Business Patterns* data, respective years.

Table 5.4 *(cont.)*

SIC	INDUSTRY	1972 TOTAL EMPLOYMENT (000s)	1972 ESTIMATED EXPORT EMPLOYMENT	1972 PERCENT EXPORT EMPLOYMENT	1964 TOTAL EMPLOYMENT (000s)	1964 ESTIMATED EXPORT EMPLOYMENT	1964 PERCENT EXPORT EMPLOYMENT	1964-1972 PERCENT CHANGE**	1972-1980 PERCENT CHANGE**
	MANUFACTURING: DURABLE GOODS								
36	Electrical/Electronic Machinery	33.1*	8,433	25.5%	23.2*	4,490	19.4%	42.7%	77.3%
37	Transportation Equipment	46.1*	20,917	45.4	40.4	19,661	49.0	14.1	0.4
35	Non-Electrical Machinery	22.4*			13.5			65.9	97.3
34	Fabricated Metal Products	14.6*			9.2*			58.7	54.1
32	Stone, Clay, Glass & Concrete Products	6.6*			4.9			34.7	34.8
38	Instruments & Scientific Equipment	3.2			1.5			113.3	65.6
24	Lumber & Wood Products	2.5			1.7*			47.1	196.0
33	Primary Metal Industries	4.0			3.3*			21.2	95.0
25	Furniture & Fixtures	6.2*			4.0			55.0	- 1.6
39	Miscellaneous Durable Goods	--			--			--	--
	MANUFACTURING: NONDURABLE GOODS								
27	Printing & Publishing	16.1*	407	2.5	11.5*			40.0	41.6
20	Food & Kindred Products	18.8			20.8	1,172	6.0	- 9.6	13.3
22	Textile Mill Products[1]	0.7*			0.7*			0.0	42.9
23	Apparel Products	20.5*			15.3*			34.0	2.0
30	Rubber & Plastic Products	4.3*	521	2.5	2.2			95.5	165.1
28	Chemicals & Allied Products	6.2			4.5			37.8	37.1
26	Paper & Allied Products	6.1			6.2			- 1.6	31.1
21	Tobacco Manufactures	0.0			0.0			----	----
29	Petroleum Refining & Related Industries	1.6			1.4			14.3	55.7
31	Leather & Leather Products	1.3*	933	71.2	0.5			160.0	107.7

SIC	Industry								
39	Miscellaneous Nondurable Goods	---			----			---	---
39	Miscellaneous Manufacturing Total[2]	3.4			2.0			70.0	79.4
	TOTAL MANUFACTURING EMPLOYMENT	217.7			166.8			30.5	43.5
	TOTAL EXPORT EMPLOYMENT		31,211			25,323			
	EXPORT EMPLOYMENT MULTIPLIER		14.3%			15.2%			
	TOTAL SMSA EMPLOYMENT	858.6			585.7			46.6	51.3

subsectors have constituted the three largest components of D-FW manufacturing. Electrical/electronic machinery (SIC 36), transportation equipment (SIC 37), and nonelectrical machinery (SIC 35) together dominate the eleven-county region's goods-producing sector and have consistently accounted for approximately one-half of the region's manufacturing employment—an estimated 46.2 percent in 1964, 46.7 percent in 1972, 47.8 percent in 1980, and 47.9 percent in 1982. The first and the third of these large industries experienced significantly greater growth during the 1972–1980 period than during the 1964–1972 period. Growth in the transportation equipment industry has been at ever-lower rates over the same periods.

Manufacturing Export Employment: Part of the importance attached to the role of manufacturing exports to locations outside the region derives from the income they draw back into the host region and the resulting local investment and employment they create. In general, export employment levels in manufacturing have fluctuated over time, although the 1982 level is about the same as the 1964 level—approximately 25,000 workers. However, the export employment share of total manufacturing employment within the region has been steadily declining from an estimated 15.2 percent in 1964 to 8.1 percent in 1982, reflecting rising levels of local demand for durable goods tied to expanding regional population and total employment.[41]

The electrical and electronic machinery (SIC 36) and transportation equipment (SIC 37) industries have long had a significant export orientation. The number and proportion of jobs tied to exports in the first of these industries increased dramatically until 1980; over the next two years both indicators fell off sharply. By comparison, the transportation equipment industry has declined almost steadily on both export indicators since 1964. In 1964, firms in this sector employed an estimated 40,400 workers in the D-FW region. Of that total, approximately half (49.0 percent) was sustained by production for export outside the region. By 1982, that proportion had declined by half to 24.9 percent, while total employment in these industries increased by 20.8 percent to approximately 48,800. There is little basis for viewing this trend negatively, however. It is common for export employment to shrink as a proportion of a growing regional economy as ever-greater proportions of goods produced go to meet rising demand within a region. In any event, as of 1982, export employment constituted nearly a quarter of total employment in both of these industries.

Approximately 30.7 percent of the region's manufacturing economy was tied to nondurable goods production in 1982. The share of manufacturing employment tied to nondurable goods has been steadily declining—39.0 percent in 1964, 36.3 percent in 1972, and 33.7 percent in 1980. Printing and publishing (SIC 27) and food and kindred products (SIC 20) are the leading nondurable goods employers in the region. The textile mill products (SIC 22) and apparel products (SIC 23) industries together have experienced relative stagnation leading to employment contraction through the 1970s, as upgraded production technologies have permitted their continued development at lower employment levels. The most rapidly expanding nondurable goods industry in the region since 1964 has been the rubber and plastic products (SIC 30) industry, in which employment increased 95.5 percent during the 1964–1972 period and 165.1 percent during 1972–1980. The only manufacturing industries to experience actual employment contraction over the 1964–1972 period were food and kindred products and paper and allied products (SIC 26). For both industries, employment rebounded during the 1972–1980 period.

The D-FW Region's Nonmanufacturing Economy

Despite the importance of manufacturing to an advanced industrial economy, in employment terms the D-FW region is dominated by nonmanufacturing industries. Moreover, regional employment has been slowly shifting toward nonmanufacturing in recent decades; its share has grown from an estimated 69.0 percent in 1964, to 71.8 percent in 1972, to 73.0 percent in 1980. Table 5.5 displays this complex and diverse economy and the steady "shift-to-services" within nine separate industry groups.[42] The data reveal that the majority of the separate industries experienced considerable employment growth over the 1964–1982 period. Overall nonmanufacturing employment increased by half during each of the two eight-year periods, 1964–1972 (52.3 percent) and 1972–1980 (54.0 percent).

The two major nonmanufacturing industry groups are retail trade (SIC 52–59) and producer services (SIC 60–67). An estimated 26.8 percent of nonmanufacturing employment is in retail trade alone. The producer services group is the fastest growing segment of the U.S. economy, and its output share (approximately 25 percent) equals that of manufacturing.[43] In the D-FW regional economy, 23.8 percent of the region's nonmanufacturing employment is tied to this industry group. Business services (SIC 73) is the major service industry in the U.S. economy hosting high-technology development. Table 5.5 indicates that this industry has experienced phenomenal regional growth since 1964. Employment in business services more than doubled during both the 1964–1972 period (137.1 percent) and the 1972–1980 period (117.3 percent). This group's share of the region's nonmanufacturing

Table 5.5
Total Nonmanufacturing and Export Employment in the Dallas-Fort Worth SMSA, 1964–1982

	1982		
SIC Industry Group	*Total Emp. (000s)*	*Est. Export Emp. (000s)*	*Percent Export Emp.*
Agriculture (SIC 01-09)	—		
07 Agricultural Services	—		
Extractive/Mining (SIC 10-14)	—		
13 Oil and Gas Extraction	31.4	17.3	55.0%
14 Nonmetallic Minerals, except Fuels	—		
12 Bituminous Coal and Lignite Mining	—		
10 Metal Mining	—		
Transformative/Construction (SIC 15-17)	—		
17 Special Trade Contractors	47.9	2.4	5.0
15 General Building Contractors	26.0	4.7	18.2
16 Heavy Construction Contractors	11.5		
Distributive Services/Transportation, Communication, and Public Utilities (SIC 40-49)	—		
42 Motor Freight Transportation and Warehousing	28.0	2.3	8.8
48 Communication	27.6		
45 Transportation by Air	17.3	8.0	45.9
49 Electric, Gas and Sanitary Services	13.3		
40 Railroad Transportation	4.4		
41 Local, Suburban, Interurban Transportation	2.8		
47 Transportation Services	—		
46 Pipe Lines, Except Natural Gas	—		
Wholesale Trade (SIC 50-51)	139.3	28.9	20.7
50 Wholesale Trade–Durable Goods	—		
51 Wholesale Trade–Nondurable Goods	—		
Retail Trade (SIC 52-59)	—		
58 Eating and Drinking Places	86.6		
59 Miscellaneous Retail	38.4		
53 General Merchandise Stores	40.7		
54 Food Stores	43.8		
55 Automotive Dealers and Gasoline Service Stations	27.8		
56 Apparel and Accessory Stores	17.5		
57 Furniture, Home Furnishings, and Equipment	16.4	4.3	26.2
52 Bldg. Materials, Hardware, Garden Supply Dealers	9.8		
Producer Services			
Finance, Insurance, and Real Estate (SIC 60-67)	—		
63 Insurance Carriers	35.4	9.5	26.9
65 Real Estate	25.1	4.7	18.8
60 Banking	25.3		
64 Insurance Agents, Brokers, and Services	11.2	2.1	18.4
61 Credit Agencies, Other Than Banks	11.7		
62 Security and Commodity Brokers	4.1		
67 Holding and Other Investment Offices	—		
66 Combination of Real Estate, Insurance, Loan Offices	—		

Table 5.5 (*cont.*)
Total Nonmanufacturing and Export Employment in the Dallas-Fort Worth SMSA, 1964–1982

1980			*1972*			*1964*				
Total Emp. (000s)	*Est. Export Emp. (000s)*	*Percent Export Emp.*	*Total Emp. (000s)*	*Est. Export Emp. (000s)*	*Percent Export Emp.*	*Total Emp. (000s)*	*Est. Export Emp. (000s)*	*Percent Export Emp.*	*1964–72 Percent Change***	*1972–80 Percent Change***
4.0*			2.0*			1.1			81.8%	100.0%
3.6*			1.8			1.0			80.0	100.0
17.8	0.5	2.8%	10.3*	1.5	15.7%	8.7*	1.0	12.0%	18.4	72.8
8.5	0.8	9.8	4.2*	1.1	25.6	4.1	1.2	28.0	2.4	102.3
1.1*			0.8*			1.1*			27.3	37.5
0.4			0.2*			—			—	100.0
0.2			—			—			—	—
102.9	25.3	24.6	65.1	14.8	22.8	44.6	11.0	24.6	46.0	58.0
57.0	16.3	28.6	33.9	7.4	21.8	24.3	6.9	29.0	39.5	68.1
29.1	7.3	25.0	16.8	2.5	14.6	12.4*	2.4	19.5	35.5	73.2
16.3*	1.4	8.9	13.5	4.5	33.3	8.2*	2.2	26.6	64.6	20.7
92.3	11.9	12.9	62.1	4.4	7.1	43.1	4.2	9.7	44.1	48.6
28.7	6.4	22.2	20.2*	4.4	22.0	13.3	2.4	18.0	51.8	42.1
20.9*			14.7*			9.6*			53.1	—
19.6	11.9	60.9	11.6	6.7	57.8	6.8	4.2	62.0	70.6	69.0
10.6*			7.5*			7.2*			4.2	41.3
—			—			—			—	—
4.2			4.0			3.1			29.0	5.0
2.8			1.0			0.6			66.7	180.0
0.6*	0.3	44.3	0.2*	0.0	10.9	0.3	0.1	37.0	33.3	200.0
116.9	26.4	24.6	81.6	21.3	26.1	59.5	17.1	28.8	37.1	43.3
71.9	20.4	28.4	—			—			—	—
36.1*	1.6	4.5	—			—			—	
254.2			174.0	1.6	0.9	112.0	2.7	2.5	55.4	46.1
72.4			37.9			21.9			73.1	91.0
36.6	3.3	9.1	22.2	3.4	15.2	12.3			80.5	64.9
33.4			40.2	7.6	18.9	26.6	6.2	23.0	51.1	16.9
31.8			20.6			15.3			34.6	54.4
26.4			22.9			16.6			38.0	15.3
14.7			10.9			7.5			45.3	34.9
8.4*			6.0*			3.3			81.8	40.0
6.1*			4.4			3.7			18.9	38.6
225.5	20.5	9.1	132.0	18.0	13.6	76.7	11.0	14.3	70.8	72.1
39.1*	17.7	45.3	27.6	12.5	45.1	22.7	11.7	51.0	21.6	41.7
23.1*	6.0	25.8	17.1*	5.0	29.3	8.2	0.9	11.0	108.5	35.1
22.2			13.1			9.4			39.4	69.5
9.9	1.8	18.5	4.7	0.4	7.6	3.2	0.3	8.0	46.8	110.6
9.5			7.0	1.4	19.7	4.5	0.7	16.0	55.6	35.7
2.8			3.1	0.2	6.1	1.2			158.3	-9.7
2.3*			2.2	1.0	48.3	1.1	0.6	57.0	100.0	4.5
0.2*			0.2			0.2			—	—

Table 5.5 (*cont.*)
Total Nonmanufacturing and Export Employment in the Dallas-Fort Worth SMSA, 1964–1982

	1982		
SIC Industry Group	*Total Emp. (000s)*	*Est. Export Emp. (000s)*	*Percent Export Emp.*
Business Services (SIC 73)			
73 Business Services	77.8	7.4	9.5
Legal Services (SIC 81)			
81 Legal Services	9.1		
Membership Organizations (SIC 86)			
86 Membership Organizations	—		
Misc. Professional Services (SIC 89)			
89 Miscellaneous Services	—		
Social Services (SIC 83, after 1974)			
83 Social Services	—		
Nonprofit Services/Health and Education (SIC 80,82)	—		
80 Health Services	72.9		
82 Educational Services	14.5		
Mainly Consumer Services	—		
70 Hotels and Other Lodging Places	20.3		
72 Personal Services	19.2		
75 Auto Repair, Services and Garages	12.8	0.6	4.9
79 Amusement and Recreation Services	12.6		
76 Miscellaneous Repair Services	4.1		
78 Motion Pictures	2.8		
84 Museums, Botanical, Zoological Gardens	—		
Total Nonmanufacturing Employment			
Total Export Employment			
Export Employment Multiplier			
Total SMSA Employment	1,549.5		

Source: Bureau of Labor Statistics, 1982.

*Adjusted for disclosure

**Data on employment are basically comparable for major SIC groups (two-digit). However, due to changes in the 1972 Standard Industrial Classification manual, data for some three- and

employment likewise doubled, rising steadily from 3.1 percent in 1964, to 4.8 percent in 1972, to 6.7 percent in 1980.

The mainly consumer services group constituted not only the smallest portion of the region's service economy by 1980, but also a relatively slowly growing industry group, especially before 1972. Many of the negative impressions associated with employment in a service economy derive from characteristics both of jobs in consumer services and of the

Table 5.5 (*cont.*)
Total Nonmanufacturing and Export Employment in the Dallas-Fort Worth SMSA, 1964–1982

1980			*1972*			*1964*				
Total Emp. (000s)	*Est. Export Emp. (000s)*	*Percent Export Emp.*	*Total Emp. (000s)*	*Est. Export Emp. (000s)*	*Percent Export Emp.*	*Total Emp. (000s)*	*Est. Export Emp. (000s)*	*Percent Export Emp.*	*1964–72 Percent Change***	*1972–80 Percent Change***
63.9	11.9	19.0	29.4	4.7	16.0	12.4	0.4	3.3	137.1	117.3
7.1			3.3			1.7			94.1	115.2
19.3			14.3			7.6			88.2	35.0
15.4			10.0	0.9	8.8	4.5			122.2	54.0
10.7*			—			—			—	—
71.5			46.1			24.5			88.2	55.1
59.5			35.4			19.1			85.3	68.1
12.0*			10.7*			5.4			98.1	12.1
63.9			43.2			34.5	2.3	6.7	25.2	47.9
17.4			8.9			6.8			30.9	95.5
18.5	2.0	11.0	15.1	1.6	10.3	14.9	3.0	20.4	1.3	22.5
9.8	0.1	0.7	6.4	0.4	6.5	4.7	0.9	19.0	36.2	53.1
9.9			6.6			3.5			88.6	50.0
5.3			3.8	0.7	17.5	2.9	0.9	31.6	31.0	39.5
2.5			2.3			1.7			35.3	8.7
0.5*	0.0	3.2	0.1*			—			—	400.0
949.0			616.4			404.7			52.3	54.0
	109.2			87.5			62.1			
		11.5%			14.2%			15.3%		
1,299.3			858.6			585.7			46.6	51.3

four-digit SIC codes may not always be strictly comparable from 1964 to 1980. Any resulting distortions are judged to be small in most cases. In one instance, SIC 48, redefinition was so complete that two-digit data are judged to be noncomparable.

low-skilled young and returning female workers who frequently hold them. These industries commonly constitute a secondary labor market in which employment is often characterized by low pay and benefits with little or no union advocacy or representation, poorly developed career ladders that offer few opportunities for advancement, high proportions of seasonal and part-time employment, tasks perceived to be dull and routine, and female-dominated occupations. The casual equation of growth in the services economy with the expansion of this industry group around the nation has led to grave concerns about the capacity of an advanced services economy to provide the standard of living to which we have become accustomed. So it is interesting to note that the employment share of nonmanufacturing tied to consumer services in the D-FW regional economy has been steadily declining from 8.5 percent in 1964, to 7.0 percent in 1972, to 6.7 percent in 1980. Clearly, the D-FW region has been witnessing the rise of what has been labeled a "new services economy," one whose linkages to manufacturing and whose capacities for generating attractive employment growth for the future have yet to be fully explored.

An Anatomy of the D-FW High-Technology Sector

Table 5.6 reports employment levels and trends for forty-eight high-technology industries within the D-FW regional economy between 1964 and 1980. It is clear that while this sector is dominated by a handful of major industries, it reflects considerable diversity, like the larger economy of which it is a part. Nearly every high-technology industry in the national economy is at least minimally represented in the D-FW region.

Three alternative definitions of what constitutes "high-technology" are employed in Table 5.6[44] Depending on the definition used, employment in high-technology industries constituted from 7.8 percent to 18.6 percent of total regional employment by 1980. Employment trends using all three definitions reveal that employment in high-technology industries expanded at a faster rate than did total regional employment during the 1964–1972 period. During the 1972–1980 period, high-technology employment levels more than doubled, and the share for the second and third definitions (Groups II and III) expanded at more than twice the rate of total regional employment. Using any of the alternative definitions, it is apparent that high-technology in the D-FW regional economy has been expanding its employment share since at least 1964. These trends clearly reveal the degree to which this sector has long been an integral and growing component of the region's economy.

At the national level, the industries in this sector that experienced the most rapid growth between 1972 and 1980 were computer and data processing services (SIC 737, 185.2 percent), communication services, NEC (not elsewhere classified) (SIC 489, 122.6 percent), optical instruments and lenses (SIC 383, 87.5 percent), and office, computing, and accounting machines (SIC 357, 66.5 percent). With the exception of D-FW's tiny and slowly contracting optical instruments industry, the rapid growth of these industries nationally is being replicated in the region. Altogether, seventeen of the forty-eight industries more than doubled their regional employment levels during 1972–1980, a feat accomplished nationwide over the same period only by the computer and data processing services and the communications services industries. Among those industries whose regional growth has accelerated to levels far above the national ones are surgical, medical, and dental instruments (SIC 384), radio, T.V., and receiving equipment (SIC 365), and agricultural chemicals (SIC 287). Even though many regional high-technology industries are still relatively small, growth across the sector has been widespread.

As noted in Chapter 2, industrial growth and development are distinct, although related, processes.[45] Frequently overlooked, moreover, is that actual employment contraction often accompanies overall industrial development. This accompaniment is certainly true for the development of the D-FW region's high-technology sector. In the U.S. high-technology sector, eight industries experienced actual employment loss during the 1972–1980 period; all of them continued to lose employment through 1982. A closer look at the evolution of this same sector in the D-FW region similarly reveals that *employment contraction, like expansion, is an integral part of the advanced industrial development of the high technology sector.* There has long been considerable diversity in industrial growth trends within the D-FW high-technology sector. Seven industries lost employment during the 1964–1972 period; eleven industries did so during 1972–1980. Two industries—industrial inorganic chemicals and noncommercial education, scientific, and research organizations—experienced extended employment declines over the entire sixteen-year period. These trends serve to illustrate that high-technology industries are not to be loosely equated with growth industries in any sense, especially as regards employment creation. Nor are those industries that are currently expanding their employment insulated in any way from the prospect that their continued development can be accompanied by little or no future employment growth.

In contrast, three industries are new additions to the region since 1964. Plastic materials and synthetics (SIC 282) and combination electric, gas, and other utility services (SIC 493) began to host very small employment levels, although they accounted for only miniscule employment shares by 1980. The third new industry, computer and data processing services (SIC 737), recorded no employment as late as 1972. However, by 1980 its estimated employment level was already 9,300.[46] Just as this high-technology service industry distinguished itself during the 1972–1982 period as being the fastest-growing high-technology industry in the United States, it likewise appears to have expanded

Table 5.6
Employment in High-Technology Industries, Dallas-Fort Worth SMSA, 1964–1980

		High-Technology Group[a]			*Employment (000s)*			*Percent Change***	
SIC	*Industry*	*I*	*II*	*III*	*1964*	*1972*	*1980*	*1964–72*	*1972–80*
131	Crude Petroleum and Natural Gas	X			2.6	2.3	4.8*	−10.1%	105.7%
162	Heavy Construction, except Highway and Street	X			3.5*	6.8*	10.9*	94.4	60.6
281	Industrial Inorganic Chemicals	X		X	0.8	0.7	0.6*	−10.0	−13.0
282	Plastic Materials and Synthetics	X		X	—	—	0.1	—	—
283	Drugs	X	X	X	0.8*	1.1*	0.9*	42.7	−19.4
284	Soaps, Cleaners and Toilet Preparations	X		X	1.1	1.4	2.0*	29.3	44.7
285	Paints and Allied Products	X		X	1.1	1.4	1.6*	29.5	10.2
286	Industrial Organic Chemicals	X		X	—	—	—	—	—
287	Agricultural Chemicals	X		X	0.1	0.2*	0.8*	47.9	285.7
289	Misc. Chemical Products	X		X	0.7	1.0	2.0	52.7	95.8
291	Petroleum Refining	X		X	0.2*	0.3*	0.2*	45.5	−37.1
301	Tires and Inner Tubes	X			0.2	0.4*	0.4*	70.1	4.5
324	Cement, Hydraulic	X			0.7*	0.9*	0.7*	32.9	−22.0
348	Ordnance and Accessories	X		X	0.3*	0.8*	0.4*	128.5	−42.3
351	Engines and Turbines	X		X	—	—	—	—	—
352	Farm and Garden Machinery	X			0.2*	0.4*	1.2*	116.3	179.5
353	Construction, Mining and Material Handling Machinery	X			4.1*	5.6*	14.6*	36.5	158.6
354	Metalworking Machinery	X			0.7	1.2*	3.3*	69.5	180.7
355	Special Industry Machinery, except Metalworking	X		X	2.4	2.1*	3.2*	−11.5	52.5
356	General Industrial Machinery	X			1.8	2.0	3.1	9.3	59.5
357	Office, Computing and Accounting Machines	X	X	X	0.6*	2.4	5.3*	313.0	119.8
358	Refrigeration and Service Industry Machinery	X			2.4*	5.0*	7.5*	106.7	49.5
361	Electric Transmission and Distribution Equipment	X		X	0.4*	0.6	0.6	64.9	−8.7
362	Electrical Industrial Apparatus	X		X	0.5*	0.5	1.1*	−2.3	120.0
363	Household Appliances	X			0.6*	0.9*	0.7*	50.7	−27.6

364	Electric Lighting and Wiring Equipment	X			0.2	1.1*	1.2*	447.0	3.2
365	Radio and T.V. Receiving Equipment	X		X	0.6*	0.2	0.9	−62.3	332.2
366	Communication Equipment	X	X	X	9.3	15.9	32.4*	71.9	103.8
367	Electronic Components and Accessories	X	X	X	1.8*	11.7*	19.7	540.2	68.0
369	Misc. Electrical Machinery	X		X	0.6	1.6*	2.1*	159.8	37.0
371	Motor Vehicles and Equipment	X			6.8	7.2*	7.3*	5.4	1.1
372	Aircraft and Parts	X	X	X	14.0*	15.5*	42.7*	11.3	175.0
376	Guided Missiles and Space Vehicles	X	X	X	—	—	—	—	—
381	Engineering, Laboratory, Scientific and Research Instruments	X		X	0.8	0.9	0.5*	17.8	−43.1
382	Measuring and Controlling Instruments	X		X	0.3	0.6*	1.7	90.8	211.5
383	Optical Instruments and Lenses	X		X	0.4*	0.6	0.6	50.1	−3.1
384	Surgical, Medical and Dental Instruments	X		X	—	0.1	1.1	—	636.1
386	Photographic Equipment and Supplies	X		X	—	0.1*	0.4*	—	230.0
483	Radio and T.V. Broadcasting	X			0.6	1.4	2.0*	169.5	—
489	Communication Services, NEC.	X			0.1	0.5*	1.9*	259.5	—
491	Electrical Services	X			3.7*	1.7*	6.2*	−53.4	255.1
493	Combination Electric, Gas and Other Utility Services	X			—	—	0.4*	—	—
506	Wholesale Trade, Electrical Goods	X			4.7	6.7	9.1	43.4	35.9
508	Wholesale Trade, Machinery, Equipment and Supplies	X			12.5	16.4	26.4	31.3	60.3
737	Computer and Data Processing Services	X		X	—	—	9.3	—	—
7391	Research and Development Laboratories	X		X	0.3	2.3*	1.0	605.1	−54.8
891	Engineering, Architecture and Surveying Services	X			2.4	4.0	9.0	65.6	125.9
892	Noncommercial Education, Scientific and Research Organizations	X			0.4	0.3	0.2	−21.8	−40.0
	Total SMSA Employment				585.7	858.6	1,299.3	46.6%	51.3%

Table 5.6 *(cont.)*

Employment in High-Technology Industries, Dallas-Fort Worth SMSA: 1964–1980

		High-Technology Group[a]			*Employment (000s)*			*Percent Change***	
SIC	*Industry*	*I*	*II*	*III*	*1964*	*1972*	*1980*	*1964–72*	*1972–80*
	Total SMSA High-Technology Employment (Percent of SMSA Employment)	Group I			85.3 (14.6%)	126.8 (14.8%)	242.1 (18.6%)	48.7%	90.9%
		Group II			26.5 (4.5%)	46.6 (5.4%)	101.1 (7.8%)	75.8%	116.7%
		Group III			37.1 (6.3%)	62.0 (7.2%)	131.2 (10.1%)	67.1%	111.6%

Source: Adapted from Richard W., Riche, Daniel E. Hecker and John U. Burgan, "High Technology Today and Tomorrow: A Small Piece of the Employment Pie," *Monthly Labor Review* (November 1983) pp. 50–58. The 1964–1980 figures are from *County Business Patterns* data, respective years.

*Adjusted for disclosure.

**Data on employment are basically comparable for major SIC groups (two-digit). However, due to changes in the 1972 Standard Industrial Classification manual, data for some three- and four-digit SIC codes may not always be strictly comparable from 1964 to 1980. Any resulting distortions are judged to be small in most cases. In one instance, SIC 48, redefinition was so complete that two-digit data are judged to be noncomparable.

[a]Group I includes industries with a proportion of technology-oriented workers (engineers, life and physical scientists, mathematical specialists, engineering and science technicians, and computer specialists) at least 1.5 times the average for all industries. Group II includes industries with a ratio of R&D expenditures to net sales at least twice the average for all industries. Group III includes manufacturing industries with a proportion of technology-oriented workers equal to or greater than the average for all manufacturing industries, and a ratio of R&D expenditures to sales close to or above the average for all industries. Two nonmanufacturing industries which provide technical support to high tech manufacturing industries also are included.

faster than any other high-technology industry in the D-FW regional economy. Nonetheless, the high-technology industries with the largest employment levels have been growing and developing for many years. In 1980, the three industries in this sector with the largest employment included two manufacturing industries—aircraft and parts (SIC 372) and communication equipment (SIC 366)—and one service industry—wholesale trade, machinery, equipment, and supplies (SIC 508). Together the three accounted for an estimated 101,500 jobs, or 7.8 percent of total regional employment. However, even more revealing is that as early as 1964, these same three industries were already the largest employers in the sector, accounting for approximately 35,800 jobs, or 6.1 percent of total regional employment.

Development Patterns Within the D-FW High-Technology Sector

In mid-1984, there were an estimated 801 high-technology firms located in the D-FW region.[47] The composition by industry sector is shown in Table 5.7. It is obvious that three of the six industry sectors have spawned the bulk of the high-technology firms in the region. Moreover, nearly half (46.7 percent) of these new firms are high-technology services. Within high-technology manufacturing, the greatest number of firms are found in the electrical and electronic equipment industry group and the instruments and related products industry group. We can expect that the sheer concentration of firms in these industry groups marks them as the principal high-technology anchors for the region. As we will see below, these groups include both new and established and large and small firms—a mix that has developed steadily over at least the past quarter-century.

High-Technology Firm Birth Cohorts: High-technology firms have filtered into the regional economy over the entire post–World War II

Table 5.7
Composition of the D-FW Region's High-Technology Sector

Industry Group		*Number of Firms*	*Percent of Total*
SIC 28 Chemicals and allied Products		15	1.9%
SIC 35 Nonelectrical machinery		20	2.5
SIC 36 Electrical and electronic equipment		222	28.1
SIC 37 Transportation equipment		8	1.0
SIC 38 Instruments and related products		156	19.8
SIC 73 Business services		369	46.7
	Totals	790	100.0%

era. A sizable portion of this sector—nearly one firm in ten (9.4 percent) of those for which information was available—is over a quarter-century old. The business formation rate has increased steadily since 1960. While 15.1 percent of the firms date to the 1960s, fully 40.6 percent are of 1970s vintage. The 1980s, however, have seen the greatest rate of growth thus far. More than one in three (34.9 percent) of the region's high-technology firms were established during the first four and a half years of this decade alone.

The shift-to-services is a major element in the advanced industrial development of urban-regional economies, as we have seen. This shift is clearly evident in the development of the D-FW region's high-technology sector. As Table 5.8 indicates, successive "birth" cohorts of high-technology firms have included progressively larger proportions of high-technology services, especially business services. Prior to 1965, one firm in five was in the high-technology services; between 1965 and 1980, the proportion rose from 39.1 percent to 50.5 percent of the respective cohorts. In the most recent cohort of firms established since 1980, 57.8 percent have been in the services.

Characteristics of the Most Recent Cohort: The 1980s have proven to be a particularly fertile period for the formation of new high-technology firms within the region, with a third of the high-technology firms belonging to this most recent cohort. Over half (57.8 percent) of these new firms were in the services, and a quarter (24.2 percent) were in the electrical and electronic equipment industries alone. The employment effect accompanying the addition of this most recent cohort has been limited, however. Together, this group created only slightly more than six thousand new jobs, with slightly fewer than half (47.7 percent) of these jobs being in business service firms. The average employee/firm ratio within this cohort of high-technology services was approximately 19.4 employees per new firm. In electrical and electronic equipment firms, the average employment per new firm was 18.7; in new instru-

Table 5.8
High-Technology Firm Birth Cohorts

Start-up Date	*Number of Firms*	*Percent of Total*	*Percent of Services*
Pre-1960	69	9.4%	(14) 20.3%
1960–1964	47	6.4	(9) 19.1
1965–1969	64	8.7	(25) 39.1
1970–1974	102	13.9	(38) 37.3
1975–1979	196	26.7	(99) 50.5
Post-1979	256	34.9	(148) 57.8
	734	100.0%	

ments and related products firms the ratio was 25.8. As might be expected, while new firms in this most recent cohort were more likely to be located in high-technology services, new employment was more likely to be in high-technology manufacturing.

Firm Size and Employment Growth: High-technology employment in the D-FW region may approach a quarter-million, although any such estimate depends largely upon what definition of the sector is used. Precise measurement is also made difficult by the presence of a handful of very large firms whose employment levels fluctuate as the regional and national economies move through the business cycle. Several dozen major high-technology employers are defense contractors and subcontractors whose employment levels are often deliberately withheld for industrial and national security reasons. Nonetheless, it is estimated that as much as two-thirds of the region's high-technology employment is located in those very few high-technology firms employing five thousand or more. It is important to note here that, as is the case throughout the entire high-technology sector, the jobs within these firms and sectors are by no means uniformly sophisticated technologically. Much traditional employment is sustained in firms throughout the high-technology sector.

Table 5.9 shows an estimate of the distribution of the region's high-technology firms by employment size. The vast majority of these firms are quite small; 28.6 percent employ five or fewer people, while 91.6

Table 5.9
Dallas-Fort Worth High-Technology Firms, by Employment Size

Firm Size	*Number of Firms*	*Percent of Total Firms*	*Number of Employees*	*Percent of Total Employment*
1–5	209	28.6%	688	0.5%
6–10	121	16.6	960	0.6
11–25	161	22.1	2,710	1.8
26–50	85	11.6	3,136	2.1
51–100	50	6.8	3,788	2.5
101–250	43	5.9	7,179	4.8
251–500	35	4.8	13,095	8.8
501–1,000	10	1.4	7,450	5.0
1,001–2,500	5	0.7	8,000	5.4
2,501–5,000	2	0.3	8,500	5.7
5,001–10,000	6	0.8	41,000	27.5
10,000+	3	0.4	52,500	35.2
	730	100.0%	149,006	99.9%[a]

[a]Total does not add to 100% due to rounding error.

percent employ fewer than 250 employees. Total employment in this sector underscores the importance of larger and older high-technology firms. Evidence of the extended development of the region's high-technology sector is found in the fact that only one job in eight in this sector is located in a firm with 250 or fewer employees. Even though the overall development of the sector may rely on the incubation of new and usually small businesses for future employment growth, *the bulk of the region's present high-technology employment is tied to older and larger established firms.*[48] There is little possibility that this pattern will change in the future, regardless of how rapidly new-firm birth rates accelerate within this sector.

International Export Orientation: A major part of the growing interest in the growth and development of the high-technology sector is related to the fact that policymakers in national and regional economies around the world are carefully assessing their competitive strengths to decide what policy options are available to retain and enhance those strengths or create them anew. Estimates of export employment are commonly used as crude indicators of the extent to which a regional economy is interdependent with the economies of other regions and nations. As we consider the implications of an emerging global economy and the growth of international trade, the extent of the international export orientation among the region's high-technology firms becomes a useful indicator of the linkage between a leading sector of the D-FW regional economy and the outside world.

Table 5.10 reports data by industry group on the international export orientation of D-FW high-technology firms. One firm in five (20.0 percent) reports exporting either goods or services in international markets. With the exception of the chemical and allied products industry, between

Table 5.10
International Export Orientation of D-FW High-Technology Firms

SIC	*Industry*	*Total Number of Firms*		*Number of Exporting Firms*	*Percent Exporting*
28	Chemical and allied products	15	(1.9%)	1	6.7%
35	Nonelectrical machinery	20	(2.5)	4	20.0
36	Electrical and electronic equipment	222	(28.1)	59	26.6
37	Transportation equipment	8	(1.0)	3	37.5
38	Instruments and related products	156	(19.8)	60	38.5
73	Business services	369	(46.7)	31	8.4
	Totals	790	(100.0%)	158	20.0%

one-fifth and two-fifths of the firms in each of the manufacturing groups report an orientation toward international markets. The rising importance of international trade involving services, especially high-technology services, is revealed in our regional data as well.[49] Fully 8.4 percent of the firms in the business services—firms that are predominantly new and small—report an international export orientation. Presumably, data on interregional domestic trade and target markets would reveal an even greater external orientation among the firms in this sector.

High-Technology Seedbed Creation: A Sociological Perspective

The creation of a high-technology seedbed is arguably more a sociological phenomenon than an economic or technological one. A main ingredient has been an expressed willingness for technically-trained people to venture out on their own and start new businesses. Yet, the prods for such risktaking may have less to do either with the entrepreneurial spirit that has been so romanticized or with related personality features that have been so thoroughly lionized in recent years. At least as important is the fact that engineers, computer scientists, related technically trained workers, and the industries they inhabit, are often extremely vulnerable to cyclical shifts in the market economy and budgetary shifts in the political economy, especially in defense-related industries. Moreover, the career prospects of such highly skilled workers, where the occupational features of their respective professions include skills that decay at a rapid rate, responsibilities that peak by the mid-30s and exhibit major salary compression, and career ladders that are relatively truncated within major companies, are hardly those that instill great loyalty to larger corporate forms.

Consider the case of engineers whose role includes the development and application of commercial technologies. This very narrow slice of the labor force is viewed as critical to the creation of new industries and the renewal of existing ones in the years ahead. Of a total labor force of nearly 107 million, engineers and ancillary scientists constitute only 1.4 percent (1.5 million) with three-quarters of these workers in industry and another 5 percent in universities. The remainder are employed by government, nonprofit organizations or are self-employed. Frequently, such technical talent is thought to be in short supply, even though the case for a national "shortage" may well be overdrawn. Willenbrock has noted that on the demand side, employer surveys consistently indicate difficulty in obtaining engineers, and hence the predictions of a shortage.[50] At the same time, however, surveys of engineers indicate that low pay, little opportunity for career advancement, and related occupational dissatisfaction are persistent conditions and may indicate that demand and supply are closer to being in equilibrium than is commonly thought. Approximately one-seventh of this workforce turns over each year with three-quarters of the new entrants to the ranks coming directly from college and the rest coming from abroad or from other occupations. At the same time, of all new positions in a given year, approximately one-fourth are tied to new business formation or secondary expansion, an equal proportion being replacements for deaths and retirees, and fully half are the result of the outmigration of engineers to nonengineering activities and occupations.

One conclusion to be drawn from these data is that with upwards of 50 thousand engineers leaving their professions each year for other careers, there must be something about the career itself that drives people out (and up) that merits consideration. Presumed specialty skill shortages may well be something that we unwittingly *create;* that is, they may well be the inevitable result of how we think about, train, and use engineers and related technical professionals. *We may not so much have a shortage of technical professionals such as engineers and computer scientists as an abundance of frustrated and dissatisfied ones.* Ironically, a high-technology seedbed may be the result of—and even be vitally dependent on—a continuing succession of cohorts of frustrated engineers "spinning off" from their old employers in search of new opportunities. As a result, high-technology seedbeds may well reflect such sources of frustration and dissatisfaction within occupations as they do the availability of other features of a "good business climate."

Spatial Patterns of Advanced Industrial Development

As discussed above, advanced industrial development invariably has a spatial dimension that holds important implications for a regional economy. This is especially true for regions like D-FW that are experiencing significant population and construction growth at the same time. The distribution of firms and employment within an industry throughout a region can and frequently does change over time. As new firms are established, the locations into which that business and employment growth filters has implications for the continued urban development of a region, as well as for the accessibility of new employment opportunities for citizens in established communities. In this section, I will explore the patterns of industrial concentration and dispersion that have unfolded within the D-FW region over the two decades since 1964. These larger spatial trends will then serve as a backdrop against which to identify and examine the spatial trends in the development of the high-technology sector within the region in recent decades.

Manufacturing: Concentration and Dispersion Trends

Table 5.11 reports data on the concentration and dispersion trends in the D-FW region's manufacturing employment.[51] These data can be used to shed light on the spatial patterns that have come to characterize the growth and development of manufacturing within the region during the past two decades. Using 1964 data, it is apparent that Dallas County was the location of considerably variable proportions of employment across the twenty manufacturing industries. In six durable goods industries and seven nondurable goods industries, Dallas County was the location of at least half of the total employment. The employment in electrical and electronic equipment was clearly the most spatially concentrated, with 95.7 percent of all jobs in the region being located in Dallas County.

The 1964–1980 period reveals *a clear sequence of increased concentration followed by dispersion within manufacturing.* Between 1964 and 1972, the proportion of all manufacturing employment located in Dallas County increased from 61.7 percent to 62.6 percent. Table 5.12 reports shifting employment shares for Dallas County across selected industries and industry groups. In contrast, by 1980 this proportion had declined once again to 58.3 percent. This same "turnaround" sequence is replicated within both durable and nondurable manufacturing industry groups. These trends are evidence not only of ways in which the region has developed, but also of ways in which manufacturing itself has changed. Clearly the continued growth and development of manufacturing in counties surrounding Dallas County, as well as in Fort Worth and Tarrant County, have served to "pull" manufacturing growth to an increasing range of acceptable and accessible intraregional locations. Rising land costs, continued residential development in suburban areas, and the increased difficulty of commuting within central portions of the county have served as "push" and "pull" factors that together have helped redirect a portion of manufacturing growth away from Dallas County. Undoubtedly, during the past two decades, progressively larger portions of the entire region have become locations into which manufacturing firms have had little difficulty filtering.

A dispersion dynamic whereby manufacturing either relocates out of central cities and large concentrated settlements or directs new and secondary expansion toward lower-density peripheral locations has been operating most dramatically in older industrial regions all through this century. In this respect, it is manufacturing itself that has changed. The

Table 5.11
Manufacturing Employment Concentration and Dispersion Trends, Dallas County, 1964–1980

		1964			1972			1980			
SIC	*Industry*	*Total Emp. SMSA (000s)*	*Dallas County Share (000s)*	*SMSA Emp. Share*	*Total Emp. SMSA (000s)*	*Dallas County Share (000s)*	*SMSA Emp. Share*	*Total Emp. SMSA (000s)*	*Dallas County Share (000s)*	*SMSA Emp. Share*	*1972–1980 Dispersion/ Concentration Index***
	Manufacturing: Durable goods										
36	Electrical/Electronic Machinery	23.2*	22.2	95.7%	33.1*	30.8	93.1%	58.7*	51.7	88.1%	95
37	Transportation Equipment	40.4	14.5	35.8	46.1*	15.5	33.6	46.3*	13.1	28.3	84
35	Nonelectrical Machinery	13.5	8.5	63.0	22.4*	15.1	67.4	44.2*	25.3	57.2	85
34	Fabricated Metal Products	9.2	6.2	67.4	14.6*	8.2	56.2	22.5*	13.2	58.7	104
32	Stone, Clay, Glass and Concrete Products	4.9	2.8	57.1	6.6*	3.7	56.1	8.9*	4.2	47.2	84
38	Instruments and Scientific Equipment	1.5	1.3	86.7	3.2	2.7	84.4	5.3*	3.9	73.6	87
24	Lumber and Wood Products	1.7*	0.8	47.1	2.5	1.8	72.0	7.4	3.5	47.3	66
33	Primary Metal Industries	3.3*	0.7	21.2	4.0*	1.4	35.0	7.8*	2.5	32.1	92
25	Furniture and Fixtures	4.0	2.2	55.0	6.2*	3.1	50.0	6.1*	2.8	45.9	92
	Manufacturing: Nondurable goods										
27	Printing and Publishing	11.5*	7.8	67.8	16.1*	10.5	65.2	22.8*	15.0	65.8	101
20	Food and Kindred Products	20.8	13.2	63.5	18.8	12.9	68.6	21.3*	12.7	59.6	87
22	Textile Mill Products	0.7*	0.2	28.6	0.7*	0.4	57.1	1.0*	0.8	80.0	140
23	Apparel Products	15.3*	11.1	72.5	20.5*	15.6	76.1	20.9*	14.5	69.4	91
30	Rubber and Plastic Products	2.2	1.4	63.6	4.3*	2.3	53.5	11.4*	3.8*	33.3	62
28	Chemicals and Allied Products	4.5	3.3	73.3	6.2	4.9	79.0	8.5*	5.6	70.0	89
26	Paper and Allied Products	6.2	4.8	77.4	6.1	4.5	73.8	8.0*	5.9	73.8	100
21	Tobacco Manufacturers	—	—		—	—		—	—		

29	Petroleum Refining and Related Industries	1.4	0.4	28.6	1.6	0.6	37.5	2.5	1.0	40.0	107
31	Leather and Leather Products	0.5	—		1.3*	0.2	15.4	2.7*	0.4	14.8	96
39	Miscellaneous Nondurable Goods	2.0	1.5	75.0	3.4	2.2	64.7	6.1	2.3	37.7	58

Source: 1964–1980 figures from *County Business Patterns* data, respective years.

*Adjusted for disclosure.

**Data on employment are basically comparable for major SIC groups (two-digit). However, due to changes in the 1972 Standard Industrial Classification manual, data for some three- and four-digit SIC codes may not always be strictly comparable from 1964 to 1980. Any resulting distortions are judged to be small in most cases.

Table 5.12
Shifting Dallas County Employment Shares: Concentration and Dispersion Trends, 1964–1980

	1964	*1972*	*1980*
High-technology sector	77.8%	77.6%	68.5%
Manufacturing			
Total	61.7	62.6	58.3
Durable goods	58.2	59.3	58.0
Nondurable goods	67.1	68.5	58.9
Nonmanufacturing (selected groups)			
Transformative (construction)	80.4	70.6	68.5
Distributive services (TCU)	77.8	79.4	62.8
Wholesale trade	77.5	79.0	75.5
Retail trade	61.7	63.9	59.8
Producer services (FIRE)	76.8	79.1	77.1
Business services	77.4	79.9	80.0
Legal services	76.5	72.7	76.1
Nonprofit services (health and education)	65.7	63.8	61.3
Mainly consumer services	66.4	67.1	65.4

[a]Each percentage indicates the proportion of all employment within an industry or industry group located in Dallas County.

location factors governing the development of new manufacturing within the D-FW region differ from those that constrained the development of early manufacturing.[52] This is especially so since the region is generally dominated by light manufacturing for which transportation costs are relatively low-and-declining proportions of total production costs. In short, the economic returns to agglomeration (that is, concentration) are now less than they once were. In the end, not only has the rapid growth of this region provided the outward momentum for manufacturing, but so also have the dynamics operating to transform industrial production within manufacturing industries.

Nonetheless, even though the dispersion dynamics of both urban-regional and industrial growth and development have operated to reinforce each other in recent years, all industries have not been swept along uniformly. The last column of Table 5.11 is a measure of the degree to which each industry's employment either dispersed away from or concentrated within Dallas County between 1972 and 1980. Among durable goods industries, only fabricated metal products (SIC 34) increased its employment share in Dallas County. Lumber and wood products (SIC 24) experienced an opposite trend. Even though total employment in the region nearly quadrupled (296.0 percent) and employment in Dallas County nearly tripled (194.4 percent), the Dallas County share declined 34.3 percent during the 1972–1980 period as the bulk of the new employment growth took place elsewhere within the region. Among nondurable goods industries, four industries either maintained their employment share between 1972 and 1980 or actually became more concentrated in Dallas County, thus countering the larger aggregate trend.

Nonmanufacturing: Concentration and Dispersion Trends

Table 5.13 reports the changing employment shares located in Dallas County for nonmanufacturing industry groups. As was the case with manufacturing above, a dispersion of growth in these industries away from Dallas County is evident but by no means uniform. Based on data reported for specific two-digit industries, 80.4 percent of the employment in the construction industries (SIC 15–17) was located in Dallas County in 1964. It is not surprising that employment in this industry has steadily dispersed ever since then, given that construction industries are commonly among the first to reflect new growth in a region. By 1972, that share had declined to 70.6 percent and again to 68.5 percent by 1980 as the rapid growth of the region was accompanied by significant residential and commercial construction in surrounding counties. Only heavy construction contractors (SIC 16) maintained their employment share during 1972–1980.

The distributive services (transportation, communication, and utilities) and the wholesale and retail trade sectors have replicated manufacturing's sequence of increased concentration followed by dispersion. Between 1964 and 1972, employment shares in Dallas County increased from 77.8 percent to 79.4 percent in the distributive services, 77.5 percent to 79.0 percent in wholesale trade and 61.7 percent to 63.9 percent in retail trade. By 1980, aggregate employment across all three industries had reversed this earlier trend and dispersed to 62.8 percent for distributive services, 75.5 percent for wholesale trade, and 59.8 percent for retail trade. In only two industry groups, SIC 41 (local, suburban, interurban transportation) and SIC 52 (building materials, hardware, and garden supply dealers) did employment continue to concentrate in Dallas County during 1972–1980.

Within the "new services economy" the producer services play a dominant role. An expanded definition of producer services includes finance, insurance, real estate (FIRE) services as well as business, legal, and assorted other minor services. In this study, we restrict producer services to the FIRE group only. The expanded role of the FIRE services is especially evident in the transformed industrial structure of many of our major urban areas. So-called "command and control" activities like

Table 5.13

Nonmanufacturing Employment Concentration and Dispersion Trends, Dallas County, 1964–1980

		1964			*1972*			*1980*			
SIC	*Industry group*	*Total Emp. SMSA (000s)*	*Dallas County Share (000s)*	*SMSA Emp. Share*	*Total Emp. SMSA (000s)*	*Dallas County Share (000s)*	*SMSA Emp. Share*	*Total Emp. SMSA (000s)*	*Dallas County Share (000s)*	*SMSA Emp. Share*	*1972–1980 Dispersion/ Concentration Index***
	Agriculture (SIC 01–09)										
07	Agricultural Services	1.0	0.7	70.0%	1.8	1.2	66.7%	3.6*	2.4	66.7%	100
	Extractive/Mining (SIC 10–14)										
13	Oil and Gas Extraction	4.1	2.9	70.7	4.2*	3.2	76.2	8.5	6.0	70.6	93
14	Nonmetallic Minerals, except Fuels	1.1*	0.6*	54.5	0.8*	0.3*	37.5	1.1*	0.2	18.2	49
12	Bituminous Coal and Lignite Mining	—	—		0.2*	0.2*	100.0	0.4	0.4	100.0	100
10	Metal Mining	—	—		—	—		0.2	0.2	100.0	—
	Transformative/Construction (SIC 15–17)										
17	Special Trade Contractors	24.3	17.4	76.6	33.9	23.3	68.7	57.0	37.3	65.4	95
15	General Building Contractors	12.4*	11.8*	95.2	16.8	12.4	73.8	29.1	21.2	72.9	99
16	Heavy Construction Contractors	8.2*	6.9*	84.1	13.5	9.6	71.1	16.3*	11.6	71.2	100
	Distributive Services/Transportation, Communication, and Public Utilities (SIC 40–49)										
42	Motor Freight Transportation and Warehousing	13.3	10.2	76.7	20.2*	16.3	80.7	28.7	21.4	74.6	92
48	Communication	9.6*	6.7	69.8	14.7*	9.6	65.3	20.9*	14.8	70.8	—
45	Transportation by Air	6.8	6.0	88.2	11.6	10.7	92.2	19.6	6.0	30.6	33
49	Electric, Gas and Sanitary Services	7.2*	5.8	80.6	7.5*	5.8	77.3	10.6	6.6	62.3	81
40	Railroad Transportation	—	—		—	—		—	—		
41	Local, Suburban, Interurban Transportation	3.1	2.4	77.4	4.0	3.5	87.5	4.2	3.7	88.1	101
47	Transportation Services	0.6	0.4	66.7	1.0	0.9	90.0	2.8	2.0	71.4	79

SIC	Industry										
46	Pipe Lines, Except Natural Gas	0.3	0.3	100.0	0.2*	0.2*	100.0	0.6*	0.4*	66.7	67
	Wholesale Trade/Durable and Nondurable Goods (SIC 50–51)	59.5	46.1	77.5	81.6	64.5	79.0	116.9	88.3	75.5	96
	Retail Trade (SIC 52–59)										
58	Eating and Drinking Places	21.9	14.1	64.4	37.9	22.9	60.4	72.4	43.3	59.8	99
59	Miscellaneous Retail	12.3	7.5	61.0	22.2	15.6	70.3	36.6	24.0	65.6	93
53	General Merchandise Stores	26.6	15.8	59.4	40.2	27.2	67.7	33.4	19.5	58.4	86
54	Food Stores	15.3	9.1	59.5	20.6	12.4	60.2	31.8	17.6	55.3	92
55	Automotive Dealers and Gasoline Service Stations	16.6	10.1	60.8	22.9	13.9	60.7	26.4	15.2	57.6	95
56	Apparel and Accessory Stores	7.5	5.5	73.3	10.9	7.3	67.0	14.7	9.2	62.6	93
57	Furniture, Home Furnishings, and Equipment	3.3	2.2	66.7	6.0*	3.9	65.0	8.4	5.4	64.3	99
52	Bldg. Materials, Hardware, Garden Supply Dealers	3.7	1.8	48.6	4.4	2.3	52.3	6.1*	3.4	55.7	107
	Producer Services/Finance, Insurance, and Real Estate (SIC 60–67)										
63	Insurance Carriers	22.7	18.3	80.6	27.6	23.6	85.5	39.1*	32.8	83.9	98
65	Real Estate	8.2	6.7	81.7	17.1*	14.0	81.9	23.1*	18.5	80.1	98
60	Banking	9.4	6.5	69.1	13.1	8.7	66.4	22.2	14.5	65.3	98
64	Insurance Agents, Brokers, and Services	3.2	2.4	75.0	4.7	3.5	74.5	9.9	7.6	76.8	103
61	Credit Agencies, other than Banks	4.5	2.9	64.4	7.0	4.9	70.0	9.5	6.3	66.3	95
62	Security and Commodity Brokers	1.2	1.1	91.7	3.1	2.8	90.3	2.8	2.4	85.7	95
67	Building and Other Investment Offices	1.1	0.8	72.7	2.2	1.7	77.3	2.3*	1.8*	78.3	101
66	Combination of Real Estate, Insurance, Loan Offices	0.2	0.1	50.0	0.2	0.1	50.0	0.2*	0.2*	100.0	200
	Business Services (SIC 73)										
73	Business Services	12.4	9.6	77.4	29.4	23.5	79.9	63.9	50.9	79.6	100
	Legal Services (SIC 81)										
81	Legal Services	1.7	1.3	76.5	3.3	2.4	72.7	7.1	5.4	76.1	105
	Membership Organizations (SIC 86)										
86	Membership Organizations	7.6	5.4	71.1	14.3	9.8	68.5	19.3	12.2	63.2	92
	Misc. Professional Services (SIC 89)										

Table 5.13 (*cont.*)

Manufacturing Employment Concentration and Dispersion Trends, Dallas County, 1964–1980

SIC	Industry group	*1964* Total Emp. SMSA (000s)	*1964* Dallas County Share (000s)	*1964* SMSA Emp. Share	*1972* Total Emp. SMSA (000s)	*1972* Dallas County Share (000s)	*1972* SMSA Emp. Share	*1980* Total Emp. SMSA (000s)	*1980* Dallas County Share (000s)	*1980* SMSA Emp. Share	1972–1980 Dispersion/ Concentration Index**
89	Miscellaneous Services	4.5	3.4	75.6	10.0	8.1	81.0	15.4	10.9	70.8	87
	Social Services (SIC 83, after 1974)										
83	Social Services							10.7*	6.1	57.0	—
	Nonprofit Services/Health and Education (SIC 80, 82)										
80	Health Services	19.1	12.5	65.4	35.4	22.0	62.1	59.5	36.0	60.5	97
82	Educational Services	5.4	3.6	66.7	10.7*	7.4	69.2	12.0*	7.8	65.0	94
	Mainly Consumer Services										
70	Hotels and Other Lodging Places	6.8	5.0	73.5	8.9	6.8	76.4	17.4	14.1	81.0	106
72	Personal Services	14.9	9.1	61.1	15.1	9.5	62.9	18.5	11.2	60.5	96
75	Auto Repair, Services and Garages	4.7	3.2	68.1	6.4	4.4	68.8	9.8	6.0	61.2	89
79	Amusement and Recreation Services	3.5	2.2	62.9	6.6	3.8	57.6	9.9	4.6	46.5	81
76	Miscellaneous Repair Services	2.9	2.1	72.4	3.8	2.8	73.7	5.3	3.9	73.6	100
78	Motion Pictures	1.7	1.3	76.5	2.3	1.7	73.9	2.5	1.9	76.0	103
84	Museums, Botanical, Zoological Gardens	—	—		0.1*	—	—	0.5*	0.1	20.0	—

Source: County Business Patterns, 1964–1980.

*Adjusted for disclosure.

**Data on employment are basically comparable for major SIC groups (two-digit). However, due to changes in the 1972 Standard Industrial Classification manual, data for some three- and four-digit SIC codes may not always be strictly comparable from 1964 to 1980. Any resulting distortions are judged to be small in most cases. In one instance, SIC 48, redefinition was so complete that two-digit data are judged to be noncomparable.

the FIRE services often benefit greatly from the economies of agglomeration and have increasingly filtered into dense and compact central city settings in major urban areas.[53] Increasingly, large, sprawling industrial activities and an expanded corporate structure are supported by the services that this group provides. Between 1964 and 1972, Dallas County's share of employment in the FIRE services increased from 76.8 percent to 79.1 percent; by 1980 that share had dropped to 77.1 percent. Despite the dispersion of employment after 1972, in 1980 FIRE services employment was more highly concentrated in Dallas County than any other nonmanufacturing group with the exception of business services.

Business services (SIC 73) plays a special role, as this group hosts a number of high-technology service industries. Of all nonmanufacturing industries, *business services alone has experienced continued employment concentration in Dallas County since 1964.* During the 1964–1972 period, the employment share increased from 77.4 percent to 79.9 percent, and by 1980, the share reached 80.0 percent. The special attractiveness of older areas within the region for newly developing high-technology services should not be overlooked. This trend constitutes evidence that central cities are continuing in their historical role as seedbeds for technological innovation, for the new industries that emerge from its commercialization, and for the renewed industries that result from its adoption and diffusion. Eventually, product and industry life-cycles may assert themselves and disperse employment in these particular industries, as they have done so dramatically in manufacturing industries throughout the nation. For the present, however, high-technology services, to a degree not found even across the larger services economy within this region, are experiencing growth and development under increasingly concentrated circumstances.

In the remaining major nonmanufacturing groups, a variety of distinct sequences are evident. Legal services (SIC 81) employment dispersed away from Dallas County between 1964 (76.5 percent) and 1972 (79.7 percent), only to begin concentrating again by 1980 (76.1 percent). This trend is a mirror image of that most frequently found within the region. Employment in nonprofit services (SIC 80 and 82), which includes both educational and health services, has continuously dispersed away from Dallas County. In 1964, 65.7 percent of the region's nonprofit employment was in Dallas County; this share declined to 63.8 percent by 1972 and again to 61.3 percent by 1980 as education and health facilities proliferated throughout the region to accommodate extensive and widespread suburbanization. Finally, the overall decline in employment share of mainly consumer services industries in the larger regional economy has been accompanied by the familiar concentration-dispersion sequence since 1964. During the 1964–1972 period, the overall employment share of these services in Dallas County rose from 66.4 percent to 67.1 percent, followed by a decline to 65.4 percent by 1980. However, employment in the hotel (SIC 70) and motion picture (SIC 78) industries actually became more concentrated within Dallas County between 1972 and 1980.

High-Technology Development: Concentration and Dispersion Trends

Not surprisingly, high-technology industrial development is very unequally distributed throughout the D-FW region. Patterns of geographic concentration can be seen at several spatial scales. As the county-level distribution in Table 5.14 reveals, an estimated 78.9 percent of the firms in this sector are located in Dallas County. Only 17.7 percent are located in Tarrant County, which includes the city of Fort Worth. The city-level distribution in Table 5.15 reveals the same essential pattern. The city of Dallas is the location of nearly half (48.9 percent) of the region's high-technology firms, while the northern suburb of Richardson has 9.1 percent and Fort Worth has 9.5 percent. Clearly, the decades-long development process in this sector thus far has been largely captured by the city of Dallas.

The spatial shifts characterizing the D-FW high-technology sector are in clear contrast to those characterizing the manufacturing and nonmanufacturing sectors and subsectors identified above. Using the most inclusive definition of the high-technology sector (Group I), Table 5.16 presents evidence of *a long-term slow dispersion of high-technology employment away from Dallas County since 1964.* Further calculations indicate that the aggregate trend of growth away from Dallas County was only

Table 5.14
County Distribution of High-Technology Firms in D-FW

County	*Number of Establishments*	*Percent of Total Establishments*
Dallas	625	78.9%
Tarrant	140	17.7
Collin	20	2.5
Denton	6	0.8
Johnson	—	—
Ellis	—	—
Parker	—	—
Kaufman	1	0.1
Rockwall	—	—
	Total 792	100.0%

Table 5.15
High-Technology Firms by City—D-FW Region

City[a]	*Number of Establishments*	*Percent of Total High-Technology Establishments*
Dallas	387	48.9%
Fort Worth	75	9.5
Arlington	38	4.8
Garland	40	5.1
Irving	52	6.6
Richardson	72	9.1
Plano	14	1.8
Grand Prairie	19	2.4
Mesquite	3	0.4
Denton	2	0.3
Carrollton	33	4.2
Hurst	2	0.3
Lewisville	4	0.5
Euless	14	1.8
Bedford	3	0.4
Other	34	4.3
Total	792	100.4%[b]

[a]Cities ranked by 1980 population size.
[b]Total does not add up to 100.0% due to rounding error.

a trickle between 1964 and 1972; Dallas County's share of high-technology employment only declined from 77.8 percent to 77.6 percent. However, the spread of high-technology employment to locations beyond Dallas County accelerated after 1972; by 1980, Dallas County's employment share had declined to 68.5 percent. Generally, this dispersion of new growth was accompanied by considerable employment expansion within existing firms in Dallas County locations, especially in the city of Dallas and Richardson, and the resulting net dispersion illustrates a difference in relative growth rates at best.

As we have seen before, aggregate trends for the entire sector tend to obscure a diversity in trends experienced by specific industries within the sector. Table 5.16 reveals that the dispersion of employment in some industries was accompanied by concentration—often considerable—in others. Among those six industries that employed more than ten thousand workers in 1980, only one—heavy construction (SIC 162)—became more concentrated in Dallas County between 1972 and 1980. Five others (SIC 353, 366, 367, 372, and 508) experienced employment dispersion. The region's largest high-technology employment industry—aircraft and parts (SIC 372)—experienced a dispersion of more than half of its employment away from Dallas County between 1972

Table 5.16
High-Technology Employment Concentration and Dispersion Trends, Dallas County, 1964–1980

		1964			*1972*			*1980*			
SIC	*Industry*	*Total Emp. SMSA 1964 (000s)*	*Dallas County Share (000s)*	*SMSA Emp. Share*	*Total Emp. SMSA 1972 (000s)*	*Dallas County Share (000s)*	*SMSA Emp. Share*	*Total Emp. SMSA 1980 (000s)*	*Dallas County Share (000s)*	*SMSA Emp. Share*	*1972–1980 Dispersion/ Concentration Index** *
131	Crude Petroleum and Natural Gas	2.6	2.1	80.8%	2.3	1.9	82.6%	4.8*	3.8*	79.2%	96
162	Heavy Construction, except Highway and Street	3.5*	3.1	88.6	6.8*	4.2	61.8	10.9*	7.8	71.6	116
281	Industrial Inorganic Chemicals	0.8	0.5	62.5	0.7	0.4	57.1	0.6*	0.3	50.0	88
282	Plastic Materials and Synthetics	—	—		—	—		0.1*	0.1*	100.0	—
283	Drugs	0.8*	0.2	25.0	1.1*	0.7	63.6	0.9*	0.2*	22.2	35
284	Soaps, Cleaners and Toilet Preparations	1.1	1.0	90.9	1.4	1.2	85.7	2.0*	1.7	85.0	99
285	Paints and Allied Products	1.1	1.1	100.0	1.4	1.4	100.0	1.6*	1.3	81.3	81
286	Industrial Organic Chemicals	—	—		—	—		—	—		
287	Agricultural Chemicals	0.1	0.1	100.0	0.2*	0.2*	100.0	0.8*	0.8*	100.0	100
289	Misc. Chemical Products	0.7	0.4	57.1	1.0	0.7	70.0	2.0*	1.4*	70.0	100
291	Petroleum Refining	0.2*	—		0.3*	—		0.2*	—		—
301	Tires and Inner Tubes	0.2	0.2	100.0	0.4*	0.4*	100.0	0.4*	0.4*	100.0	100
324	Cement, Hydraulic	0.7*	0.4*	57.1	0.9*	0.4	44.4	0.7*	0.2*	28.6	64
348	Ordnance and Accessories	0.3*	0.2	66.7	0.8*	0.6	75.0	0.4*	0.4*	100.0	133
351	Engines and Turbines	—	—		—	—		—	—		
352	Farm and Garden Machinery	0.2*	0.2*	100.0	0.4*	0.4*	100.0	1.2*	0.4*	33.3	33
353	Construction, Mining and Material Handling Machinery	4.1*	2.4	58.5	5.6*	3.9	69.6	14.6*	9.3	63.7	92
354	Metalworking Machinery	0.7	0.7	100.0	1.2*	0.6	50.0	3.3*	1.0	30.3	61
355	Special Industry Machinery, except Metalworking	2.4	2.2	91.2	2.1*	1.7	81.0	3.2*	2.1	65.6	81

356	General Industrial Machinery	1.8	0.7	38.9	2.0	0.6	30.0	3.1	1.9	61.3	204
357	Office, Computing and Accounting Machines	0.6*	0.4*	66.7	2.4	2.4	100.0	5.3*	4.4	83.0	83
358	Refrigeration and Service Industry Machinery	2.4*	1.3	54.2	5.0*	3.0	60.0	7.5*	2.8	37.3	62
361	Electric Transmission and Distribution Equipment	0.4*	0.2	50.0	0.6	0.4	66.7	0.6	0.6	100.0	150
362	Electrical Industrial Apparatus	0.5*	0.5*	100.0	0.5	0.5	100.0	1.1	0.8	72.7	73
363	Household Appliances	0.6*	0.2*	33.3	0.9*	0.5	55.6	0.7	0.2	28.6	51
364	Electric Lighting and Wiring Equipment	0.2	0.2	100.0	1.1	0.5	45.5	1.2*	0.7	58.3	128
365	Radio and TV Receiving Equipment	0.6*	0.6*	100.0	0.2	0.2	100.0	0.9	0.9	100.0	100
366	Communication Equipment	9.3	9.3	100.0	15.9	15.4	96.9	32.4*	27.9	86.1	89
367	Electronic Components and Accessories	1.8*	1.4	77.8	11.7*	11.6	99.1	19.7	18.9	95.9	97
369	Misc. Electrical Machinery	0.6	0.6	100.0	1.6*	1.4	87.5	2.1*	2.0	95.2	109
371	Motor Vehicles and Equipment	6.8	2.4	35.3	7.2*	1.7	23.6	7.3*	1.8*	24.7	104
372	Aircraft and Parts	14.0*	12.0	85.7	15.5*	13.3	85.8	42.7*	18.0*	42.2	49
376	Guided Missiles and Space Vehicles	—	—		—	—		—	—		
381	Engineering, Laboratory, Scientific and Research Instruments	0.8	0.8	100.0	0.9	0.9	100.0	0.5*	0.1	20.0	20
382	Measuring and Controlling Instruments	0.3	0.1	33.3	0.6*	0.3	50.0	1.7	1.1	64.7	129
383	Optical Instruments and Lenses	0.4*	0.4*	100.0	0.6	0.6	100.0	0.6	0.6	100.0	100
384	Surgical, Medical and Dental Instruments	—	—		0.1	0.1	100.0	1.1	0.9	81.8	82
386	Photographic Equipment and Supplies	—	—		0.1*	—		0.4*	0.2	50.0	—
483	Radio and TV Broadcasting	0.6	0.2	33.3	1.4	1.0	71.4	2.0*	1.7	85.0	—
489	Communication Services, NEC.	0.1	0.1	100.0	0.5*	0.1	20.0	1.9*	1.8*	94.7	—
491	Electrical Services	3.7	3.1	83.8	1.7*	1.1*	64.7	6.2	3.8	61.3	95
493	Combination Electric, Gas and Other Utility Services	—	—		—	—		0.4*	0.2*	50.0	—

Table 5.16 *(cont.)*
High-Technology Employment Concentration and Dispersion Trends, Dallas County, 1964–1980

		1964			*1972*			*1980*			
SIC	*Industry*	*Total Emp. SMSA 1964 (000s)*	*Dallas County Share (000s)*	*SMSA Emp. Share*	*Total Emp. SMSA 1972 (000s)*	*Dallas County Share (000s)*	*SMSA Emp. Share*	*Total Emp. SMSA 1980 (000s)*	*Dallas County Share (000s)*	*SMSA Emp. Share*	*1972–1980 Dispersion/ Concentration Index***
506	Wholesale Trade, Electrical Goods	4.7	4.2	89.4	6.7	5.9	88.0	9.1	7.6	83.5	95
508	Wholesale Trade, Machinery, Equipment and Supplies	12.5	10.4	83.2	16.4	13.2	80.5	26.4	20.3	76.9	96
737	Computer and Data Processing Services	—	—		—	—		9.3	8.1	87.1	—
7391	Research and Development Laboratories	0.3	0.3	100.0	2.3	1.7	73.9	1.0	1.0	100.0	135
891	Engineering, Architecture and Surveying Services	2.4	1.8	75.0	4.0	3.0	75.0	9.0	6.2	68.9	92
892	Noncommercial Education, Scientific and Research Organizations	0.4	0.4	100.0	0.3	0.3	100.0	0.2	0.1	50.0	50

Source: The 1964–1980 figures are from *County Business Patterns* data, respective years.

*Adjusted for Disclosure.

**Data on employment are basically comparable for major SIC groups (two-digit). However, due to changes in the 1972 Standard Industrial Classification manual, data for some three- and four-digit codes may not always be strictly comparable from 1964 to 1980. Any resulting distortions are judged to be small in most cases. In one instance, SIC 48, redefinition was so complete that two-digit data are judged to be noncomparable.

and 1980, reflecting the fact that the military-industrial complex in this region is increasingly a suburban phenomenon.

Spatial Features of High-Technology Development Within the City of Dallas

Table 5.17 and Figure 5.9 display the distribution of high-technology firms throughout the city of Dallas. While 6.2 percent of these firms are located in the downtown area and another 11.4 percent are located adjacent to it, the vast majority are located in the northern tiers of the city. Once again, in this cross-sectional data for mid-1984, there is evidence of a dispersion dynamic at work within the high-technology sector at even this smallest of spatial scales.

Spatial Filtering of Successive Firm Birth Cohorts: The historical importance of urban concentration and agglomeration economies to the D-FW region's high-technology industrial development is undeniable. However, even though high-technology remains distributed extremely unequally, there is evidence that *at the intracity level,* just as was the case relative to Dallas County discussed earlier, *a deconcentration of high-technology is underway.* For the most part this is due to the locational decisions accompanying new firm start-ups, rather than to the relocation of existing firms per se.

Table 5.17
Geographic Distribution of High-Technology Firms in the City of Dallas

Geocoded Subsectors	*Number of Establishments*	*Percent of Total Dallas Establishments*	*Service Industry Share*	*Percent of Firms Established Since 1980*
Oaklawn—Downtown	24	6.2%	70.0%	20.8%
Love Field/Industrial	44	11.4	50.0	20.5
Southwest Dallas	12	3.1	—	25.0
Redbird	2	.5	—	—
East Oak Cliff	3	.8	—	—
Fair Park	4	1.0	—	—
Pleasant Grove	4	1.0	—	—
East Dallas	9	2.3	66.7	11.1
White Rock/Greenville	61	15.8	55.7	33.7
North Dallas I	40	10.3	40.0	35.0
North Dallas II	52	13.5	57.7	36.5
North Dallas III	56	14.5	82.1	48.2
Vickery-Lakewood	9	2.3	66.7	44.4
Farmer's Branch	67	17.3	47.8	32.8
Totals	387	100.0%		

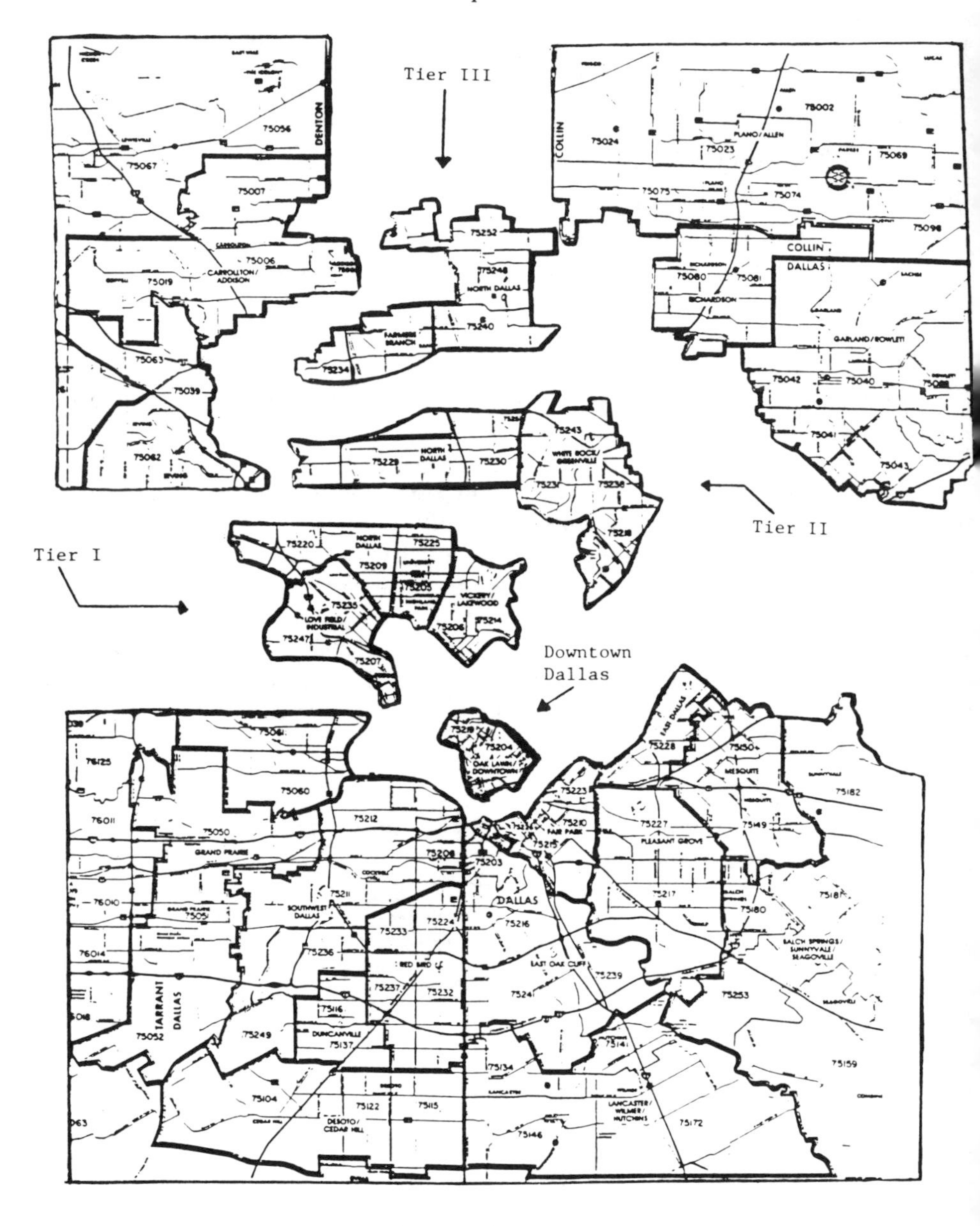

Figure 5.9. *City of Dallas: development tiers.*

Figure 5.10 illustrates how recent high-technology growth has steadily migrated away from the downtown area to newer parts of the city. It does so by reporting the spatial distributions of a succession of high-technology firm birth cohorts across a downtown central core area and three tiers of the city of Dallas extending northward from the central business district. High-technology firms have been entering the city's economic base for more than a quarter-century, and it is not surprising that the smallest proportion of firms from each of the cohorts is located in downtown Dallas, given the outward thrust of industrial development that was already building by the 1960s. Nevertheless, since then, consistently declining proportions of successive cohorts have located in the city center. Of the firms established before 1960, 15.8 percent were located in the city core. The extent to which inner-city areas have progressively "lost" their unique capacities to nurture new high-technology industrial development is illustrated by the trend whereby only 3.8 percent of the firms in the most recent 1980s cohort chose to locate in this area. The turning point for downtown Dallas seems to have come in the mid-1970s. While just under a third of the firms in the cohorts established prior to 1975 located there, after that date the proportion dropped to near one in five. The northernmost part of Dallas—Tier III—has consistently outdrawn the downtown area. However, beginning in the 1970s, the relative attractiveness of Tier III began to increase dramatically. Fully 37.7 percent of the high-technology firms established in the 1980s were located in Tier III.

Interpreting Spatial Patterns of Development

Must these spatial patterns of industrial development necessarily constitute evidence of diminishing returns to scale and therefore a waning of the traditional advantages of concentration in large urban areas? Put more simply, is this dispersion of high-technology firms better viewed as a flight away from or avoidance of inner-city locations rather than as an accommodation to a wider range of peripheral locations? In this case, probably not. The bulk of the recent physical development in the city of Dallas has been located north of the city center. While the greatest portion of the region's growth is tied to the incubation of new business development and secondary expansion in-place, for large central cities like Dallas, encircled as it is by a ring of incorporated suburban jurisdictions, much growth has either avoided locating within its borders or has taken root initially in locations peripheral to it. The emergence of such patterns can slowly shift to once-peripheral locations like North Dallas and the northern suburbs the *economic centrality* that has been gradually displaced and no longer coincides with *geographic* and *jurisdic-*

Figure 5.10. *Shifting locations of high-technology firms.*

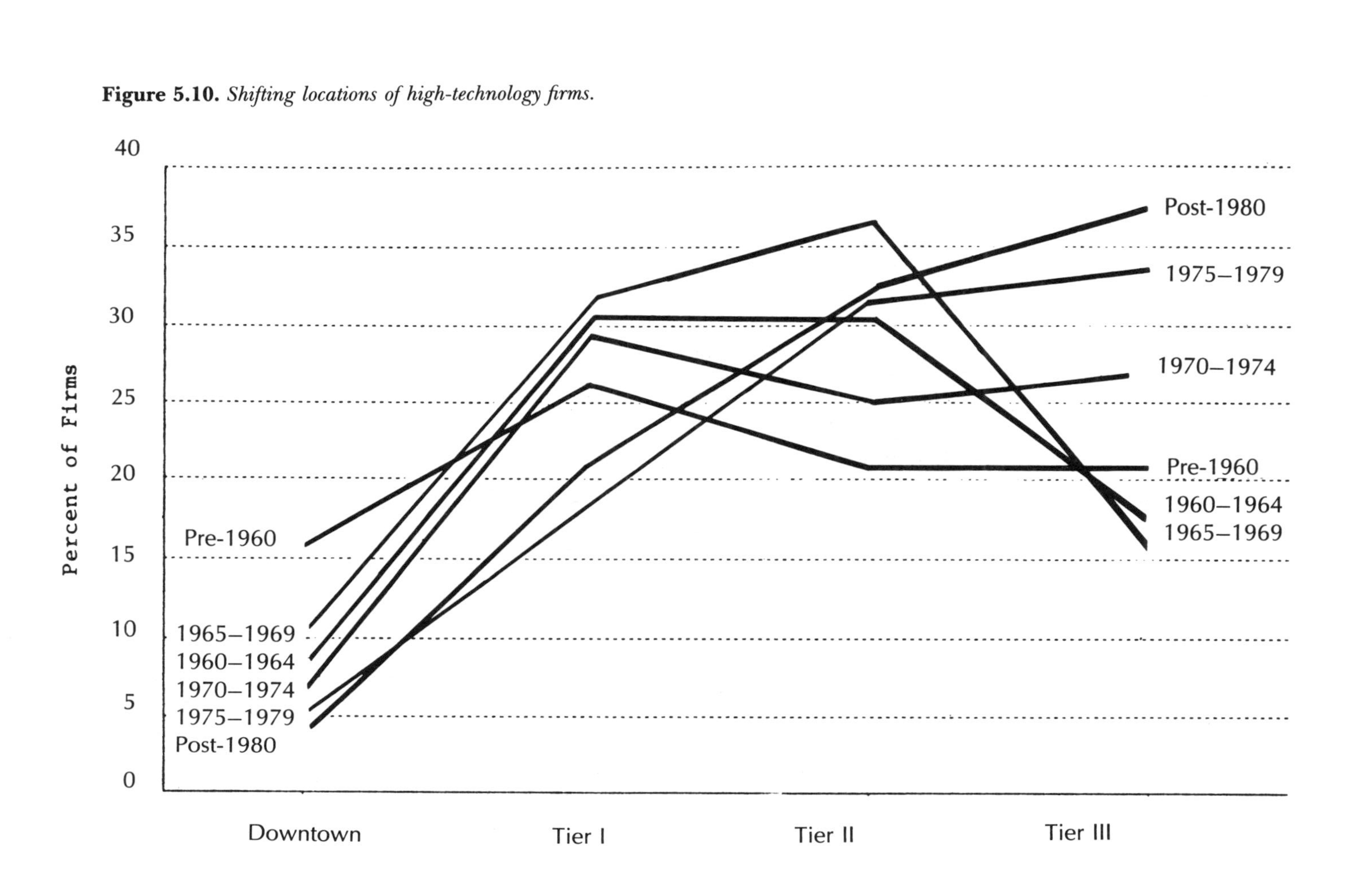

City of Dallas Locations

Hi-Tech Birth Cohorts	*Downtown*	*Tier I*	*Tier II*	*Tier III*
Pre 1960	15.8%	26.3%	21.0%	21.0%
1960–1964	8.3	30.4	30.4	17.4
1965–1969	10.5	31.6	36.8	15.8
1970–1974	6.3	29.2	25.0	27.1
1975–1979	4.8	19.3	31.3	33.7
1980 +	3.8	20.8	32.3	37.7

Downtown: Downtown (Central Business District)
Oaklawn

Tier I: North Dallas I
Love Field-Industrial
Vickery-Lakewood

Tier II: North Dallas II
White Rock-Greenville

Tier III: North Dallas III
Farmer's Branch

tional centrality within the region. Therefore, it is not so much true that high-technology development has broken its original bonds of dependence on urban concentrations, thus permitting it to migrate northward away from the city center; rather, high-technology development appears simply to have accompanied the expansion of concentrated urban development northward to the edge of the city. As a result, high-technology development has not so much moved away from the center of Dallas as the favorable conditions that originally attracted such development have themselves dispersed to new locations, and high-technology development along with them.[54]

It is likely that were we to look in on this spatial shift at the end of the century, we would find that the trajectory of development would by then have carried the critical mass of high-technology firms northward and westward out of the city of Dallas altogether. Any shifts in the location of employment growth would undoubtedly lag well behind those of new business growth, since the older and larger firms in the sector are responsible for the greatest part of its employment, and they are generally found in closer-in locations. However, should the pace of business relocation and secondary expansions to suburban and exurban locations elsewhere in the region begin to quicken, then the spatial gap between business and employment growth could well narrow significantly in the future.

It is in this way that industrial and urban development influence each other and confirm the importance of a spatial perspective on advanced industrial development. While large cities like Dallas have in the past provided an attractive setting for what has become today's high-technology sector, there is ample evidence to suggest that what currently looks like a loosening grip, first on new firm growth and then on new and existing employment throughout the sector, may in the future become a loosened grip on the entire sector altogether.

6

Agriculture and Land Redefined

The capacity of old-line industries—especially basic manufacturing—to make adjustments to new economic circumstances and technological possibilities and to eventually prosper is an important determinant of the ability of localities, regions, and the nation at large to preserve existing competitive strengths and create new ones. As we have been reminded in recent years, the ease with which industrial change in these sectors unfolds translates directly into changing patterns of employment opportunities and income distributions, shifting investment patterns, and the restructuring of local and regional economic bases and the settlement patterns tied to them. Ultimately, living standards and the quality of life depend on the direction and pace of these adjustments. Yet, as enormously consequential as these impacts of industrial change may be, some would suggest that they pale in significance compared to those commonly associated with the restructuring of agriculture. Major corresponding shifts in the capacity of agriculture to produce are related not just to how well we will live but to whether we will live.

In 1983, Americans spent over $300 billion for domestically produced agricultural products. Despite the inflation that accompanied the most recent recessions, the real price of food has declined since 1970. Agriculture is today the nation's largest industry, employing over 15 million people, the majority of whom perform support roles through a massive agricultural services sector. The enormous stakes tied to agricultural change tend to complicate and distort analyses of it. It becomes difficult to distinguish between agriculture as an economic activity undergoing changes in its production arrangements and agriculture as the fragile buffer between life and death for hungry millions around the world. Objective and seemingly detached analysis often appears to blur what some view as the moral distinction between the capacity to produce corn and the capacity to produce cars. The motivations that keep alive such a distinction are entirely understandable, but they could result in reinforcing outmoded perceptions and policy stances that actively seek to preserve traditional agricultural arrangements rather than to foster and promote the emergence of a "new agriculture." The results may be as disastrous as they are ironic. The inability to anticipate and appreciate the adjustments underway within agriculture today may well reduce our ability to meet the needs of an expanding world population tomorrow.

In this chapter an attempt is made to understand agriculture as one industry among many others undergoing major changes in an advanced industrial economy. Modern farming and agribusiness are characterized by lengthy and complex linkages in both production and distribution. The agricultural economy has evolved in such a way that only 5.9 percent of those employed in the sector produce final products. Agricultural production is supported by a vast services system that receives 72 percent of consumer expenditures for food in the United States. The importance of agriculture to advanced economies is especially evident when examining its major role in world trade, and so I begin there. I go on to explore the existence of instructive similarities between the longer-term restructuring of agriculture and that of older basic industries—especially manufacturing. The chapter also examines the way in which productivity and output are decoupled from direct employment in agriculture, the expanded importance of the agricultural services industries, the locational shifts that redistribute agriculture within and between regions, the way new corporate arrangements accompany advanced industrial development, and the eclipse of traditional views of land and land-based natural resources as the heart of agriculture.

Next, the chapter introduces categories of land-policy controversies that have developed in recent years and color our interpretations of agriculture in transition. There is concern that with the spread of industrial and residential land uses at all spatial scales, urban land uses must necessarily collide with the land and related natural resource requirements of agriculture; this concern has prompted increased vigilance and advocacy aimed at the preservation of agricultural land in the face of threatening accompaniments of urbanization. Similarly, ecological concerns have prompted the renewed "neo-Malthusian" fear that food resources may be losing a race with population growth. Finally, a section is devoted to understanding the rationale underlying a general industrial policy for agriculture, which is often proposed to bolster its competitiveness in increasingly important global markets.

A Second Agricultural Revolution Takes Shape

Agriculture in the United States offers a far more instructive model of industrial change than is commonly realized. Ironically, the developments within the land-based economy that shape and redefine it have been largely invisible in an advanced industrial economy increasingly organized around economic activity within a system of metropolitan or regional subeconomies. Yet the ways in which the sector has come to be organized, the ways in which land and related natural resources are redefined and used and their economic value determined, the ways in which commercial crop and related production take place, and the ways in which scientific research and technological innovations are adopted and diffuse throughout the industry have quietly moved agriculture to the frontier of our advanced industrial society. At a time when the stimulative role of new technology in industrial change has attracted widespread interest, agriculture reveals itself to be an industry that has been continuously reshaped by technological innovation from the preindustrial era through the postindustrial one.

The productivity of the farm economy has been high and increasing all through the post–World War II era.[1] During the 1972–1982 decade alone, while nonagricultural productivity languished, agricultural productivity surged at an average annual rate of 4.4 percent. There is growing sentiment, however, that continued increases in agricultural productivity may no longer be forthcoming. Agriculture as an industry is commonly judged to be losing momentum and adaptability as well as to be increasingly constrained by the specter of certain "natural" limits on its development. The evidence includes the diminishing prospect of new technological breakthroughs and the fixed and inherent features of the natural environment that during the past 10,000 years have been so crucial to the land-based economy. These features are beginning to define what some consider to be an outer bound of productivity for the industry.[2]

Nonetheless, in the face of these concerns, evidence is also accumulating that may signal a new revolution slowly gaining momentum. In the process of this revolution, agriculture may well reveal itself more fully as the increasingly urbanized and industrialized activity that it has become. The past century has witnessed quantum leaps in agricultural productivity, derivative principally from new mixes of traditional factor inputs of land, labor, and capital. A major implication of these developments is that agriculture is in the process of being restructured even as its essential products, production processes, and traditional relationships to "natural" resources are being substantially redefined. Despite its origins and traditional location in nonmetropolitan and rural areas, the land-based economy is no longer economically or locationally peripheral. Indeed, U.S. metropolitan areas have provided ideal and improving platforms for agriculture throughout this century. Wolf has noted that only 10 percent of the average metropolitan area is devoted to urban uses, 24 percent to cropland, 32 percent to woodland, 19 percent to pasture, and 15 percent to miscellaneous uses.[3] The nation's metropolitan system has increasingly dispersed across the landscape; yet, with its mobility, flexibility, and new structural features, modern agriculture may be seen to have thoroughly and successfully invaded metropolitan-based industrial and service economies as much as they have invaded it. As Table 4.1 showed, between 1951 and 1979, the share of U.S. agricultural employment located in metropolitan areas increased from 71.0 percent to 81.1 percent, with the bulk of the increase coming during the 1970s. Urban and rural—land-based and nonland-based— economies, then, have not so much collided in conflict as they have converged in cooperation.

The Role of U.S. Agriculture in a Global Economy

I begin this reconsideration of the land-based economy by exploring its role in foreign trade, for it is in this role that agriculture deserves to be appreciated as a foremost link to the emerging global economy. Steadily during the 1970s, as the United States was drawn into world markets, the trade balance became an important indicator of national economic functioning. By far the world's largest economy was rapidly losing its insularity as import and export trade dependencies developed. Over the 1970–1980 period, exports as a share of GNP more than doubled, from 4.4 percent to 8.5 percent, and they are expected to rise to 15 percent by 1990. Concurrently, imports increased their share of GNP from 4.1 percent to 9.5 percent.

On the surface, the overall trade picture steadily deteriorated as deficits with other developed nations persisted throughout the 1970s. However, beneath the surface important, if usually obscured, trends reflected the changing ways in which the U.S. economy was tied to the rest of the world. We are accustomed to assigning greater significance to trade involving goods than to that involving services. Understandably, then, the mounting trade surpluses in sectors other than traditional manufacturing have seldom been judged sufficient to dispel building concern over the threats to the nation's industrial status posed by soaring trade deficits in merchandise. Likewise, as we shall discover below, the extra margin of deficits triggered by our heavy dependence on foreign oil largely obscured the fact that our comparative advantage in global trade had fragmented into several distinct pieces. During the 1970s, the principal trade deficit sectors included automotive products, consumer goods, low-technology manufactures, and petroleum. While the cumulative deficit for these four sectors ballooned from $17.6 billion in 1970 to $140.1 billion by 1980, it was the petroleum sector in which deficits swelled faster than any other—from $2.3 billion in 1970 to $75.8 billion a decade later.

Over the same decade, the principal trade surplus sectors included agriculture, crude materials and fuels (excluding petroleum), high-technology manufactures and services (including investment earnings), all of which together registered an increase from $18.7 billion in 1970 to $127.4 billion by 1980. Of these, agriculture displayed the most explosive growth—from a $1.6 billion surplus in 1970 to $24.3 billion in 1980.[4] These balances are based on agricultural exports valued at $7.3 billion in 1970 and $41.2 billion in 1980.[5] By 1982, agricultural products accounted for approximately one-fifth of all U.S. exports, with an agricultural trade surplus for the 1972–1982 period of approximately $166 billion. A recent analysis of U.S. agricultural policy notes that foreign sales as a percent of gross farm income increased from less than 5 percent in 1940, to only 11 percent in 1958, to nearly 30 percent in 1981.[6] Our building orientation toward export production was registered quite literally on our farmland. While in 1960 only 19.8 percent of our acreage had been dedicated to exports, that proportion had grown to 39.2 percent by 1980.[7]

Increasingly, then, our strengths in international trade developed in two very different sectors. Agricultural exports—from our oldest economic sector—mushroomed to become a mainstay of U.S. export trade. High-technology goods and services—from what came to be widely regarded as the world's newest coherent sector—had become the other. In the end, as we shall see below, it is not the great differences, but the great similarities between these sectors that demand recognition.

The U.S. Agricultural Advantage in Flux. Since 1980, the picture has changed rather significantly. As we have noted in previous chapters, the pace and substance of the restructuring of U.S. agriculture were influenced significantly by the recessions of the early 1980s. Recession created an environment conducive to the reorganization of the financial, organizational, and technological arrangements that characterize the industry. Shifts in the structure of supply of many farm commodities have accompanied shifts in the structure of demand. Farm exports have steadily declined from 163.9 million metric tons in 1980 to an estimated 140.0 million metric tons in 1984. From a supply perspective, while production has continued to pace domestic consumption by a wide margin, this downturn in exports during the 1980s has generated enormous surpluses that have severely depressed domestic prices both for farm products and the land on which they are grown; as well, it has caused the cost of federal farm programs to escalate dramatically. While the prolonged initial effect of the Soviet grain embargo is now largely spent, in the interim, Canada, Argentina, and Australia, among other nations, have stepped up their competition with the United States for lucrative export trade in agricultural goods.

From a demand perspective, other nations have scaled back their purchases of American commodities for a variety of reasons. Some former importing countries, such as those in the Third World and the Fourth, are chronically poor and in debt; some are tied to hopelessly inefficient command economies, such as those in Eastern Europe; some are still slowly emerging from a protracted recession throughout the industrialized world; and finally, some, such as Mexico and some of its Latin American neighbors, have had to slash their imports due to severe declines in the purchasing power of their currencies. The continuing strength of the dollar in foreign markets has hindered the ability of American farmers to retain the niches they carved out in the 1970s. Increasingly coherent and aggressive agricultural policies—complete with export targeting, domestic subsidies, tariffs, and related trade barriers—have helped competing food and fiber producers gain deepening footholds in markets that the United States once dominated.

The Land-Based Economy in Transition: A Model for Manufacturing?

Agriculture is commonly viewed as a sector largely incapable of exhibiting the full range of adjustments that are commonly found among other industrial sectors in an advanced industrial economy. This perception is rooted mainly in the assumption of the relatively immutable ways in which agriculture must always remain tied to so-called "natural" resources. Yet a closer examination of agriculture as an economic activity suggests that this assumption may be too rigid. The distinctions between domestic- and foreign-trade issues and appropriate policy responses have become blurred in recent years as agriculture's role has expanded and contracted. The factors that enabled U.S. agriculture to be so resilient in international trade ultimately derive from those that have allowed it to become so productive at home in the first place.

In the broadest terms, urban-industrial transformation has gradually dislodged agriculture as the backbone of the U.S. economy. From an employment perspective, at least, industrial activity—especially factory-centered manufacturing—steadily eclipsed farm-centered agriculture as the core economy. The restructuring of agriculture, unlike that of manufacturing, for instance, seems relatively familiar to us, if only because it has been going on for so long. We have come to associate agriculture with bounteous yields from abundant amounts of "prime" farmland. The mechanization and resulting labor substitution within the industry have long been regarded as politically acceptable because the widespread social welfare gains of life in an industrialized and urbanized society have been judged to more than compensate for what was left behind. The living standards of the historical farm economy were found to be no match for those of the new factory/office economy.

Today, another shift is underway: office and home-centered economic activity in a broad range of services and manufacturing activity in a variety of redefined factory settings are slowly displacing the traditional factory-centered economy best illustrated in older basic industries. A new and diversified farm-factory-office-home economy is taking shape that no longer preserves traditional distinctions between urban and rural economies. What, if anything, does agriculture have in common with older industrial sectors such as manufacturing as their respective restructuring continues?

Decoupling Productivity and Employment. No sector reveals more clearly the extent to which the traditional indicators of economic functioning have been decoupled from one another than does agriculture. Even though agriculture has steadily contracted as an employment sector, its output as a contribution to GNP remains large. By comparison, manufacturing employment, while growing slowly in absolute terms, is expected by some to shrink to as low as 10 percent of total employment by the year 2000.[8] Meanwhile, its output share is likewise expected to stay high.[9] This single similarity between these two sectors encourages one to search for other respects in which these key economic sectors are restructuring along common paths that share common landmarks.

The 1920s held special significance for U.S. industrial development, since during that decade for the first time in our history the majority of Americans resided in settlements officially defined as "urban." Still, in 1920, 32 million people—30.2 percent of the nation's population—lived on farms.[10] Since then, not only the countryside's predominant economy, but its demography as well, has continued its historical restructuring. From 1930 to 1980, the nation's farm population has declined from 30.5 million to less than 6 million—that is, from 24.9 percent to 2.7 percent of total population.[11] Since 1980, the farm population has declined to 5.6 million or 2.4 percent of total population. Concurrently, farm employment has declined from 12.5 million to 3.7 million.[12] This, then is the latest installment in a long history of outmigration from the agricultural sector that throughout the twentieth century until the mid-

1970s was accompanied by net outmigration from the countryside itself.

However, not all those who reside on farms participate directly in the agricultural economy. It is estimated that "farmer," as a description of occupation rather than residence, applies to only 1 percent of the population. Since the 1970s, many people who came to reside on farms were primarily employed in nonfarm industries in expanded nonmetropolitan and extended metropolitan economies. The demography of this sector and its capacity to make dramatic adjustments over time is especially noteworthy given the relative inability of other advanced industrial economies, such as those in West Germany, France, and Japan, to shift population and labor out of agriculture as rapidly as they would like.[13]

Industrial "Mobility" and Regional Shifts. In recent years, we have come to associate the current restructuring of the industrial factory/office economy with rapid capital mobility and somewhat less rapid demographic redistribution. In the past quarter-century, as we have seen, manufacturing establishments and employment have filtered out of central cities and into suburbs and exurbs, out of older industrial regions and into the South and West, and even out of the country itself.[14] Given the locationally fixed productive inputs in agriculture, it is commonly assumed that a similar filtering or aggregate mobility has no corollary there. However, there is evidence suggesting that agriculture, too, has a capacity to redistribute itself locationally—both within and between regions—as it adjusts to changing economic and technological circumstances.

Shifts in agricultural employment and demography have not proceeded uniformly across multistate regions; in the resulting patterns there is clear evidence of an often overlooked repositioning of the industry at large. Historically, the South was the region most dominated by agriculture. However, as the South industrialized and thus diversified its economic base, agriculture was redistributed among the regions. Today, the Midwest, not the South, is the agricultural heartland of the nation. As the nation's farm population declined steadily during the post–World War II period, its regional distribution shifted significantly as well. Consequently, over the 1950–1982 period, the regional division of labor in agriculture became less marked. Gradually, from an employment perspective, agriculture filtered into the North Central and West regions, while shifting away from the South. The proportion of total farm population in the North Central region increased from 32.3 percent to 44.8 percent, and in the West from 8.3 percent to 13.0 percent. In the South, the proportion declined steeply from 51.6 percent to 34.8 percent, while it dropped only slightly from 7.8 percent to 7.5 percent in the Northeast.[15]

Industrial Structure and the Organization of Farms, Land, and Land-Based Resources. In manufacturing, common features of modern industrial development include new more complex corporate structures defined by the decentralization of production centers through "branching," the centralization of administrative and control functions, and the rise of conglomerates through corporate mergers.[16] While not to the same extent or following precisely the same track, agriculture has evolved a business structure that resembles that found in other restructuring sectors. Especially evident are the ways in which the industry has expanded its scale and sorted itself into new corporate arrangements. Even though production is still tied principally to the family farm, the sector is increasingly influenced by the development of corporate agriculture or agribusiness. The signs of agricultural concentration are clearly found in the way farmland is now organized for production.

A half-century ago, in 1930, the nation's agricultural sector was organized into 6.5 million farms that used 987 million acres; the average farm size was 151 acres. The increasing scale of agriculture is illustrated by the fact that by 1982, the number of farms had dropped to 2.4 million with the average size rising to 433 acres. While small farms of less than 50 acres have increased since 1978, reflecting the increased popularity of rural residence among those whose income is largely derived from nonfarm employment, the number of very large farms of over 2,000 acres likewise continued to grow. As the 1980s began, slightly more than a quarter of all farms and all farmland were devoted to cash grains including wheat, rice, corn, and soybeans. Subsectors devoted to cotton, tobacco, dairy, poultry and eggs, fruit, and nuts and vegetable production are all considerably smaller. Approximately four farms in ten and slightly more than half of all farmland were devoted to livestock production.[17]

By 1982, despite the enormous increases in productivity, only 5.3 percent more acres were under cultivation than in 1930.[18] Few statistics indicate so clearly the "plastic" nature of land as a factor in agricultural production.[19] Gradually, a traditional conception of land as a wholly natural input began to yield to one that gave new consideration to its many man-made attributes.

Modern U.S. agriculture is characterized by a structure of land holdings and uses that reveal at once both great variability and great stability. As might be expected, a few large farms control the largest share of cultivated land. In 1978, only 2.6 percent of all farms were larger than 2,000 acres, yet these few large farms accounted for nearly half (45.5 percent) of all land in farms.[20] In contrast, while the survival of the family farm has been a persistent concern within the agricultural community for decades, its role in the industry has not declined substantially. Although the total number of farms has dropped dramatically over the past half-century, the pattern of farm ownership has not changed. In 1983, as in 1930, only 2 percent of all farms were corporate, and an estimated 85 percent of these were family corporations.[21]

Agriculture has become increasingly capital-intensive and accessible to wave after wave of technological innovation intended to enhance productivity and operating efficiency; land values have responded accordingly. Over the 1969–1978 decade alone, and for reasons relating as much to how the organization of land uses reflects shifting demand pressures as to inflation, the total value of land in all farms soared from $206.8 billion to $650.0 billion.[22] Between 1975 and 1982, the total value of farmland and farm real estate increased from $358.6 billion to $821.5 billion.[23] While land values have been increasing steadily over the half-century since 1933, the inflation of the 1970s exaggerated the upward trend. In contrast, the 1980s have been marked by declines in the value of U.S. farmland.[24] In 1980, a record high of $795 per acre was reached; declines in price over the next five years reduced the average price to $744 per acre by 1983 and to less than $700 per acre by 1985.[25]

It appears, then, that as the scale and organization of agriculture continue to change, the sector may be in the process of becoming a factory/office–centered sector in the same way and for the same reasons, if not yet to the same extent, as other basic industries have. Similarly, agricultural production is in the process of revealing considerable industrial mobility, not only between regions but also between scales of economic activity. Likewise, agriculture appears to have acquired a new set of corporate arrangements reflected in new patterns of landholding. In the end, the range of similarities between agriculture and manufacturing as each restructures is broader than may commonly be realized.

New Spatial Arrangements and Land Policy Controversies

In the beginning, all land was not created equal. And the way agriculture as an industry distributed itself across the landscape has typically followed the contours of important resource inequalities. As the original land-based economy, agriculture has been strongly influenced by the distribution of natural features of the land and by related environmental factors. It is estimated that it takes a millenium to create a single inch of top soil. Understandably, then, the production factors for agriculture commonly have been viewed with a reverence that is unequaled today in the manufacturing and services sectors of an industrialized economy.

One dimension of growth in an advanced economy is expressed through the urbanization of traditionally low-density peripheral—and often rural—areas. Urbanization has less to do with the growth of population and employment per se than with the way both have come to be redistributed relative to one another across the landscape. There is a common view that in an advanced industrial economy the general deconcentration patterns in residential settlement and industrial location—and the new concentration and reconcentration that accompany them—must necessarily conflict with the functioning of the agricultural sector, especially as the latter competes for land, water, and other resources. Modern agriculture is still significantly dependent on resource inputs for which there are thought to be few easy substitutes. Nonetheless, access to "prime" farmland that has sufficiently rich topsoil and exposure to proper amounts of water and sunlight has been viewed as a relatively fixed constraint on food—including livestock—and fiber production. Not just any land, then, is suitable for agriculture; and so, to a degree far greater than that which characterized the early stages of manufacturing development, the "agricultural heartland" of the nation has come to be associated with natural boundaries that reflect the distribution of natural resources.

From the perspective of industrial location, agriculture has traditionally been viewed as relatively tightly anchored to specific locations within these set boundaries. Agriculture as an industry has been assumed to lack the mobility found in more advanced sectors of an industrial economy such as services—and even manufacturing itself—as we have seen in the past quarter-century. As a result, the resource needs and factor inputs for agriculture have claimed, if not always actually received, priority in any conflicts for land and space with other more urban-industrial uses.

Today, there are pressures building to topple such a rigid conception. The heated debate that has ensued has been a classic example of the way in which information and data are capable of generating more controversy than consensus, and the way in which "zero-sum" politics continues to guide thinking about agriculture as an economic activity. The conventional wisdom suggests that the gains of a developing industrial economy must henceforth come increasingly at the expense of a far more fragile and vulnerable agricultural economy. It was not until the 1970s, however, that the encroachment on traditional rural land by urban uses generated significant alarm. There was abundant evidence that specific parcels of farmland were indeed being converted to nonagricultural uses, and warnings began to appear concerning the impending and inevitable collision involving the land use patterns of an advanced industrial economy and those of modern agriculture. It was widely assumed that in such a contest, agriculture would be at a natural disadvantage and unable to compete effectively for the land and other resources it would require in the future.

Urbanization and Farmland Preservation. Since the late 1970s, a number of factors have conspired to draw attention to a variety of agricultural land policy issues. Among them are the recent leveling off of U.S. agricultural productivity rates, a continuation of topsoil loss through water and wind erosion, the building concern over pesticide use and the chemical upgrading of lower-grade soil, the energy intensivity of agriculture, and the increasing vulnerability of irrigation systems to the geopolitical distribution of water and its allocation among competing industrial and residential uses.[26] Even the dim but troubling specter of hunger in the United States at some future date, as population and food production increase at unequal rates, has helped focus concern on the quantity and quality of the nation's "platform" for agriculture—land and land-based resources.

The data supporting these concerns appear formidable indeed. Crosson and Brubaker have reported estimates that in the next quarter-century, the expansion of agriculture will require 60–70 million acres more than is required today—an increase of roughly 12 percent over the approximately 540 million acres that are either currently being used as cropland (413 million acres) or have the potential for being converted into cropland in the future (127 million acres).[27] This general view has spawned a number of interrelated agricultural land policy debates in recent years, including controversies surrounding the need for agricultural land preservation, open-space preservation, growth controls to prevent urban encroachment on rural land, and topsoil erosion protection.

In 1981, the Cornucopia Project report on the effects of erosion and development on U.S. land resources estimated that 26 square miles of cropland is lost daily to soil erosion, while half again as much is claimed by encroachment on rural land by urban uses.[28] The Worldwatch Institute has similarly warned that the productivity of a third of U.S. cropland is declining due to topsoil erosion. Alarm over the incursion by competing alternative rural uses is likewise evident in the findings of a United States Forest Service survey, which reported that commercial forest land had expanded onto marginal cropland in New York State to the point that by 1982 timber covered three-fifths of the state.[29] And, also, in 1981 ominous and foreboding headlines were generated by the conclusions of the National Agricultural Lands Study—an interagency effort commissioned in 1977—which not only estimated that approximately 3 million acres of agricultural land were being lost annually to residential and industrial (urbanization) uses, but also that the federal government was an active, if unwitting, promoter of this encroachment.[30]

Reinforcing Demographic Trends. The complex demographics of the 1970s seemed to buttress the concern over statistics on soil erosion and urban encroachment on rural lands. Multiscale patterns of population deconcentration resulted in the reinforcement of a century-long suburbanization trend with a new larger-scale pattern whereby nonmetropolitan areas began to grow faster than metropolitan ones. By 1974, the smallest places in the country were discovered to be growing the fastest.[31] Moreover, this dynamic was not confined to specific regions; between 1970 and 1978, three-fourths of all nonmetropolitan counties in the United States experienced net population increases. These net shifts of population growth to the periphery of metropolitan areas and beyond did much to increase the pressure from essentially urban demands on traditionally rural—and especially agricultural—land parcels whether or not they were in actual use.[32] Population growth rates in the one hundred counties for which agricultural outputs were the highest were found to be nearly twice those of the nation at large during the decade.[33]

Reinforcing Government Policies. As the National Agricultural Lands Study correctly indicated, the spreading out of our advanced industrial economy has indeed benefited from a variety of public sector reinforcements. Ninety federal programs were found to encourage directly or indirectly the displacement of rural land uses.[34] Highway construction programs, rural development programs, subsurface infrastructure subsidies for water and waste-water treatment projects also have had the direct effect of converting rural land to urban uses. The same array of indirect impacts can be seen from homeownership subsidies such as mortgage interest deductions and FHA-VA home loans; accelerated depreciation schedules for business that tilt investment decisions in favor of new construction as opposed to rehabilitation of older physical capital; and even revenue sharing, which has extended an intergovernmental fiscal lifeline to new and small incorporated communities. As with urban development policy, the federal government has countless programs and policies that reinforce with public sector money flows—if they do not actually set the trends into motion initially—the deconcentration and mobility associated with patterns of private investment and disinvestment.

Policy Activity in Response. As a result of these several trends, since the 1970s policy discussion and activity has commenced at all levels. In 1980, the Agricultural Land Protection Act (H.R. 2551) sought both to build state and local capacities to prevent farmland conversion and to ensure that federal actions no longer hindered those efforts; it only narrowly failed to gain passage in the House. In addition, all but two states have adopted some sort of agricultural land retention programs, usually involving differential tax assessment strategies to encourage the insulation of land currently in agricultural use. And there is widespread support at the local level for systematic efforts to reduce farmland conversion.[35]

The Urbanization Threat to Farmland Reexamined. The clash between urban and rural land uses may well be more apparent than real, however. Two traditional dimensions of land—the quality of the soil and its location—may increasingly be viewed as growing independent of one another. Despite all the seemingly compelling reasons proposed for why a farmland "crisis" is at hand, a reexamination of the evidence can do much to defuse concern about the issue, as well as to illuminate the methodological and valuative reasons that account for why the crisis was declared in the first place. Simon and Sudman have reviewed the methodological reasons showing that the crisis interpretation rested on an unreliable empirical basis.[36] New evidence indicates that the amount of arable land in the United States is increasing rather than decreasing, and that this trend is evident around the world.[37] Moreover, Crosson and Stout have concluded that even the high rate and widespread nature of soil erosion afflicting U.S. farmland is unlikely to have a significant effect on productivity.[38] Overall, the impact of new technologies on productivity can be expected to continue to outpace the deleterious effects of soil erosion on specific land parcels, while the costs of responding to the soil erosion that does take place continue to outweigh those of the erosion itself.

While urbanization poses no demonstrable negative net consequences for farmland, Vining et al. and Dillman and Cousins have concluded that within counties, the highest-grade soils are indeed in the path of urbanization trends, despite the absence of such a trend at larger regional and nationwide scales.[39] This is so, they say, principally because of the tendency for settlements to begin and prosper near the best-quality land. Like Hart, Brown et al, Luttrell, and Brown however, they argue that the conversion process threatens such a tiny proportion of our land resources as to dismiss the farmland loss as an issue.[40] Urban and rural land uses accommodate and reinforce each other more than they conflict. Even where actual encroachment occurs at the edges of growing settlements, the net impacts on arable land resources, and ultimately on agriculture as an economic activity, are insignificant.

Simon has used this controversy to develop the argument that economic growth is ever-less constrained by physical features of the environment.[41] In general, this perspective is critical of the view that in an advanced industrial society the agricultural subeconomy competes with the urban-industrial economy for the same platform—land. Resources like land, water, or related natural inputs to the economic activity of food production are not so rigidly fixed by nature—either quantitatively or qualitatively—that they cannot be altered by technological innovations or augmented by new waves of economic demand.

Moreover, Schnidman has argued that agricultural land preservation policies are questionable to the extent that they focus more on land than on the continuing viability of agriculture as an economic activity.[42] Policies that seek to preserve land—even so-called "prime farmland"—risk the preservation of increasingly outmoded and obsolete production arrangements as well as excessively narrow conceptions of what land is or can be. To zone out or otherwise discourage competition among alternative uses of land may actually be dangerous in the long run. Land and land-based resources are fixed in only the most basic physical sense. Yet other land-based resources like water can also be the focus of explicit policies whose continuation might well be reconsidered. It has been argued that the federal role in subsidizing water projects all through the twentieth century has hindered the overall development of agriculture.[43] For example, water priced below its true cost delays the eventual economic diversification of parts of the West and Southwest whose crop land is of marginal quality and whose industrialization might be more widely beneficial to the communities and residents in the long run. Productivity increases in the future will come more from the ways in which the "natural" qualities of land can be developed or better organized, if not bypassed altogether, rather than simply preserved.

The critical data sets that could resolve these issues have not yet been analyzed, and it may be premature to assume that the organization of an advanced industrial economy will necessarily be threatening to agriculture. The remarkable flexibility that has characterized agriculture to date would seem to argue against a pending "agricultural Armageddon." Agricultural production is less threatened by spreading urban and industrial arrangements than it is an integral part of them. As such, agriculture may well be in the process of adjusting to a mix of factor inputs that are both unfamiliar and unlikely by historical standards. Precisely this adaptability has characterized agriculture throughout the industrial-urban era thus far. And technology, in its various guises, has played an important part of that continuous process of adjustment and restructuring.

Technology and the Redefinition of Agriculture

The reigning conception of agriculture places great significance on humankind's dependence on the natural resources around us. From this perspective, agriculture is viewed as more of a trust relationship among participants in the ecosystem than as an industry or economic activity per se. Vivid illustrations of this perspective abound; for instance, the estimate that the earth's biological diversity may well sustain the loss by the end of the century of a million existing species, or the observation by an agricultural feed company spokesman that "... about 15 species of cultivated plants literally stand between man and starvation."[44]

While it would be folly to suggest that humankind is somehow exempt from the dependency relationships with other living things around us, it is perhaps also the case that agriculture as an economic activity has continuously redefined those relationships where it has not rendered them obsolete altogether. In the process, modern agriculture bears little resemblance to the industry it was only a few short years ago. Today, agriculture can no longer be equated simply with food production. Rather, agriculture must also be considered an energy industry, given the potential for diversion of plant biomass and agricultural wastes into methane and grains into gasohol and other fuels. Likewise, agriculture has prompted a redefinition of what "food" itself is or will be, as is evidenced by research into ways to tap the nutrition of plant proteins found in tobacco or to cultivate seaweed as a food source.[45] Finally, the usefulness of the distinction between plant and animal life has recently been blurred by efforts to reprogram plants genetically to produce meat-like proteins.[46] The continued development of hydroponics and mariculture illustrates the dispensibility of land itself. In short, recent history has shattered a wide variety of inadequate and outdated definitions of what agriculture is supposed to be and how or where it is supposed to produce. Gradually, agriculture has shifted its dependence away from land and land-based resources and toward a succession of new technologies that permit new factor mixes in production.

In large part, this redefinition of agriculture has been due to the way successive waves of technology have filtered into and through the land-based economy. Indeed, science and technology have caused the role of knowledge, information, and other artificial factors to eclipse those of land and related natural factors in the original land-based economy. The increased capital intensity of agriculture is reflected not only in how farms and ranches are organized on the land and how they are run from day to day, but also in the new and expanded definitions of what the industry is all about. In 1982, capital investment in farming reached $110 billion.[47] Of that, just over a quarter was devoted to capital expenses including land, buildings and structures, farm machinery, and livestock. The rest was devoted to expenses tied to ways of actually organizing for production and operating more efficiently.[48]

Technological inroads into agricultural production—through the employment of either new products or new processes—is by no means new. Substitutions of newer for older technologies are common and desirable, as is illustrated by the long search for nonchemical means of controlling pests such as insects, fungus, and weeds, which claim nearly a third of food crops grown.[49] At the same time, the continued use of latter-day versions of older technologies is also still common. Between 1968 and 1979, for instance, the proportion of all land in farms that was irrigated increased modestly from 3.7 percent to 4.9 percent.[50] Consequently, the flexibility of the industry is tied in large part to the increasing diversity of techniques that can be tailored to specific locations, soil characteristics, and other environmental or topographical features.

The Diffusion of Biotechnology into Agriculture. The diffusion of biotechnology to and through agriculture illustrates well the life-cycle features of an industry as it steadily evolves to higher stages of technology intensity. For decades now, the traditional means of creating plant and animal species with new desirable traits has involved a time-consuming process of selective and guided breeding. Modern tissue culture and genetic engineering techniques eliminate the intermediate stages of successive approximation across generations by altering the genetic communication from one generation to another, which takes place at the molecular level.[51]

As is the case with the technology underlying forms of factory and office automation, the core capabilities of genetic engineering have existed since the 1950s. In the past three decades, however, these techniques of molecular biology have diffused out of their nearly exclusive use in medicine and into more commercial application in agriculture.[52] While the new biotechnology industry has had a troubled and lackluster infancy, there is already clear progress toward application goals that promise not only to boost productivity but also to continue the redefinition of agriculture itself. A new agriculture can be expected to do new things in new ways and in new locations.

Biotechnology's goals would loosen the constraints on *how* agriculture produces. These goals already include, for example, both the "creation" of fruits and vegetables that contain less water so that processing and canning costs can be reduced and the development of plants that can absorb mineral supplements directly through the leaves ("foliar nutrition"), rather than indirectly through the roots. Animal life, too, is a frontier for genetic engineering. Artificial insemination and embryo transfer are already used widely, and eventually these techniques, together with cloning, promise to revolutionize the reproductive process in the livestock industry, even as they redefine what is meant by the term *livestock* itself. The tripling of milk production from "supercows," the boosting of egg production from "superhens," and the improvement of the health and reproductive capacities of animal stock generally suggest the myriad ways in which science and technology have altered, if not come close to severing, the tie between agriculture and nature.

Technology can also relax the limitations imposed by nature on where the industry would be possible and profitable. A number of technological innovations have the potential to loosen the constraints imposed on locations by the environment. Substituting the bacteria on fruit so that crops can resist frost damage would increase yields in areas subject to unpredictable bouts of freezing weather. Chemical treatment to reduce plant transpiration—a process whereby plants use water to keep cool—would permit more continuous cultivation in areas where the availability of water is uneven or is actually declining. Engineering plants able to fix nitrogen more efficiently would appreciably expand the amount of land that can produce acceptable yields while simultaneously reducing our dependence on chemical fertilizers.

Technology and the Pace of Adjustment. While the impact of successive technological innovations has had the effect of altering the cost structure for mixes of land, labor, and capital as agriculture has industrialized, the role of technology as it diffuses through modern agriculture is far from straightforward. The direction of adjustment is more predictable than the pace. Long lag times between innovation and diffusion/adoption are as commonplace in agricultural technologies as elsewhere. While the likelihood is high that future and continuing technological breakthroughs will accelerate the rate of agricultural productivity, there can be no assurance that their timing will make the adjustment process orderly. It may be decades before we feel the full range of impacts of frontier technologies like genetic engineering that define the new agricultural revolution. Smith, in a study of the prospects for reversing, through an increase in agricultural research, the decline of agricultural productivity in Texas brought on by higher energy and water costs, suggests that such a strategy is unlikely to outpace the trends leading to productivity decline.[53] While the domestic food supply is not in jeopardy, the role agriculture will play in U.S. export trade may very well fluctuate, and the dependence of individual states on agriculture may decline as agriculture continues to evolve toward a more advanced industrial status.

The Limits-to-Growth Debate and Agricultural Change

The presumed threat to agricultural lands posed by urbanization, as we saw above, is tied to the changing patterns of population redistribution in an advanced industrial economy. At a more basic level, growth understood as sheer population increase is also commonly viewed as a threat to all aspects of economic growth including agriculture. As the twentieth century began, global population had not yet expanded to 2 billion. Today, this level exceeds 4.5 billion, and by the turn of the century it will have surpassed six billion.[54] Inevitably, the background against which sectoral changes in agriculture are viewed includes widespread concerns for the capacity of our global agricultural apparatus to feed the millions who are added each year. Estimates by the World Bank indicate that one in nine people in the world—or a total of more than 500 million—have access to less food than is necessary to sustain normal activity.[55]

Despite the difficulties of even defining hunger in a world where significant proportions of the populations of even wealthy nations may be simultaneously overfed and undernourished according to prevailing nutritional guidelines, the sheer magnitude of such estimates of hunger heightens the urgency of discussions about how agriculture as an economic activity has and can be expected to develop. Furthermore, since something as basic as food is the essential product of the agricultural industry and the stakes are perceived to be so high, the acceptable margin for error in policy debate is generally smaller than it might be elsewhere. Consequently, there is often little patience for those who would offer bold and optimistic scenarios of a future that deviates markedly from the past.

Expressions of grave concern for the biological carrying capacity of the earth have continuously accompanied this population expansion. The perils of growth, whether focused on rates of population increase or on the capacity of global resources to meet the demands placed on them, frequently lead to the prediction of major catastrophes.[56] In response, more optimistic scenarios predicated on the view that resources of past importance to agriculture are fixed neither in definition nor amount, and therefore are "created" as opposed to simply inherited, have challenged this "limits-to-growth" thesis.[57] The implication is that the production involved in agriculture is not fundamentally different from that in other industrial sectors. Mixes of inputs are not fixed, and economic competition with other, more urban activities for land, water, and even location is something that increases the efficiency of agriculture rather than threatens or diminishes it.

In the end, the current stock of "natural" resources available to agriculture is probably not fixed and is in little danger of exhaustion. Recent trends involving slackened rates of increase in U.S. agricultural productivity, dampened U.S. food exports in spite of growing demand for food from those nations ever less able to afford it, and the unstable politics of world trade help underscore the essential confusion of two separate issues. The issue of whether or not there will be significant hunger and even famine is quite distinct from whether or not agriculture has the capacity to feed six, twelve, or even twenty billion people, or whether the United States will remain competitive with other agricultural producers. As the two issues are confounded, the continuing adaptability of agriculture to new mixes of land, labor, and capital has come to be associated mistakenly with an unpredictability that some see as threatening the survival of millions of people.

A more accurate assessment would probably be that tragic outbreaks of famine in the world will probably continue to occur in the future. However, their occurrence will not be evidence of the incapacity of agriculture, as one aspect of advanced industrial development, to accommodate increased population growth. Neither will it indicate the existence of inherent limitations on how agriculture as an economic activity can arrange resources for use. Rather, any such famine will have more to do with the inadequacy of trade mechanisms between and political will within nations and food distribution infrastructures within nations than with agricultural production per se. To that extent, nations collectively and unwittingly choose the levels of such calamities they experience. In the end, their political and cultural relations with one another more profoundly distort the distribution of food—like that of all wealth—than does their capacity for its creation. That children may starve arguably tells us less about agriculture than about relations among nation-states in the global community in which it functions.

An Industrial Policy for Agriculture?

A familiar response to those who argue that we need national policies of one sort or another is that while often not explicit, coherent, or free of contradictions, the sum total of all direct and indirect impacts of the federal government's actions and inactions does indeed constitute a national policy. A case has been made, however, that with regard to agriculture, federal actions have been both reasonably compatible and successful. Industrial policies have as their ultimate aim the enhancement of the competitiveness of an industry in international markets. Thurow, for example, has argued that federal actions in agriculture constitute a successful national agricultural policy, including as they do federal sponsorship of agricultural research and development, provision of physical infrastructure, establishment of farm financial institutions, and a system of price supports and acreage controls[58] Taken together, these key features have been responsible for restoring productivity and competitiveness to U.S. agriculture since World War II.

It is quite likely that aside from the land and natural resource policy debates that have arisen recently, agriculture viewed as an industry and as an important export link to the international economy will increasingly be at the center of debates concerning the need for a sector-specific industrial policy. While agriculture may serve as a useful model of industrial restructuring, however, there are serious doubts that existing federal policies of demand—and supply—management directed toward agriculture likewise constitute a test of the merits of an industrial policy for agriculture.[59] The record of federal interventions into agriculture has been mixed at best. Furthermore, it is debatable whether these interventions have been the *sine qua non* of the successes that have been realized. Present policies sustain and mask much economic inefficiency and are responsive to the political pressures of well-defined interest groups and existing arrangements for stability and predictability within the industry.

In the end, the trade-off between policy-induced industry stability and less constrained industrial change is difficult to assess. However, the importance of U.S. agriculture to the nation and the world argues for making certain that as both an economic activity and as an industry, it retains the flexibility to nurture and respond to new resource and technological opportunities in a timely manner.

7

Localism, Governance, and Public Finance Options[1]

> By far the most significant of recent developments in the field of American taxation is the new temper of the taxpaying public. This new temper scarcely requires statistical demonstration. Its day to day manifestations are amply recorded in the public press. Among these manifestations may be mentioned the birth of hundreds of new taxpayer associations, the resolutions and activities of farm and trade organizations, the multiplication of tax investigating bodies, both public and private, and the growing volume of reports, studies and recommendations which represent the results of their findings. The new mood of the taxpayer is reflected in the current platforms of candidates for political office, in the long drawn out and acrimonious sessions of legislative bodies and in the wave of budget slashing which has recently become epidemic.[2]

While the above quotation strikes a responsive chord in many today, it was originally intended to tap a mood formed over a half-century ago when it was used to introduce a chapter on trends in taxation and public finance in President Herbert Hoover's *Report of the President's Research Committee on Social Trends*. In 1932, no less than today, the fundamental role of the nation's tax system was clearly to act as a template for how responsibilities were to be delegated among levels of government and between the private, public, and nonprofit sectors. That these patterns of assignments continue to engender as much controversy as they dispel is likewise clear.

But it is not the mood of the taxpaying public that interests us here so much as the emergence of new forms of industrial capital, a new industrial structure complete with new urban settlement patterns, and the politics that derive from the pace and patterning of these changes.[3] Specifically, systems of local public finance built around the property tax and land-value taxation, both of which matured in step with an earlier industrial-urban system, are now exhibiting signs of being increasingly inappropriate, inadequate, and obsolete. They are inappropriate because they reinforce a notion of "localism" that is itself unraveling in the wake of deconcentration and counterurbanization trends that have been building throughout the latter decades of an evolving era. They are inadequate because they are sometimes difficult to export into newly decentralized and multinodal settlements that define our industrial economy. A property tax base is weakened through the attenuation and attrition that accompany widespread mobility and lead to lower-density growth and development. And such taxes are obsolete to the extent that they are unable to assimilate, either conceptually or administratively, new forms of property that derive from the creation, communication, and commercialization of knowledge and information—the new critical capital of advanced industrial economies. Computer software, cable television, and communication satellites, as Woolery has noted, are just a few examples of the new kinds of property that dominate an expanding information sector and for which valuation and geographical siting pose enormous challenges.[4] In the end, a tax instrument like the property tax, whose base is undergoing such redefinition and structural change, is a tax whose "fit" to an emerging economic order deserves careful reexamination.

The property tax as a fiscal tool has long been compatible with the structural and spatial dynamics of the industrial-urban era and the expanded variety of localized wealth that could be used as a tax base. Today, the property tax cannot escape being affected by the restructuring of local economies. Increasingly, this restructuring is creating value in intangible and historically unfamiliar forms and redistributing employment and population growth across the land in new ways. Primarily a local tax, the property tax will continue to be buffeted by this nation's passage from an older industrial era into an advanced industrial one. The ravaging of older local tax bases and the construction of new ones in the system of older industrial settlements will be accompanied not only by the reconstitution of new and smaller tax bases in previously remote locations outside metropolitan areas, but also by a redefinition of "localism" itself.

The Eclipse of Land and Traditional Patterns of Land Use

The property tax is the oldest tax in the nation's fiscal armamentarium. Over the centuries, it has exhibited considerable flexibility in adjusting to a nation whose fiscal requirements have changed dramatically. During the nineteenth century, there was a tremendous expansion of the bases of wealth—and a broadening of the notion of private property—as the nation began to rationalize its agrarian economy, to urbanize rapidly, and to assemble a burgeoning capital plant in its new industrial cities. The simple classification scheme based on land and livestock grew obsolete as these developments continued. The property tax base expanded to include not only land itself but also selected tangible resources on it and beneath it that were increasingly relevant to an industrial and commercial economy. Ironically, however, this flexibility has had the look of a process of contraction because in most locations so many other forms of tangible and intangible property have remained exempt from the property tax altogether.

It is not surprising that the early property tax focused primarily on land, since land was the principal source of wealth in a preindustrial economy. As the land-based agricultural sector gave way to an emerging industrial sector, it was inevitable that the property tax would change to reflect society's more complex and indirect relationships with the land. The conception of "land," understood in the narrow sense of soil and land-based resources, expanded to reflect the increasing importance of "location," understood in terms of access to labor pools and other factor inputs, transportation routes for growing industries, and mass markets for consumption.

During the half-century between the New Deal and the New Federalism of the first Reagan term, the rise of income taxation superseded the taxation of wealth and other tangible property. The fiscal system gradually reoriented itself toward tapping newly developed income streams, via income taxes, and tapping patterns of consumption, via sales, gross receipts, and excise taxes. This reorientation reflected the increasing importance of wage labor and the expanded market for consumer goods. In the end, these new revenue sources became better articulated with the industrial-urban economy of the early twentieth century. One feature, in particular, indicates the local character of the property tax and the constitutional reality that localities are the creatures of states: as economic and social life in the country burst the bonds of localism, extralocal levels of government—especially state government—appropriated the new taxing instruments as their own. The federal government has no property tax provision, and the states underemploy property taxes where they have them. This underuse has occurred in part because state-level government expanded most dramatically during a period that offered opportunities for the adoption of new taxing instruments less dependent on land and the specific uses to which it was put.

The property tax, which is both old and local, may therefore be turning into an anachronism. Although it is in little danger of extinction, its base within jurisdictions has been systematically transformed by a powerful dispersion of households and businesses to new locations.[5]

Likewise, its incidence frequently involves the shifting of burdens to those often least able or willing to accommodate them—the immobile elderly whose incomes may be mismatched to the assessed values of their homes—or to those who have the widest range of mobility options to circumvent them. This latter group would include newly forming businesses and households. Cities try desperately to attract and retain individual and corporate citizens in their tax bases, yet both have been reluctant to locate in urban locations that impose high property tax burdens. Finally, property tax administration suffers from the inevitable comparisons, made across jurisdictions, that reveal a pattern of considerably varied tax rates and administrative apparata for classification and assessment. Within any single jurisdiction, taxation's frequent role as a political tool clearly lays bare the fact that ultimately no fiscal tool can be administered apolitically.

Property tax systems not only exhibit great variability in administrative standards and practices nationwide, but they often tend not to represent equity standards that prevail even at more local levels. It has been accepted that interjurisdictional differences are ipso facto undesirable as increasingly extralocal standards encourage uniformity at all costs. At the local level, the inevitable tendency to shift taxation away from existing homeowners for the political expedience of cultivating or maintaining a political base has—usually without it ever being stated explicitly—shifted the property tax burden toward businesses large and small, as well as toward new homeowners, whose rates are relatively higher because their assessments more closely reflect current market values than do those of longtime residents for whom reassessments are an infrequent event. Consequences of this shift may well be registered in the patterns of industrial, commercial, and residential growth that now shape older industrial cities in particular.

Historical Property Tax Trends in Advanced Nations

Advanced industrial nations reveal much variability in the degree to which they rely on property taxation to finance their public sectors. However, data displayed in Table 7.1 suggest either a common dynamic or a diffusion or imitation process at work among nations that share common cultural, if not geographical, characteristics. The property tax contributes a relatively large portion of total revenues in English-speaking nations; it accounts for fully 10.0 percent of the total tax revenue in the United States. The United Kingdom is the only other advanced industrial nation that relies more heavily on the property tax. The tax treatment of property in predominantly Anglo-Saxon cultures may well illustrate the way in which a shared cultural heritage and shared legal

Table 7.1
Distribution of Tax Revenues in Selected Countries, by Source, Fiscal Year 1980
(percent of total taxes from each source)

Country	Property Taxes[b]	Taxes on Income and Profits[a]		Social Security Taxes	Taxes on Goods and Services			Other Taxes[d]
		Individual Income	Corporate Profits		Total[c]	Consumption taxes		
						General	Specific	
United Kingdom	12.09	30.02	7.69	16.87	28.81	14.09	13.11	4.52
United States	10.04	36.80	10.13	26.41	16.61	6.62	7.80	.01
Canada	9.22	34.02	11.34	10.41	32.82	11.66	12.72	2.19
Australia	8.34	43.67	11.95	—	31.18	5.23	22.38	4.85
Japan	8.20	24.22	17.28	29.02	16.35	—	14.15	4.93
New Zealand	7.85	61.67	7.77	—	22.30	10.22	11.19	.41
Greece	7.73	13.91	4.19	28.70	44.15	22.08	19.87	1.32
Switzerland	7.32	35.68	5.80	30.77	20.43	9.12	9.73	—
Denmark	6.17	51.40	3.21	1.77	37.32	22.15	13.36	.13
Turkey	6.16	49.52	4.71	5.76	29.14	—	28.69	4.71
Luxembourg	5.66	27.36	16.45	30.31	19.50	10.72	8.21	.72
Ireland	5.19	32.04	4.55	14.32	43.55	14.79	28.05	.35
Spain	4.53	20.22	5.56	48.21	20.55	6.69	18.45	.93
Italy	4.28	24.69	8.34	35.71	26.97	15.78	10.38	.01
Netherlands	3.89	26.32	6.61	38.15	24.76	15.86	6.86	.27
France	3.55	12.92	5.04	43.16	29.98	21.05	8.10	5.35
Austria	2.81	23.20	3.43	31.40	31.32	20.00	10.39	7.84
Belgium	2.59	35.25	5.72	30.71	25.59	16.43	8.11	.14
Germany	2.63	29.92	5.51	34.10	26.99	16.79	8.99	.85
Finland	2.00	44.80	4.45	8.46	40.11	20.11	19.61	.18
Norway	1.74	34.01	13.22	15.16	35.37	18.27	16.58	.50
Portugal	1.45	—	—	26.99	40.12	16.20	23.18	31.44
Sweden	.93	40.97	2.45	28.63	24.31	13.34	9.52	2.71

Source: Basic data from Organization for Economic Cooperation and Development; *Facts and Figures on Government Finance*, Tax Foundation, 1983.

[a]Includes taxes on capital gains.

[b]Includes taxes on movable and immovable property, net wealth taxes, and estate and gift taxes.

[c]Includes import duties, profits of public fiscal monopolies, licenses, and other business taxes.

[d]Includes general and selective taxes on payrolls that are not earmarked for social security purposes, and other taxes not elsewhere specified.

traditions rooted in English common law transcend national boundaries and comingle to shape images of the public sector and the fiscal arrangements devised for its maintenance. In a similar manner, although with a contrasting result, the Scandinavian countries have come to rely very little on the property tax as a revenue source.

While general sales taxes, social security taxes, and assorted nontax revenue instruments such as user charges have increased as important U.S. government revenue sources since World War II, reliance on the corporate income tax, selected sales and excise taxes, and the property tax has declined substantially. The property tax contribution to total government revenue dropped by a third from 10.7 percent to 7.0 percent between 1960 and 1981, rising only slightly to 7.5 percent by 1983.[6] All local governments in the United States employ the property tax, and many localities rely heavily on its revenue. Yet, like the localities that it serves, the local property tax itself has been beset by several influences stemming from the continued industrial development of the nation. As a result, there has been a long-term lessening of dependence on the property tax.[7] In the early decades of this century, the property tax share of total locally raised revenues remained fairly steady; in both 1902 and 1932 it was 73.1 percent. By 1950 it had shrunk to 50.2 percent, and by 1980 the share had declined to only 28.2 percent, thereafter rebounding slightly to 28.8 percent by 1983. The bulk of this long-term decline came during the 1970s.[8]

Retreat of the Local Property Tax. In the late 1970s, after a long period of dormancy, the combative mood of the American taxpayer reemerged once again as a significant factor in any analysis of the nation's tax system. Nowhere has this factor been more evident than with respect to the local property tax and its highly visible role in local public finance. In recent years, a spate of fiscal-containment measures have been used to convey local citizens' exasperation with the size, cost, and perceived growth rate of the public sector at all levels.[9] This uprising has occurred despite the fact that while spending has continued to increase at all levels since the mid-1970s, it has done so at declining rates. Further, after sequential surges in government employment at the federal, local, and then state levels in the past quarter-century, public sector employment growth rates have flattened across the board, with actual shrinkage in the federal civilian labor force since 1969.[10]

Engineered reductions in growth in California and Maryland in 1979 notwithstanding, the post-war period has witnessed uninterrupted growth in property taxes. Between 1970 and 1978, the rise occurred at an average annual rate of 8.4 percent. During FY 1979–1980 alone, property tax revenues increased 5.5 percent to $68.5 billion. Yet the inflation-enhanced visibility of the property tax easily distorts evidence of its long-term retreat in several respects. Over the 1960–1980 period, property tax revenues for all governments declined from 14.5 percent to 11.9 percent of all tax revenue.[11] During the 1970–1980 period alone, property taxes declined in relation to personal income by 22 percent. Furthermore, the Tax Foundation has reported that while total collections rose 134 percent, from $27.7 billion to $65.6 billion, between 1969 and 1981, and per capita collections rose 112 percent, from $139 to $295 per $1,000 of personal income, property taxes actually declined 15 percent over the same period. Property taxes accounted for 40 percent of the $380 per capita state and local taxes in 1969 but only 32 percent of the $934 in 1979, owing to the 180-percent rise in other state and local taxes over the same period.[12]

Local taxes account for approximately 96 percent of all property tax revenues. Intergovernmental aid constituted 45 percent of local general revenue in 1979, up from 36 percent after a decade of increasing local dependency on extralocal revenue sources. During that same period, the property tax share of local general revenue actually declined from 41 percent to approximately 29 percent.[13] While property taxes continue as the primary source of revenue at the local level—approximately 76 percent of the total of $86.4 billion in local general revenues in 1980—localities are having to shift their dependence to new revenue sources as their local tax bases either fail to grow at historical rates or even decline. The question might well be raised whether or not the property tax system as it has evolved through the industrial era retains enough flexibility to serve as a key finance tool in the local economies around the United States in the future.

Localism and Local General Revenue

In our federal system, states were originally viewed as sovereign; both cities and the federal government were to be derivative and subordinate. Yet, following the Civil War, on the strength of their new economic roles, a national urban system of large industrial cities gradually usurped the primacy of state government. Shifting patterns of local public finance all through this century reflect the changing patterns of intergovernmental relations that resulted. In 1902, local governments raised nearly all (93.4 percent) of the revenues they required; this proportion was still 85.7 percent in 1932. Gradually, however, as the scale of economic activity expanded, the primacy of local economies began to recede, and with it their ability to finance the full range of public activities launched in or directed at central cities. By 1950, locally raised revenues had declined to 68.4 percent of local expenditures with state governments accounting for nearly all the rest (30.1 percent).

The pendulum gradually swung toward the federal government, the only level of government perceived capable of handling the new scale of economic and political activity in industrialized society. During the 1960s and 1970s, the role of the federal government expanded dramatically, with its contribution to local revenues nearly tripling to 10 percent by the late 1970s. By 1979, the dependence of localities on their own resources reached a low of 55.3 percent.[14] While this local share rebounded slightly to 58.5 percent by 1982, constituting for some observers evidence that the long slide had ended, the decades-long slide into dependency at the local level remains the more important trend.

Declining Local Autonomy and the Fiction of Localism

The forces shaping local economies around the nation are increasingly insulated from the impacts of policy tools traditionally at the disposal of local governments. As a result, an array of institutions that in an earlier time reflected well the localism and insularity of the local residential and business sectors are now increasingly ill-suited to the expanding scale of an advanced industrial economy and its reworked settlement patterns.[15] The local property tax is one such institution. A continued reliance on the local property tax as a major revenue source for local governments reinforces an outdated conception of localism that has gradually lost both its relevance and its appropriateness.

Political jurisdictions—especially local and state governments—with their relatively fixed political boundaries, are steadily declining in their ability to keep pace with the more fluid forces defining as well as locating the key features of an advanced industrial economy. The full range of problems, from traffic congestion to varieties of environmental pollution, are ones that local governments are only marginally able to influence in most instances. Local governments exercise little influence over decisions by large national and multinational corporations to shift capital, production, and jobs and thereby to open, expand, close, shrink, or relocate plants and offices. Planning and zoning authority exercised by local officials nearly always threatens to discourage new growth or displace existing development. A system of intergovernmental arrangements capable of responding to the increased scale of urban economic and social life has yet to be devised and successfully implemented.

While economic and social life appear to accommodate themselves to increasing spatial scales with relative ease, that adjustment invariably has led to reduced allegiances to traditional local political arrangements. Suburbanization and related forms of capital mobility have dispersed residences, employment, income, and ultimately tax bases. As a result, many localities have watched their jurisdictional boundaries become decoupled from their economic and social ones. Until recently, they have responded politically by becoming increasingly dependent on higher levels of government for revenues. The political impulse to compensate localities for what has been called the "historical decline of the significance of place" has been especially apparent since the Depression.[16]

Property Taxation Patterns and Advanced Industrial Development

The private economy of the United States dwarfs that of the public economy, although government has expanded dramatically in recent decades. In FY 1983, spending by all levels of government reached $1.3 trillion, with the share spent by local and state governments at 33.5 percent. Between 1950 and 1983, government expenditures soared from $1,614 to $15,345 per household—from $466 to $5,587 per capita.[17] The major taxing instrument for all nonlocal levels of government since World War II has been the individual income tax. Despite fluctuations over the years, in 1981 it provided approximately 30.9 percent of total government revenues—the same as in 1944.

In 1981, Americans paid nearly $824 billion—or approximately $3,600 per capita—in taxes to all levels of government. Of this total the federal burden has grown at a more rapid rate than have either the state or local burdens, despite the fact that revenues from several taxing instruments declined and revenues from state and local instruments generally increased. Clearly, the expansion of the federal government in the 1930s and again in the 1960s, as well as the subsequent expansion of local and state government purviews in the past fifteen years, are largely registered in the form and function of the nation's nested tax structure and resulting fiscal federalism.

The overall tax system, along with the fiscal federalism that it reflects, is correctly seen as a political compact on the assignment of responsibilities throughout the public sector. However, it is also properly viewed as an elaborate system of signals and stimuli whose envisioned outcomes in the private and nonprofit sectors are intended to be both sectoral and spatial. Predictably, as government has grown large and complex, its revenue base has done the same. In a pluralistic political system, a public finance system becomes an elaborate mosaic of compromises and trade-offs whose impacts are complex, contradictory, and often incoherent. During the past decade, both the net impact and the internal architecture of the current tax system have come under considerable scrutiny and have been attacked for contributing to the flattening of national economic growth; the weak propensity to save and the difficulties of accumulating capital necessary for industrial expansion; urban "sprawl" and high land vacancy rates in central cities; population and industry shifts within and between regions; and a wide variety of other outcomes commonly considered to be economic "problems."

Taxation Trends and Industrial-Urban Development

After half a century of vigorous expansion in the size, scope, and resource requirements of the public sector, the responsiveness of the tax system to larger trends is amply demonstrated. As it emerged from the Depression, the nation's tax structure was having its role reshaped by a combination of economic, demographic, and technological influences. The industrialization of the United States was by then almost a century old; the nation's economy had transformed from one based on the primacy of the farm and land-based resources to one tied to the factory in densely compact central cities whose residential, commercial, and industrial roles secured their new economic dominance—the engines of the new industrial economy had moved to town. The nation's cities were now completing their transition from market appendages of an agricultural economy to centers of production as well as distribution and consumption.

The cityward drift of population during the century was a necessary feature of industrialization. Advancements in architectural engineering and transportation—especially the elevator and the automobile—did much to dictate the vertical and horizontal redistribution of population and jobs in these industrial centers. Communication advancements like the telephone and radio compensated for the expanded physical distances that separated urban areas and parts of towns and linked the administrative functions of city centers to increasingly dispersed production and related activities.

The inherited tax structure, however, was relatively rigid and ill-suited to the new scale of economic activity and the mobility at the nation's disposal. As an unwitting accomplice, government substantially amplified the impact of new technological capabilities by assuming responsibility for the construction of a vast and costly network of public infrastructure. This facilitated the dispersal of population and employment that, as we saw in previous chapters, would eventually contribute to the unraveling of the physical arrangements of the early urban-industrial era. The new economy that was made possible by centralization and concentration now extended its reach back across the countryside and was transformed in the process through decentralization and dispersion.

A long cycle of concentration and centralization had not only brought jobs to cities and people to jobs, but also had expanded enormously the wealth of the nation's cities. The sorting of residences, offices, retail outlets, factories, and public buildings gradually differentiated central business districts, as well as industrial and residential areas, from other functional areas of now-complex cities. The new order in land uses that market processes provided through ecological and technological changes, as well as through cultural preferences, soon became the object of official attempts to preserve and protect them. The widespread adoption of zoning ordinances and related land use restrictions and building regulations beginning in the late 1920s signaled the beginning of this trend. As a result, the social and spatial aspects of local tax burdens shifted as well. As industrial cities developed, newly mobile citizens, as well as physical and financial capital, began to spread out ways that eventually rendered the fixed jurisdictional boundaries of urban areas and their tax bases increasingly obsolete.

Regional Patterns in Property Taxation

The relationship between taxation and economic growth is complex. Invariably, industrial development is characterized by large-scale spatial shifts among regions both in population and in business growth. At least part of this dispersion reflects industry-technology life-cycles, the mobility of capital, and the continuing search for lower-cost environments, initially for industrial growth but also for the population growth that accompanies it. The bulk of this century has witnessed the slow spread of industrial growth and development from the heartland of the Northeast and Midwest to the West and, later, to the South. The declining industrial attractiveness of this heartland and the building attractiveness of the South as the new growth region of the nation is reflected at least in part by regional patterns of total and property taxation.

Table 7.2 reports state tax capacity (TCI) and tax effort indices (TEI) for selected tax bases. An examination of trends in the two indices for total taxes reveals that while the Northeast, relative to the nation as a whole, has the least capacity of any region for taxing its citizens, it nonetheless does so to a greater extent than does any other region. In contrast, tax effort lagged behind tax capacity in the Midwest, South, and West. Patterns of historical development of local public sectors and a diversity in cultural tastes for government activity, as well as real differences in potential tax bases, are revealed across regions. For example, southern states in aggregate imposed an even lighter total tax burden on their citizens than would have been dictated by their relatively low tax capacity alone.

Table 7.2
State Tax Capacity and Tax Effort for Selected Tax Bases

	Total Taxes (1979)		*Property Taxes*						
			Total		*Res.*	*Farm*	*Comm.*	*Public Utility*	*Vacant Land*
	TCI[a]	*TEI*[b]	*TCI*	*TEI*	*TCI*	*TCI*	*TCI*	*TCI*	*TCI*
I. *Northeast*									
New England									
Connecticut	105.6	102.5	113.4	130.5	135.5	11.7	106.9	93.4	92.5
Maine	79.8	110.7	81.9	130.3	99.3	27.2	65.6	87.9	72.8
Massachusetts	90.7	145.3	89.1	220.6	104.4	5.8	86.0	82.5	72.2
New Hampshire	97.1	78.3	91.1	168.2	113.2	19.7	77.6	83.2	45.9
Rhode Island	83.5	122.8	82.5	148.4	97.8	5.4	79.2	47.6	123.9
Vermont	85.6	109.9	86.7	146.5	104.2	87.1	68.3	79.2	2.8
Mid Atlantic									
New Jersey	101.0	117.4	100.8	180.0	112.6	11.3	103.6	89.0	108.9
New York	86.9	171.6	78.4	221.1	79.8	13.7	94.2	73.2	48.6
Pennsylvania	92.5	105.2	94.1	87.3	93.6	35.5	106.3	119.8	99.0
II. *Midwest*									
East North Central									
Illinois	111.5	98.8	115.6	107.1	109.9	171.5	116.4	115.8	71.8
Indiana	97.4	84.2	97.5	83.9	76.7	176.1	109.9	124.5	67.1
Michigan	101.9	113.8	96.2	135.0	92.3	41.3	116.9	102.3	54.4
Ohio	99.2	86.3	102.6	91.3	98.7	86.5	112.3	107.6	90.0
Wisconsin	95.9	118.8	106.0	116.9	113.8	120.1	93.4	91.0	113.7
West North Central									
Iowa	106.3	93.4	129.2	94.5	94.6	641.8	83.9	103.9	72.1
Kansas	106.9	86.5	106.0	105.6	78.5	334.0	99.1	134.1	47.9
Minnesota	101.7	116.5	108.0	103.0	102.3	239.5	97.6	88.6	57.1
Missouri	94.8	82.9	92.8	72.9	81.6	168.1	90.7	100.1	112.2
Nebraska	96.3	98.4	101.1	124.0	59.6	536.3	76.4	57.0	168.1
North Dakota	106.1	77.3	111.3	82.1	61.8	729.5	76.5	54.9	28.3
South Dakota	92.4	84.4	102.3	113.7	65.7	637.0	55.3	71.6	58.1
III. *South*									
South Atlantic									
Delaware	110.9	95.1	116.1	51.3	103.9	69.1	146.7	109.0	122.6
District of Columbia	107.2	132.8	101.5	114.7	120.5	0.0	88.2	125.9	96.8
Florida	103.9	78.8	101.2	87.4	119.5	59.4	66.3	108.3	255.2
Georgia	83.1	96.7	76.1	88.9	67.6	70.2	80.6	121.4	73.2
Maryland	98.0	110.0	98.5	93.8	121.2	45.7	71.2	98.4	127.3
North Carolina	81.6	92.3	79.9	60.3	72.4	68.1	85.0	107.3	111.0
South Carolina	77.3	92.3	75.6	60.2	72.7	52.9	72.8	124.5	84.7
Virginia	92.6	88.8	94.1	74.3	105.7	60.6	83.1	88.6	105.8
West Virginia	95.0	80.9	102.6	36.4	71.8	39.5	139.5	232.3	56.3
East South Central									
Alabama	75.8	86.6	67.0	34.1	57.4	67.8	76.6	103.6	33.6
Kentucky	86.2	86.4	80.6	44.2	60.8	121.5	106.0	86.7	28.9
Mississippi	71.1	96.3	65.2	64.2	51.7	116.6	67.1	95.5	79.9
Tennessee	81.5	87.4	71.7	76.3	67.2	78.6	85.8	41.7	55.4

Table 7.2 *(cont.)*
State Tax Capacity and Tax Effort for Selected Tax Bases

Personal Income Tax		*General Sales Tax*		*Property Taxes as Percent of Personal Income*				*Property Taxes Percent of Local General Revenue*
					Percent of U.S. Avg.		*Percent Change*	
TCI	*TEI*	*TCI*	*TEI*	*1980*	*(1980)*	*1967*	*1967–80*	*(1978–79)*
130.0	12.5	89.0	125.9	4.7%	(134)	4.8%	−2.1%	59.0%
66.3	93.8	90.3	94.6	4.7	(134)	5.1	−7.8	45.2
100.9	170.0	92.2	64.0	6.2	(177)	5.9	5.1	50.1
91.5	6.9	105.3	0.0	5.6	(160)	5.9	−5.1	61.5
89.8	111.6	78.7	102.5	5.0	(143)	4.5	11.1	54.3
69.1	148.3	95.3	38.6	5.3	(151)	5.0	6.0	57.6
111.0	64.7	98.9	71.8	5.1	(146)	5.4	−5.6	45.7
104.6	193.8	86.9	140.0	5.5	(157)	5.2	5.8	29.3
97.6	128.0	91.1	84.1	2.9	(83)	3.2	−9.4	26.9
123.7	76.1	105.9	106.6	3.8	(109)	4.2	−9.5	37.7
100.7	71.8	96.7	118.9	2.9	(85)	4.7	−38.3	32.4
111.9	127.3	107.8	81.3	4.4	(126)	4.3	2.3	39.9
105.2	84.4	95.8	72.3	3.2	(91)	4.3	−25.6	29.1
89.8	196.9	97.2	84.7	4.2	(120)	5.1	−23.8	30.6
90.9	128.5	101.7	65.2	4.1	(117)	5.6	−26.8	36.7
94.8	80.4	101.5	82.8	3.9	(111)	5.5	−29.1	42.0
94.9	197.7	108.9	65.7	3.7	(106)	6.0	38.3	27.8
96.4	82.3	102.8	92.8	2.6	(74)	3.8	−31.6	28.7
87.7	91.6	98.8	86.8	4.6	(131)	6.7	−31.3	41.5
76.3	59.6	112.4	70.0	3.3	(91)	5.9	−44.1	33.1
64.3	0.0	102.9	98.1	4.7	(132)	7.0	−32.9	49.6
116.6	202.0	98.7	0.0	1.8	(51)	1.9	−5.3	22.1
133.3	165.7	99.5	125.1	3.2	(91)	2.9	10.3	10.6
99.4	0.0	119.3	87.4	2.9	(95)	4.2	−31.0	26.5
76.4	113.1	93.1	104.9	2.8	(82)	3.0	−6.7	24.1
116.7	183.2	100.9	79.2	3.1	(89)	4.2	−26.2	26.6
73.6	146.5	84.9	82.4	2.4	(69)	2.6	−7.7	22.2
66.3	129.7	82.7	102.8	2.4	(69)	2.0	20.0	26.1
98.6	114.4	95.4	68.8	2.8	(80)	2.8	—	30.4
80.0	87.7	83.1	130.9	1.9	(54)	2.7	−29.6	21.1
66.2	94.4	82.5	112.7	1.2	(34)	1.7	−29.4	9.5
73.3	145.1	87.0	92.6	2.6	(54)	2.6	−26.9	14.3
56.5	85.5	74.4	158.3	3.1	(69)	3.1	−22.6	19.5
78.7	4.6	91.7	141.1	2.8	(63)	2.8	−21.4	22.8

Table 7.2 (*cont.*)
State Tax Capacity and Tax Effort for Selected Tax Bases

	Total Taxes (1979)		*Property Taxes*						
			Total		*Res.*	*Farm*	*Comm.*	*Public Utility*	*Vacant Land*
	TCI[a]	*TEI*[b]	*TCI*	*TEI*	*TCI*	*TCI*	*TCI*	*TCI*	*TCI*
West South Central									
Arkansas	78.3	82.1	74.7	53.3	57.1	200.1	70.4	95.8	76.6
Louisiana	108.4	79.2	101.2	34.2	78.2	73.5	139.6	128.3	78.9
Oklahoma	113.1	71.2	107.9	45.7	71.8	201.0	138.8	139.3	86.3
Texas	121.5	62.9	106.6	82.3	75.4	137.9	143.7	137.0	93.4
IV. *West*									
Mountain									
Arizona	95.0	115.8	94.3	130.3	96.3	83.2	79.2	81.1	281.8
Colorado	14.4	95.8	116.1	98.4	117.1	171.0	108.4	85.9	128.6
Idaho	90.8	92.3	100.1	85.3	87.3	310.1	74.3	76.4	165.3
Montana	111.4	87.6	123.3	114.3	89.6	551.2	85.5	124.1	103.2
Nevada	163.7	65.3	117.0	85.7	136.5	50.0	83.5	171.5	165.5
New Mexico	105.4	84.2	94.3	45.8	74.7	141.4	89.6	163.1	196.8
Utah	88.4	99.0	95.6	79.7	101.8	93.2	89.6	83.6	91.2
Wyoming	178.7	79.0	173.4	99.4	98.7	348.7	232.8	302.0	80.7
Pacific									
Alaska	215.2	126.4	147.4	99.3	139.3	10.9	174.0	38.0	591.7
California	115.9	95.0	131.1	60.5	168.6	50.1	100.5	76.3	146.0
Hawaii	105.1	128.3	119.5	57.8	154.1	73.1	70.3	74.7	277.2
Oregon	104.8	94.1	110.8	111.8	122.4	91.5	96.7	80.3	185.9
Washington	102.6	97.3	108.3	87.0	123.1	90.3	93.1	59.5	182.0

Sources: Advisory Commission on Intergovernmental Relations, "Tax Capacity of the Fifty States: Methodology and Estimates" (Washington, D.C., March 1982), and ACIM, *Significant Features of Fiscal Federalism,* 1980–81 Edition, December 1981.

[a]TCI-Tax Capacity Index: A tax capacity index score of greater than 100 indicates that a state

With respect to property taxation in particular, some of the same patterns are revealed. The West leads and the South trails all other regions in property tax capacity. However, both regions impose lighter tax burdens than do other regions. In contrast, while the property tax capacity of the Northeast exceeds that of the South only slightly, the property tax burdens borne by northeastern households and businesses are by far the heaviest in the nation. Predictably, a history of reliance on the property tax to finance local government is considerably more evident in the older industrial regions of the United States. In aggregate, property taxes constitute 44.3 percent of local general revenue in Northeastern states, 35.5 percent in the Midwest, and only 29.3 percent in the West and 20.9 percent in the South. These variations reflect the relatively recent expansion of local government in the West and South as well as leaner urban service packages provided by localities in these

Table 7.2 (*cont.*)
State Tax Capacity and Tax Effort for Selected Tax Bases

Personal Income Tax		*General Sales Tax*		*Property Taxes as Percent of Personal Income*				*Property Taxes*
					Percent of U.S. Avg.		*Percent Change*	*Percent of Local General Revenue*
TCI	*TEI*	*TCI*	*TEI*	*1980*	*(1980)*	*1967*	*1967–80*	*(1978–79)*
62.2	102.3	84.0	89.8	2.6	(57)	2.6	−23.1	21.7
88.8	40.9	86.5	159.8	2.4	(43)	2.4	−37.5	13.2
87.1	80.5	92.0	84.2	3.4	(54)	3.4	−44.1	22.0
112.0	0.0	111.6	82.4	4.1	(97)	4.1	−17.1	34.9
91.1	73.4	100.5	162.5	4.6	(131)	5.8	−20.7	29.9
108.9	91.8	111.0	122.8	3.8	(109)	5.4	−29.6	31.9
68.7	139.9	98.1	69.4	3.1	(89)	4.4	−29.5	33.2
77.0	141.8	110.2	0.0	5.9	(169)	6.4	−7.8	42.9
143.3	0.0	279.3	49.3	2.8	(80)	4.4	−36.4	27.0
75.3	44.5	97.2	152.5	2.0	(57)	2.6	−23.1	14.0
73.1	137.2	90.3	136.4	3.5	(100)	5.0	−30.0	30.3
127.5	0.0	128.8	134.4	5.8	(166)	6.9	−15.9	37.4
173.3	100.0	122.5	34.2	7.9	(226)	2.3	243.5	22.5
112.9	112.6	113.6	126.8	2.8	(80)	6.2	−54.8	19.4
102.6	172.5	123.1	183.0	2.2	(63)	2.7	−18.5	41.6
98.5	196.6	114.4	0.0	4.5	(129)	5.2	−13.5	35.4
119.3	0.0	102.0	198.3	3.2	(91)	3.5	−8.6	20.0

has a capacity to tax its total tax base (or a specific tax base) at a rate greater than that for the average state.

[b]TEI-Tax Effort Index: A tax effort index score of greater than 100 indicates that a state is taxing the overall specific tax base at a rate greater than the average for all states.

regions. Finally, from the perspective of property taxes as a percentage of personal income, once again the data reveal that the burden imposed in the Northeast greatly exceeds the U.S. average, while that imposed in the South is dramatically lower than the average.

Trends among regions differ considerably from those within regions. Table 7.3 reports population and property tax trends for the nation's twenty-five largest cities. While large cities in all regions experienced population loss during the 1970s, the experience was all but universal for the large cities of the older industrial regions. Nevertheless, large cities in the South reported considerably greater dependence on real property tax to finance local budgets than did large cities in other regions. Moreover, a majority of large southern cities reported a property tax increase in 1983. On both counts, large southern cities were more similar to large northeastern cities than they were to large cities

Table 7.3
Population and Property Tax Trends for the 25 Largest U.S. Cities

	City Population (000s)	*1970–80 City Population Change*	*Change in Central City Population as Percent of SMSA, 1970–1980*	*Real Property Tax as a Percent of City Budget*	*1983 Tax Increase?*
Northeast					
New York (1)[a]	7,071.0	−10.4	9.7	24%	Yes
Philadelphia (4)	1,688.2	−13.4	−4.5	16	Yes
Boston (20)	562.0	−12.2	−2.9	38	No
Midwest					
Chicago (2)	3,005.1	−10.8	−5.6	19	No
Detroit (6)	1,203.3	−20.5	−8.3	10	No
Indianapolis (12)	700.8	−4.9	−6.8	33	Yes
Milwaukee (16)	636.2	−11.3	−5.4	25	Yes
Cleveland (18)	573.8	−23.6	−8.0	14	No
Columbus (19)	564.9	4.6	−16.4	4	No
South					
Houston (5)	1,594.1	29.2	−7.0	47	No
Dallas (7)	904.1	7.1	−23.8	37	No
Baltimore (10)	786.8	−13.1	−7.4	37	Yes
San Antonio (11)	785.4	20.1	−2.3	15	Yes
Memphis (14)	646.4	3.6	−10.0	24	Yes
Washington, D.C. (15)	633.7	−15.7	−5.5	19	Yes
New Orleans (21)	557.5	−6.1	−9.6	6	No
Jacksonville (22)	540.9	7.3	−26.5	24	Yes
Nashville (25)	455.7	7.0	−27.9	53	No
West					
Los Angeles (3)	2,966.8	5.5	−0.2	19	Yes
San Diego (8)	875.5	25.5	−4.2	15	No
Phoenix (9)	789.7	35.2	−9.5	18	No
San Francisco (13)	679.0	−5.1	−2.0	27	No
San Jose (17)	636.6	38.4	7.4	15	No
Seattle (23)	493.8	−7.0	−6.5	19	No
Denver (24)	491.4	−4.5	−11.5	16	No

Source: Census Bureau, NJ 11-12-83:2359.
[a]The number in parentheses is the rank of each city based on 1980 population.

in the West, where dependence on the property tax is relatively low and increases were generally not made in 1983.

Recession, Recovery, and the Property Tax

The changing role of the property tax has been partially obscured by the severity and timing of recent business-cycle downturns. Not surprisingly, the existing property tax base in many localities expanded very slowly, if at all, during the most recent recessionary period in the early 1980s. Not only were home construction retarded and household formation rates slowed, but interest rate–sensitive industrial development of all kinds was dampened considerably. As a result, the tax bases of many localities were not augmented by new construction. Where rate hikes were not politically feasible, a further reduction in the contribution of the local property tax to local revenues resulted.

Understandably, given the long period of high inflation, this trend remained largely invisible. As property values in many locations steadily inflated, revenue available to localities increased, often substantially. In the quarter-century between 1956 and 1981, the assessed value of taxable property in the United States increased tenfold, reaching nearly $3 trillion. However, two-thirds of the increase took place in the five-year period from 1976 to 1981. The dramatic impact of inflation on households living in owner-occupied single-family dwellings in metropolitan areas throughout the country can be traced as follows. Locally assessed real estate accounted for over 80 percent ($2.5 trillion) of this national total; three-quarters (76.0 percent) of this amount was located in metropolitan areas. Of all local realty value, residential value accounted for approximately $1.5 trillion or approximately 79.0 percent. Finally, of this total, 87.3 percent was accounted for by single-family residences.[18] As a result, even though localities were gradually shifting away from their historical reliance on the local property tax, by the early 1980s property tax collections were soaring. Local governments actually collected 13 percent more in property taxes in 1982 than in 1981, after 10-percent and 6-percent increases in the previous years.[19]

Central Business District Transformation. As we have seen, cyclical downturns promote selected development processes while retarding others. Many large central cities continued to expand their local property tax bases through new nonresidential construction during the extended recession of the early 1980s. Many of the nation's largest cities, in slowly growing and rapidly growing regions alike, experienced a boom in high-rise office construction that continued even during the deepest trough of the business cycle. This reflects that once begun, such massive construction projects have an inertia that carries them forward through a sequence of cyclical shifts in the economy. The vigor with which central cities built accommodations to house the expanding office-based services and information sector components of their rapidly restructuring local economies is evident in the way structural shifts proceeded undeterred amidst cyclical downturns in the early 1980s. Once again, the role of recessions as filters of industrial growth and development is revealed.

State-Local Tax Policy Influences on Advanced Industrial Development

> There is, in fact, no economic advantage of one place over another which may not be over borne by an artificial advantage.[20]

The creation or restoration of comparative advantage has been a policy goal of local, regional, and national economies for a very long time, as the above statement, made by E.A. Ross in 1896, indicates. As the economy and patterns of settlement have slowly restructured and repositioned themselves at several spatial scales since World War II, much speculation has arisen again concerning the influence of a wide variety of policy-manipulable factors in stimulating and steering new and renewed industrial development. Against the concern that the adjustment of the economy to fluctuating energy assumptions and new technological innovations is somehow inexorable, a search has ensued for means of intervention to slow the pace, if not alter the direction, of the changes. The response is a quintessential political one that seeks to alleviate the problems afflicting both localities defined by relatively fixed political boundaries and residents for whom mobility options either in-place or to new places may be few or constrained.

Among the factors that have been scrutinized in recent years for their ability to lure economic development and attract additions to a local tax base has been the state-local tax burden. After World War II, growth gravitated first to the West and then to the South, which is characterized by relatively low state-local tax burdens. Yet the conclusions of half a century and more of research on the matter indicate that state and local tax-incentive strategies aimed at deferring, shifting, or lessening this burden have only a marginal impact on a firm's investment decisions.[21] This is due principally to the relatively insignificant contribution of taxes to the cost of production and also to the relative ease with which they may be passed on.

Tax Policy and Business Climate Reexamined

Recently, however, there has been an attempt to resensitize economic development efforts to the fact that differences in the rate of growth among states may indeed be explained more than previously thought by state and local tax and expenditure policies. A report by the Joint Economic Committee of the U.S. Congress on a supply-side perspective on state and local economic development strategies indicated that (1) the fastest-growing states are those with relatively low tax rates; (2) states with rapidly rising tax burdens tend to grow more slowly than states with a stable or falling burden; and (3) states that place relatively greater reliance on the taxation of capital (for example, taxation of income or property) tend to grow more slowly than states that rely relatively more on consumption-based taxes (for example, sales taxes or user charges).[22] While these findings hold for the state level of analysis, a similar pattern of findings also holds across metropolitan areas. From a sample of thirty large U.S. cities, those that demonstrated the greatest economic growth between 1969 and 1978 were those that had much lower tax burdens than cities experiencing lower rates of growth. The property tax burden in low-growth cities was on average 6.04 percent of the total income ($15,000) for a family of four in 1975; the average property tax burden for high-growth cities was only 3.29 percent. In an exploration of the impact of business costs on business formations and employment growth in thirty-five SMSAs during the 1976–1980 period, local taxes were discovered to have a significant depressant effect in low-technology industries, but no discernible impact on high-technology industries. These effects were particularly pronounced on the formations among very small firms, thus indicating the possible role of large size as an insulator against the effects of high business costs.[23]

The issues surrounding the contribution of state-local tax burdens to a climate for economic growth and development are far from settled, but in the interim it should not be assumed that tax burdens are of little or no consequence. This may be especially unwise as the nation continues its transit from an older to a newer industrial base. The "demise of manufacturing" tenet often woven into the post-industrial scenario is clearly refuted in the JEC study discussed above, which reports that those states whose employment structures were relatively heavily weighted toward manufacturing experienced greater economic growth than those states for whom manufacturing played a relatively lesser role. Economic prosperity continues to be related to a healthy goods-production sector, no matter how complex that sector's relationship to an advanced services economy may have become in recent years.

Thus, a newly emergent industrial structure and the tax policies accompanying it appear to be especially relevant to the nature and distribution of future economic growth.

Skilled Labor Pools and Tax Policy. These findings do little to challenge the accumulated wisdom of a decade and more of research, which suggests that on the whole, factors other than state-local tax burdens effectively steer investment across an economic landscape. However, there is increasing justification for examining the mix of tax instruments used, as well as the rates at which their cumulative burdens change over time, to understand better the geography of business investment and economic development. The generally negative relationship between economic development and property taxation, for instance, properly directs our attention beyond tax burdens that may inhibit the establishment or secondary expansion of firms and toward tax burdens on the households of firms' employees.[24] A study of the locations of high-technology firms and regional economic development indicated that the premier factor influencing the future interregional and intraregional location choices of these companies was the availability of appropriately skilled labor.[25] Yet the same study noted that the third most important steering factor among or between regions, and the second most important factor within regions, is likely to be the structure of state-local tax burdens. These findings indicate the need to be especially attentive to the burden of taxes on households; the burden of taxes on workers and their families, in certain instances perhaps more so than the tax burden on business, may impede the crucial job-labor matches required for localities seeking to nurture or sustain rapidly growing technology-intensive goods and services sectors.

The traditional sequence of job growth following population growth in the suburbanization process has long attested to the greater geographic mobility of the household. As the population has suburbanized, potential consumer and labor markets lured business from the cities and nurtured new business. The same dynamic operated on the regional level. In the 1970s, as we have seen, that sequence was often reversed. In general, the spatial or locational requirements of firms in growing new industrial and service sectors are much less stringent than they were for the older mainline industrial sectors. Indeed, it would be difficult to discover a major technological innovation whose net impact on production, distribution, and consumption has necessitated that production inputs be physically closer together than before. At the same time, despite growing speculation that the mobility of modern households has declined, we have seen in Chapter 4 that the available evidence does not completely support this view.

Residential deconcentration is likely to continue relatively unabated as it redefines local and regional economies. Consequently, the economic development problems of the coming years are more likely to be those that arise from the mismatch of mobile capital and immobile labor amid deconcentration than from the dynamics of rapid, if uneven, urbanization in the 1950s and 1960s. Yet there are still magnets for economic growth, including the availability of skilled labor, moderate tax burdens, and the amenities that have nurtured significant population growth in nonmetropolitan areas in all regions during the past decade. The extent to which they are policy manipulable, however, varies from factor to factor.

Looking Ahead: Growing Threats to Property Taxation

This chapter has viewed the property tax against the background of restructuring regional and local economies and changing settlement patterns. New and renewed production capacities, developing in tandem with rapid employment growth in services, invite speculation about a whole range of industrial-urban arrangements that may be in the process of significant redefinition. Land and its uses will necessarily continue to be inputs to production in the central cities, the suburbs, and beyond. However, there is reason to anticipate the rise of new mixes of factor inputs, with the result that the high ratios of nonland to land inputs once confined to high-density central cities may well push outward to once-peripheral areas and accompany the redefinition of the production activities that once dominated those places.[26] In that light, concern for the viability of the property tax system raises a series of questions concerning how it meshes with social, political, and economic developments that are taking place at the national, regional, and local levels.

Physical Capital and Demographic Trends

In recent decades local tax bases have been diminished not only by the inability of older industrial jurisdictions to compensate for the outmigration of businesses and households, but also by high inflation rates that curtailed residential and nonresidential construction and inflated the value of existing stock. Nonetheless, the shift to a services-based economy—a development more easily seen in the local economies of older industrial cities than in the larger national economy—has been augmented by a boom in the construction of high-rise offices and hotel-convention facilities, as well as by the dwindling supply of affordable housing in and near central business districts around the nation. Even so, this exchange of older for newer economic activities that continue to benefit from urban concentration may not necessarily provide a steady and reliable tax base for localities in the future. Already there are signs that many central city nonresidential markets are overbuilt. Proposed tax reforms—such as abolition of the deductibility of state and local taxes and the elimination of tax exemptions from the bulk of industrial development bonds—promise to dampen urban physical development in a clear reversal of the surge of urban development triggered by the 1981 tax bill.[27] The long-term net impacts of telecommunications development will likely lessen the necessity of separate and specialized meeting and working places for an information-based central city economy.[28] The same basic revolution that swept through the fields and factories of the goods-producing economy and rationalized their production processes is now making yet another sweep through the offices of the services economy.

Likewise, the residential sector has been transformed as low birth rates and high household formation rates dramatically alter traditional conceptions of homeownership and perhaps even reduce the desirability of ownership and maintenance of real estate.

Households and Housing Reconfigured. On the surface, it would appear that little settlement change should be expected from net replacement activity of the nation's housing stock. Only an average annual expansion of 2 percent in housing stock is expected in the near future. Moreover, it has been estimated that 75 percent of Americans in the year 2000 will be living in dwellings already built as of 1980.[29] Yet beneath the surface, the adaptation of housing options to a new mix of households reveals more volatility. The very notion of housing appears to be undergoing significant change in response to shifting demographics. During the 1970s, housing stock increased annually an average of 2.7 percent, although Sternleib has estimated that nearly half of this increase was in response to the fragmenting of existing households rather than to population growth.[30] In recent decades, household formation rates have commonly exceeded overall population growth. In the 1970s alone, 2.3 million people sorted into 1.7 million new households were added, on average, each year. The average annual rate of household formation (2.7 percent) far exceeded the average annual growth rate of total population (1.1 percent). Moreover, the average household size has been steadily shrinking from 5.79 in 1790, 4.93 in 1900, 3.67 in 1940 to 2.75 in 1980.[31]

Inevitably, these trends had quite different impacts across localities. It was not unexpected for overall population declines to be accompanied by household growth in certain localities. Yet, the erosion of the local tax bases through population decline was usually not compensated for by the stable or rising number of households, thus placing special burdens on cities whose service responsibilities did not decline in step with their total population. The income selectivity of these trends exacerbated the impacts on localities as a disproportionate amount of suburbanization and central city-avoidance behavior was tied to middle- and upper-income residents. In aggregate terms, these trends permitted relatively poorer households to filter into better housing stock in many neighborhoods. However, the attractiveness of such housing for many nonpoor households was blunted by its architectural and space-use characteristics, which were originally designed to meet the demands of larger traditional families who occupied the structures during an era of more forgiving energy assumptions. The poorer "fit" between new households and old houses did little to stem the flow of population away from older cities.

Household formation rates are sensitive to business-cycle activity. The early 1980s witnessed a temporary decline in the formation rate for new households. With the economic recovery, this rate rebounded, although only to a more modest level because of the onset of a longer-term decline tied to the baby boom cohorts' passage through and beyond their peak household formation years. New young households' difficulties in affording housing is further blunting demand. In response, not only has the average size of new houses built since 1978 shrunk nearly 10 percent, but the rapid conversion of nonresidential structures will continue to affect the volume of construction activity in coming years.[32]

Were it not for the tax advantages that have made homeownership the principal means by which the broad middle classes can amass an estate, changing work and leisure patterns reflecting expanded options for women and the redefinition of the traditional family would probably have reduced the demand for ownership of land and housing in many locations throughout the United States. Only recently has it become evident that the demand for single-family detached housing, like that for automobilies, can no longer be automatically extrapolated from past demand trends. Low birth rates, shrinking household sizes, and rising proportions of elderly living alone, all despite continued suburbanization, are demographic trends that, together with advanced industrial development patterns noted previously, might well result in dramatically lower levels of demand for land and in alterations in traditional distributions of its uses in the future.

After the current economic recovery stabilizes, a new mix of households will likely fuel a demand for housing and transportation forms so fundamentally different from what was seen only a few short years ago as to offer little solace to those in both the housing and automobile industries who pin their hopes to a return to the past. Smaller houses will require less land. Such trends could conceivably diminish further the expected revenues of a local property tax that is accompanied by a politically reasonable rate. Where tax-rate increases occur, they may be especially burdensome for the 23 percent of all households composed of single individuals, even though two-earner households will remain more insulated from their impacts.

Finally, the blurring of the distinction between the homeplace and the workplace through the rise of telecommuting and self-employment heralds the tendency for more and more workers to work out of their homes. This tendency may well unfold in newly settled areas where new construction has either not yet begun or will never take place on the scale known in an era during which separate structures for workday and residential occupancy had to be built. Further, it may proceed in older settled areas where the costs of new construction and the congestion associated with commuting encourage the tendency to work at home. These general trends in and around the residential sector could well result in a declining demand for land and a slower pace of transit for land parcels between uses.

The Prospects for Property Tax Reform

Stop-gap measures designed to salvage the equity of the property tax have become as increasingly complex and poorly articulated as they are politically predictable. Homestead exemptions, statewide circuit-breakers, and the like, do indeed respond to the reality that the burden of property taxes is falling increasingly on a central city-based residential sector that is often older and poorer than ever before. Nonetheless, these reforms fall short of solving the fundamental dilemma of localities whose taxation schemes no longer are able to capture the scale of the urban economy in and around them. Further, popular efforts, such as sizable homestead exemptions, that shift the incidence of tax burdens to present and future business are a shortsighted political effort to buy residential peace. They will likely inhibit the kind of new urban economic development that is crucial to a restructuring local economy and that would permit industrial communities to play as dominant a role in the next industrial era as they did in the receding one.

One must be open to the possibility that the local property tax may not make the transit between industrial eras and still retain its historical form and function.[33] What was once the most important tax in the nation's tax structure through the 1930s had already been much diminished in importance as state and federal levels tapped more fluid bases through sales and income taxes. For the past half-century, the property tax has endured as the most important local tax source, thus being perceived by many as serving as a hedge against ever-greater tax and, therefore, political centralization. Yet as manufacturing growth shifted from central cities to small and often remote localities and as localities grew more dependent on extralocal revenue sources up to 1980, many of the myths about local autonomy in social and economic life have been laid bare.

The local property tax is likely to diminish even more as it fails to accommodate the ever-larger scale of economic and social life. The deconcentration of people, jobs, capital, and older political constituencies and coalitions will continue to redefine bounded municipal economies as far into the twenty-first century as one can see. Persistent calls for local property tax "reform" increasingly reveal the widening gulf between the local nature of a fading industrial-urban economy with its settlement patterns and those new industrial patterns and settlement forms that are emerging to take their place. Does the property tax reform movement honestly face the reality that localities are no longer self-contained entities whose tax bases can be harnessed to serve local needs in accordance with local wishes? May this reform movement not finally be recognized as a myth that no longer serves a constructive purpose? Indeed, the property tax may preserve a certain fiction of localism; to express local preferences may no longer mean much when life is lived at greater spatial scales than ever before.

The Rate-Setting Ritual and Rethinking Urban Services

Not only is the local property tax the most unpopular of taxes, since it is associated with annual or semiannual lump payments; its complicity in a traditional political ritual at the local level also often stirs up suspicion and animosity. As with a restaurant tab after a business lunch, the size and substance of a local budget are usually determined first, and only then is the tax rate calculated that will bring in the necessary funds. Despite the procedural scrutiny that may be exercised at this point, such a political-fiscal sequence does not lend itself easily to budget discipline or to the reconsideration of whether or not a task should be—or should continue to be—assigned to the local public sector rather than to the private or the voluntary sector. As a result, the tendency among tax scholars and local officials has been to spend a great deal of time viewing the local property tax as an independent variable—a tool in the local fiscal system that needs to be recalibrated to achieve greater efficiency and equity—rather than as a dependent variable whose form, function, and ultimate efficacy are the consequence of other forces at work in society.

Approximately half of the revenue from property taxes comes from residential uses of land, and the other half comes from business property. The difficulty of deciding how to apply the homestead exemption without jeopardizing business growth or risking voter wrath and outmigration illustrates the volatile politics accompanying its use. Yet the discussion of property tax reform has continued as if some combination of the property tax base, rate, and assessment procedures and overall administration can cement households, businesses, and their respective loyalties to localities and thus maintain the local economies in historically familiar ways. The reform approach to the property tax, while laudable in the short run, may miss the crucial ways in which this fiscal tool may no longer be used.

The Continuing Preoccupation with Land and Land Use

Land and its uses figure prominently in the series of redefinitions accompanying the present restructuring of the national and local economies and the settlement patterns to which they are tied. This role of land may well be too prominent to insulate the local property tax—essentially a real estate tax—from significant reconceptualization and reevaluation. Ultimately, the local property tax is registered on the land, since the economic activity of businesses and the residential activity of households are believed to have locational features that are sensitive to the influence of taxes. The property tax can obstruct the "career" of a specific parcel of land and prevent its transit from one use to another. Dual-rate or split property tax systems that shift the tax burden to the land and away from the development on it have made only modest inroads throughout the United States. Such a system has the potential for reducing the incentives facing developers to hold land for speculation rather than develop it and face higher property tax assessments. However, an outcome widely (if not wisely) perceived as undesirable is illustrated by farmland that if taxed at its highest and best use rather than at its actual use, often becomes the target of more aggresive urban development.[34] A reverse image of this process is illustrated by the underdevelopment of central city land parcels that is believed to result from the disincentives to development that flow from heavily taxing the improvements on the land rather than just the land itself. The continuing emphasis on land and even location, then, whether understood as urban "sprawl" or land vacancy and speculation in central locations, preserves a preoccupation with land and the uses and abuses that are associated with it. Whether this preoccupation with land and what happens to it continues to be justified as it once may have been, however, is another matter.

Shifting Focus: From Land to Location. The waning years of the older industrial era were the setting for a thorough redefinition of the uses and significance of land. The accomodation to technological innovations has rendered the primary sector of the economy (especially agriculture) ever-less dependent on land as it has been traditionally defined. Accompanying the rise and dominance of manufacturing and construction, land was redefined to mean little more than location itself. Today, the ascendency of advanced manufacturing in new and renewed goods-producing industries and services—especially administration and information exchange—appear to be further reducing the significance of land in the economy and the larger society. Harnessing the latest advances in transportation and telecommunications, the newer sectors of the economy have rendered electronic proximity of greater importance than physical proximity, thus loosening even more the bond between specific land parcels and specific land uses. People and businesses simply do not have the same locational constraints now that they had during the older industrial era. Does it not follow that a tax system historically based on land and its uses will likewise be transformed?

Perhaps a case can be made that at least as much effort should be spent on redefining the task of local governance as on salvaging the local property tax as an instrument of local public finance. Certainly, questions concerning why money is raised at the local level for purposes that may no longer be strictly "local" merit as much consideration as how it is raised. As the society and economy restructure and reposition themselves, as the intangible bases of wealth such as information and knowledge migrate to the core of an advanced industrial economy, and as new forms of property defy easy assignment to specific locations for purposes of assessment, reconceptualizations of land, land use, and even location per se lead irrevocably to the redefinition of land-value taxation and the property tax. To the extent that these processes continue, the reform of the property tax system not only may be increasingly caught up in efforts that cancel each other out but may also soon be so inadequate that the property tax will be seen as ever-less relevant to local public finance.

Prominent among the local functions funded by the property tax is a whole series of municipal services that have migrated onto the public sector agenda during the twentieth century. The steady elaboration and expansion of the municipal services package through the past century reflected not only the political opportunity to maintain voter contentment and develop a power base in the public work force, but also the economic fact that the engines of the private sector were located in the municipality. Today, as both the economy and the population continue to deconcentrate and public responsibilities continue to decentralize, the rationale is no longer intact for using local fiscal instruments to maintain traditional service packages. The ultimate worth of many of these services can be questioned, given the tendency in recent decades for households and businesses to leave them behind.[35] Where specific services continue to be justified, a search for alternative delivery and/or financing arrangements may well be a useful local exercise that should be undertaken periodically.[36]

In the past, lower local taxes were not necessarily desirable when quality municipal services were at stake. Today, however, higher local taxes may not be necessary to ensure services of high quality. A local budget discipline that capitalizes on the widespread reassessment of the scope and substance of municipal services packages promises a bottom-up reform of the system that generates property tax revenues, if only by reducing the need for them. The privatization trend offers guarded guidance for curbing costs, especially in younger growing communities before large public sectors become entrenched. Likewise, the planned disinvestment possibilities open to contracting communities whose service packages can be better tailored to the needs, if not the demands, of the remaining population offer ways to lighten the burden placed on local property taxpayers. Already the shift to user charges for many services with a private consumption character demonstrates that communities have indeed adapted in ways that relieve the historical burden placed on the local property tax system.[37]

Notes

Chapter 1

1. Charles P. Kindleberger and Bruce Herrick, *Economic Development* 3rd ed. (New York: McGraw-Hill, 1977); Roger J. Vaughan and Peter Bearse, "Federal Economic Development Programs: A Framework for Design and Evaluation," in *Expanding the Opportunity to Produce,* ed. Robert Friedman and William Schweke (Washington, D.C.: Corporation for Enterprise Development, 1981), pp. 307–321; Philip Shapira, "Economic Development Analysis and Planning in Advanced Industrialized Economies: A Bibliography" (Chicago: Council of Planning Librarians, May 1983).

Chapter 2

1. For general discussions that illustrate the ways in which secular and cyclical productivity trends both during and following recessions are commonly blurred, see "The Revival of Productivity," *Business Week,* February 13, 1984, pp. 92–100; "Still Waiting for a Breakthrough in Productivity," *Business Week,* March 5, 1984, p. 20; and "U.S. Productivity Gains Are Still on a Roll," *Business Week,* September 17, 1984, p. 28. In an international context, a growing wage gap between U.S. and foreign manufacturing industries makes it all the more likely that productivity increases alone cannot restore U.S. competitiveness. See Alfred L. Malabre, Jr., "Gap Between U.S., Foreign Wages Widens: Productivity Gains Fail to Offset Disparity with Other Nations," *Wall Street Journal,* July 17, 1985, p. 6.
2. As the United States emerged from back-to-back recessions in late 1982, a distinct gap between productivity trends in manufacturing and nonfarm nonmanufacturing industries opened up. According to data provided by Robert J. Gordon (cited in "Potential GNP May Still Be on a Slow-Growth Track," *Business Week,* December 24, 1984, pp. 13ff.), rates of productivity growth for the two sectors were relatively high and similar during the 1960s. During 1970–1974, an enormous gap developed wherein cyclically adjusted annual rates of increase in manufacturing exceeded 3.5 percent while average rates for nonmanufacturing fell below 0.5 percent. The gap closed somewhat during 1974–1979, principally as a consequence of decline in manufacturing productivity. During 1979–1984, however, the gap began to open up once

again. While productivity rebounded in manufacturing to the 1960s level, nonmanufacturing productivity has remained stagnant.

3. See Bureau of Labor Statistics (BLS), 1981, *Productivity Chartbook,* for details on productivity measurement and recent trends.
4. For a discussion of the implications of industrial concentration within the United States, see Steven Lustgarten, *Productivity and Prices: The Consequences of Industrial Concentration* (Washington, D.C.: American Enterprise Institute for Public Policy Research, 1984).
5. For a view of productivity stagnation as symptomatic of the adjustments required by decades-long cycles of capital construction and replacement, see Jay W. Forrester, "More Productivity Will Not Solve Our Problems," *Business and Society Review,* Fall 1980, pp. 10–18.
6. Data on productivity trends are from BLS file data and from information provided by the American Productivity Center in Houston, Texas.
7. See BLS, *Productivity Chartbook,* 1981, Charts 8 (p. 17) and 9 (p. 19), for industry-specific productivity trends. For related discussion, see a staff study conducted by the Joint Economic Committee, U.S. Congress, *Productivity: The Foundation of Growth* (Washington, D.C., November 1980).
8. Committee for Economic Development, *Productivity Policy: Key to the Nation's Economic Future* (New York, 1983).
9. Congressional Budget Office, "Federal Support for R&D and Innovation" (Washington, D.C., April 1984).
10. Gregory Schmid, "Productivity and Reindustrialization: A Dissenting View," *Challenge,* January–February 1981, pp. 24–29.
11. See Morgan O. Reynolds, *Power and Privilege* (New York: Universe Press, 1984).
12. Eight studies that demonstrate the beneficial influence of unions on productivity are reviewed in Richard B. Freeman and James L. Medoff, *What Do Unions Do?* (New York: Basic Books, 1984).
13. See Senator Lloyd Bentsen, "Chairman's Introduction," in Joint Economic Committee, *Productivity,* pp. i–ii.
14. Capital spending trend data are from Bureau of Economic Analysis (BEA) and BLS data reported in Charts 24 (p. 53), 26 (p. 57), and 28 (p. 61) of the BLS *Productivity Chartbook.*
15. Based on BLS data (cited in "Capital May Not Be the Key to Productivity," *National Journal,* April 16, 1983, p. 820).
16. Congressional Budget Office, "Federal Support for U.S. Business," (Washington, D.C., January 1984).
17. See Charles R. Hulten and James W. Robertson, "Taxation: Our De Facto Industrial Policy," *The Urban Institute Policy and Research Report,* Vol. 14, October 1984, pp. 1–5. Nonetheless, as suggested above, there is no guarantee that increased capital investment will result in productivity increases. Heavy investment in the face of limited market growth or profit potential can result in redundancy and accelerated obsolescence. Commerce Department data illustrate that since 1965, capital intensity—plant and equipment expenditures per worker—outpaced productivity gains (cited in "Why Investment Yields Less Bang for the Buck," *Fortune,* March 7, 1983, p. 42.
18. See Thomas M. Humbert, "Understanding the Federal Deficit, Part 3: The Unproven Impact," Heritage Foundation, *Backgrounder,* No. 330, January 27, 1984.
19. See National Science Foundation, *Science Indicators, 1980,* (Washington, D.C., 1981), Appendix, Tables 1–4; and also BLS, *Productivity Chartbook,* Charts 30 (p. 65), 32 (p. 69), and 33 (p. 71). For a discussion and overview of R&D spending trends across five advanced industrial nations, see Richard R. Nelson, *High-Technology Policies: A Five-Nation Comparison* (Washington, D.C.: American Enterprise Institute for Public

Policy Research, 1984), pp. 19–22. See also The Report of the President's Commission on Industrial Competitiveness, *Global Competition: The New Reality,* Vol. II. (Washington, D.C., January 1985), p. 97.

20. Bureau of Labor Statistics, *Productivity Chartbook,* p. 6.
21. Ibid., Chart 3 (p. 7).
22. Office of U.S. Trade Representative, *U.S. National Study on Trade in Services* (Washington, D.C., December 1983), p. 1.
23. See R. Kutscher and J. Mark, "The Service-Producing Sector: Some Common Perceptions," *Monthly Labor Review,* Vol. 106, No. 4, 1982, pp. 21–24.
24. Cited in "Service Industries Aren't to Blame for Low Productivity," *Business Week,* June 13, 1983, p. 20.
25. See discussion in Thomas M. Stanback, Jr., Peter J. Bearse, Thierry J. Noyelle, and Robert A. Karasek, *Services: The New Economy* (Totowa, N.J.: Rowman & Allanheld, 1981), pp. 6 ff.
26. See "A Productivity Revolution in the Service Sector" *Business Week,* September 5, 1983, pp. 106 ff.
27. Of the 630 months between August 1929 and November 1982, the U.S. economy has been in recession in 152 (23.8 percent). For a discussion that distinguishes "classic cycles" characterized by absolute expansion and contraction from "growth cycles" as fluctuations around generally rising trends, see Philip A. Klein, *Business Cycles in the Postwar World: Some Reflections on Recent Research,* (Washington, D.C.: American Enterprise Institute for Public Policy Research, 1976).
28. See Henry F. Myers, "Economists Say Slump Hastened Some Trends but Spawned Very Few," *Wall Street Journal,* May 25, 1983, p. 1. For a somewhat different perspective, see the discussion by Freeman et al. of G. Mensch's view that depressions can function as accelerators of technological innovation. Christopher Freeman, John Clark, and Luc Soete, *Unemployment and Technical Innovation: A Study of Long Waves and Economic Development* (Westport, Conn.: Greenwood Press, 1982), pp. 51 ff.
29. See Joseph Schumpeter, *Business Cycles: A Theoretical Historical and Statistical Analysis of the Capitalist Process* (New York: McGraw Hill, 1939), for a classic discussion of capitalist economies in dynamic disequilibrium.
30. Peter Drucker has underscored the remarkable capacity of the U.S. economy to create new employment by noting that between 1965 and 1984, job growth of 45 percent outpaced the 38 percent growth of the adult population between the ages of 16 and 65. Moreover, as I shall emphasize later in this and the following chapter, the manufacturing sector has experienced an expansion through the past quarter century. Between 1960 and 1984, U.S. manufacturing employment alone grew 14.2 percent. See Peter Drucker, "Why Has America Got So Many Jobs?" *Wall Street Journal,* January 24, 1984. p. 32.
31. The 1981–1982 recession lasted sixteen months with employment reaching 10.7 percent.
32. BLS data (cited in *National Journal,* June 18, 1983, p. 1304).
33. Data on the changing structure of unemployment are from Paul Manchester, "Are Business Cycles Symetric?" Working Paper 1444, and "The Changing Cyclical Variability of Economic Activity in the United States," Working Paper 1450, (Washington, D.C.: Joint Economic Committee, September 1984).
34. This conclusion is substantiated in a paper by J. Bradford Delong and Lawrence H. Summers presented at a National Bureau of Economic Research conference on business cycles (cited in *Business Week,* April 23, 1984, pp. 17–18.)
35. See George Gilder, "The Numbers Tell a Supply-Side Story," *Wall Street Journal,* June 13, 1983, p. 22.
36. From a regional perspective, Bluestone and Harrison, also using Dun and Bradstreet

data, have shown that between 1969 and 1976, the number of jobs created and destroyed by openings, relocations, expansions, and contractions of private business establishments were quite substantial. The net employment change for individual regions understood as a ratio of jobs destroyed to jobs created varied from 0.68 in the West South Central region to 1.02 in the Northeast. See Barry Bluestone and Bennett Harrison, *The Deindustrialization of America: Plant Closings, Community Abandonment and the Dismantling of Basic Industry* (New York: Basic Books, 1982), Table A.1, pp. 266–271.

37. Data reported in "Is the Economy to Blame for Unemployment?" *Wall Street Journal,* June 13, 1983, p. 1.
38. See Anne Swardson, "Malady in Prosperity: Growth Comes at a Cost" (citing unpublished Federal Reserve Board data), *Dallas Morning News,* March 22, 1984, p. 10.
39. Data reported in "A Sudden Rush to Spend on America's Factories," *Business Week,* January 30, 1984, pp. 83–84.
40. See Congressional Budget Office, "Dislocated Workers: Issues and Federal Options," July 1982; also see M. Bendick and J. Devine, "Workers Dislocated by Economic Change: Do They Need Federal Employment and Training Assistance?" in *Seventh Annual Report: The Federal Interest in Employment and Training* (Washington, D.C.: National Commission for Employment Policy, 1981.
41. See Daniel Bell, *The Coming of Post-Industrial Society* (New York: Basic Books, 1973), p. 133.
42. See Victor R. Fuchs, "Economic Growth and the Rise of Service Employment," Working Paper No. 486, National Bureau of Economic Research, June 1980; Thomas M. Stanback, Jr., *Understanding the Service Economy: Employment, Productivity, and Location* (Baltimore: Johns Hopkins University Press, 1979); Stanback et al., *Services: The New Economy;* Joachim Singelmann, *Agriculture to Services: The Transformation of Industrial Employment* (Beverly Hills, Calif.: Sage Publications, 1978); Herman Kahn and B. Bruce-Biggs, *Things to Come: Thinking About the 70s and 80s* (New York: Macmillan, 1972); Bobbie H. McCrackin, "Services: Key to Current Stability and Future Growth," *Economic Review,* Federal Reserve Bank of Atlanta, July 1983, pp. 36–52; Lora S. Collins, "The Service Economy," *Across the Board,* November 1980, pp. 17–22; Alfred Malabre, Jr., "Service Jobs Keep Expanding in Recession, Make up Even Larger Share of Work Force," *Wall Street Journal,* January 15, 1982, p. 44; Eli Ginzberg, "The Mechanization of Work," *Scientific American,* Vol. 247, No. 3, September 1982, pp. 39–47; Jerome A. Mark, "Measuring Productivity in the Services," *Monthly Labor Review,* June 1982, pp. 3–8; Eli Ginzberg and George J. Vojta, "The Service Sector of the U.S. Economy," *Scientific American,* Vol. 244, March 1981, pp. 48–55; Ronald K. Shelp, "The Service Economy Gets No Respect," *Across the Board,* February 1984, pp. 49–54; and Ronald K. Shelp, "Myth Vs. Reality—A Service Economy," *The Journal of the Institute for Socioeconomic Studies,* Autumn 1983, pp. 26–38.
43. See Bluestone and Harrison, *The Deindustrialization of America.*
44. Colin Clark, *The Conditions of Economic Progress* (London: Macmillan, 1940).
45. Victor Fuchs, "Economic Growth, and Kahn and Bruce-Biggs, *Things to Come.*
46. See Stanback, *Understanding the Services,* Stanback et al., *Services: The New Economy,* and Singelmann, *Agriculture to Services.*
47. See U.S. Department of Commerce, Bureau of Economic Analysis, *The National Income and Product Accounts of the United States, 1929–74 Statistical Tables,* 1977, Table 6.2; "Gross National Product by Major Industry, Workfile 1205–02–03," and "U.S. National Income and Product Accounts: Revised Estimates, 1975–77," *Survey of Current Business,* July 1978, Table 6.2, p. 52.
48. Moreover, while the share of national output as measured by percent of real GNP for services increased by only 5.8 percent over the 1948–1978 period, the growth

pattern for manufacturing revealed a major internal inconsistency. Over the same period, nondurable goods, which essentially dominate the output of the manufacturing sector, experienced a contraction from 34.4 percent to 26.5 percent. Simultaneously, durable goods industries increased their combined output share from 15.9 percent to 19.4 percent. See Collins, "The Service Economy."

49. BLS data (cited in *National Journal,* April 16, 1983).
50. The glacial quality of these shifts over time is further revealed in a study by the New York Stock Exchange, Inc., of employment projections by sector between 1977 and 1995; employment in goods production is expected to decline from a 47.9 percent share to a 42.8 percent share, while employment in services is expected to increase from 52.1 percent to 57.2 percent over the same period. This translates into less than a 0.3 percent average annual shift in each case. (Data cited in *National Journal,* "Laboring Under an Illusion of More Service?" September 9, 1984, p. 1694). Over the longer sweep of time between the two most severe cyclical downturns of the century—1929 to 1977—the share of employment in the extractive (mining) and transformative (manufacturing and construction) industries only declined from 36.5 percent to 29.7 percent, while the employment share in services of all kinds increased from 55.1 percent to 68.4 percent. (Reported in Stanback et al., *Services: The New Economy,* pp. 12–13). See also Amitai Etzioni, "Prematurely Burying Our Industrial Society," *New York Times,* June 28, 1982, p. A15. Finally, it may well be worth remembering that causal analysis in the sciences builds upon the concept of variance. It is generally agreed that constants or circumstances that vary only minimally and within a restricted range make poor candidates for serving as causes of changes in other variables. Demonstration of their effects is hindered by the fact that the linkages cannot be empirically isolated and identified. From this perspective, it does not appear that the shift-to-services is sufficiently pronounced to be useful in accounting for major changes in other parts of the economy and society. I will attempt to develop this point more fully below.
51. Stanback et al., *Services: The New Economy,* p. 44, likewise conclude that in a consumption sense, we remain a goods oriented economy, even though the nature of those goods and consumption patterns are changing.
52. For instance, the nonprofit services including health and education registered important gains on both indicators over the same period. Between 1948 and 1977, the health sector tripled its employment share from 1.7 percent to 5.2 percent. Likewise, retail services expanded from 12.5 percent to 14.2 percent. In contrast, the distributive services, which includes the linking industries of transportation, communication, and utilities as well as wholesale trade, experienced a contraction in employment share over the same period from 13.5 percent to 11.4 percent. (Data reported in Stanback et al., *Services: The New Economy,* Table 1.2, pp. 12–13.)
53. Stanback et al., *Services: The New Economy,* and Ginzberg and Vojta, "The Service Sector of the U.S. Economy."
54. Further, over the 1948–1977 period the employment share for manufacturing shrank from 32.3 percent to 24.1 percent while that for producer services nearly doubled from 6.1 percent to 12.0 percent. (Data reported in Stanback et al., *Services: The New Economy,* Table 1.2, pp. 12–13.)
55. For a brief introduction to "economic base" and export base theory, and how to study them, see "Economic Base: What Our Jobs Are Tied To," in Eva C. Galambos and Arthur Schreiber, *Economic Analysis for Local Government* (Washington, D.C.: National League of Cities, November 1978), pp. 13–24.

    For discussions arguing for a reappraisal of the role of services in a regional economy, see John M.L. Gruenstein and Sally Guerra, "Can Services Sustain a Regional Economy?" *Business Review,* Federal Reserve Bank of Philadelphia, July–August,

1981, pp. 15–24; L. Hirschhorn, "Toward a National Urban Policy—Critical Reviews, The Urban Crisis: A Post-Industrial Perspective," *Journal of Regional Science*, Vol. 19, No. 1, 1979, pp. 109–118; and Stanback et al., *Services: The New Economy*, p. 6.

56. See Joseph Schumpeter, *Capitalism, Socialism and Democracy* (New York: Harper & Row, 1976; originally published 1942), pp. 65 ff., for an analysis and interpretation of data focused on the same question. He discovered that in the face of rapid industrial development, the structure of the English income distribution as reflected in income pyramids did not change during the entire nineteenth century.

57. See Bluestone and Harrison, *The Deindustrialization of America*, and Stanback et al., *Services: The New Economy*. Also see Harry Braverman, *Labor and Monopoly Capital: The Degradation of Work in the Twentieth Century* (New York: Monthly Review Press, 1974), and David M. Gordon, Richard Edwards, and Michael Reich, *Segmented Work, Divided Workers: The Historical Transformation of Labor in the United States* (New York: Cambridge University Press, 1982).

58. See Bennett Harrison, "Rationalizing, Restructuring, and Industrial Reorganization in the Older Regions: The Economic Transformation of New England Since World War II," Working Paper No. 72. (Cambridge, Mass.: Joint Center for Urban Studies of MIT and Harvard, February 1982). Also see Bennett Harrison, "Regional Restructuring and 'Good Business Climates': The Economic Transformation of New England Since World War II," in William K. Tabb and Larry Sawyers (eds.), *Frostbelt-Sunbelt: The Political Economy of Urban Development and Regional Restructuring* (New York: Oxford University Press, 1983), and Lester C. Thurow, "The Disappearance of the Middle Class," *New York Times*, February 5, 1984, p. F3; Robert Kuttner, "The Declining Middle," *Atlantic Monthly*, Vol. 252, July 1983, pp. 60–72.

59. Moreover, this occupational restructuring has been accompanied by a dramatic shift in business investment patterns. Jonscher has reported that between 1977 and 1982, investments in productivity-enhancing tools such as computers and related office automation equipment for use by white-collar employees rose from a level of one-half that invested in equipment for use by blue-collar workers to an amount equal to that invested in equipment for use by blue-collar workers. See Charles Jonscher, "The Impact of Information Technology on the Economy: Problems of Modelling and Measurement," paper prepared for conference on Impact of Information Technology on the Service Sector, University of Pennsylvania, February 1985 (cited in Mitchell L. Moss, "Telecommunications and the City," paper presented at the Landtronics Anglo/American Conference, London, England, June 1985).

60. Robert Z. Lawrence, "Sectoral Shifts and the Size of the Middle Class," *The Brookings Review*. Fall 1984, pp. 3–11. See also Kenneth McLennan, "A Low-Wage America?" in *The Journal of the Institute for Socioeconomic Studies*, Vol. 9, No. 3, Autumn 1984, pp. 43–58; and Frank Levy and Richard C. Michel, "Are Baby-Boomers Selfish?" *Policy and Research Report*, The Urban Institute, Vol. 14, No. 3, December 1984, pp. 12–14.

61. See Greg J. Duncan, *Years of Poverty, Years of Plenty* (Ann Arbor: Survey Research Center, Institute for Social Research, University of Michigan, 1984).

62. See Steven F. Benz, "High-Technology Occupations Lead Growth in Services Employment," *Business America*, September 3, 1984, pp. 19–20.

63. While shifts in the nation's industrial and occupational structures are not viewed in this chapter as setting the stage for a declining middle class, there is evidence that changing distributions of total family income and per capita family income have been registered as a 15.7 percent contraction of the "middle class" between 1968 and 1982. See McKinley L. Blackburn and David E. Bloom, "What's Happening to the Middle Class?" *American Demographics*, January 1985, pp. 19–25. Part of the explanation for this is the rise of the two-earner family, which has likely had the effect of shifting a portion of previously middle-class families into a higher-income stratum in much

the same way and for much the same reasons that new families formed through divorce, single parenthood, and widowhood jettison another portion into lower-income strata. The 1970s saw the building of a tidal wave of new households that may have been sociologically sound yet were also economically tenuous. The impact of this diversity of new household forms was amplified by the sheer size of the baby-boom cohort entering household formation life stages and the rise of elderly households. It is entirely consistent with the data available to date that any contraction of the middle class that has taken place is much more likely to be linked to the new scales and diversity of the household formation process than either industrial or occupational shifts per se. How available employment opportunities and wages are pooled across this greater diversity of household and family forms is arguably more consequential for the formation and retention of a broad middle class than are the larger processes of an evolving economy and industrial development. For a discussion of the enduring wage distribution in the computer and business equipment manufacturing industries, see Peter Behr, "This Idea's Time Hasn't Come," *Washington Post National Weekly Edition,* April 16, 1984.

64. See Robert J. Samuelson, "Middle-Class Media Myth," *National Journal,* December 31, 1984, pp. 2683–2678; also see Robert J. Samuelson, "Job Fears and Facts," *National Journal,* June 25, 1984, p. 1348.
65. See Fabian Linden, "The Myth of the Disappearing Middle Class," *Wall Street Journal,* January 23, 1984, p. 22.
66. See Marc U. Porat, *The Information Economy* (Washington, D.C.: Office of Telecommunications, May 1977).
67. See Geza Feketekuty and Jonathan D. Aronson, "Meeting the Challenges of the World Information Economy," *The World Economy,* Vol. 7, March 1984, pp. 63–86. The authors note that the information-related share of total employment was only 17 percent as recently as 1950. According to Jonscher, as measured by the proportion of white-collar workers, the proportion of information workers in the United States is estimated to have increased from 18 percent in 1900 to 55 percent (and 64 percent of total wages) in 1984. Nonetheless, this growth may well have peaked by now. While "production sector" productivity has grown 3 to 5 percent annually since 1950, productivity in the information sector has lagged at only ½ to 1 percent per year. Consistent with a projection made by Leontieff, Jonscher anticipates a decline in the productivity of the information economy between now and the year 2000 as an expanded white-collar labor force results from the shift of labor into a relatively low-productivity sector. See C. Jonscher, "Information Resources and Economic Productivity," *Information Economics and Policy,* Vol. 1, No. 1, 1983. William Baumol likewise has suggested that the expansion of the information sector cannot be expected to continue uninterrupted. He notes that many activities within the information sector have resisted productivity growth. Consequently, as productivity growth is stimulated elsewhere, the costs of laggard sectors like information production are driven up. As information costs rise, a substitution in favor of other relatively cheaper inputs which may be registered as a slowdown in the replacement of selected labor inputs such as clerical workers can be expected to result. (Oral remarks entitled "Technology and Service Industry Structure: an Overview," made at a conference, The Impact of Information Technology on the Service Sector, The Fishman-Davidson Center for the Study of the Service Sector, The Wharton School, University of Pennsylvania, Philadelphia, February 8, 1985.)
68. Estimate provided by Electronic Services Unlimited and reported in *Wall Street Journal,* February 5, 1985, p. 1. At this time, however, measurement of the dimensions of this phenomenon remains extremely difficult. In recent research by the National Research Council, it was estimated that the number of U.S. households in which at

least one worker would be linked full-time or part-time to his or her employer via a computer would increase from only 15,000 today to nearly 10 million by 1990. See National Research Council, *Office Workstations in the Home* (Washington, D.C.: National Academy Press, 1985) (cited in *National Journal,* July 6, 1985, p. 1954).

Chapter 3

1. Joseph Schumpeter, *Capitalism, Socialism and Democracy* (New York: Harper & Row, 1976; originally published 1942), p. 68. Italics added.
2. Roger Schmenner, "Every Factory Has a Life Cycle," *Harvard Business Review,* Vol. 62, No. 2, March–April 1983, pp. 121–129. Also see Roger Schmenner, "Choosing New Industrial Capacity: On-Site Expansion, Branching and Relocation," *Quarterly Journal of Economics,* Vol. 95, No. 2, August 1980, p. 105. The increasing age of U.S. manufacturing facilities relative to those of major competitors has been identified repeatedly as a principal reason for the United States recording the lowest productivity rates during the 1970s of all major industrial nations. As a result of the recent surge of capital investment in the United States, by 1984 the average age of U.S. facilities had declined to levels lower than those in Japan. (Data cited in *U.S. News and World Report,* December 3, 1984, p. 46).
3. For discussions of the trend toward increasingly automated manufacturing, see Office of Technology Assessment, *Computerized Manufacturing Automation: Employment, Education and the Workplace,* (Washington, D.C.: OTA-CIT, April 1984). Also see Gene Bylinsky, "The Race to the Automated Factor," *Fortune,* February 21, 1983.
4. PRODUCTION Research, "Computers in Manufacturing Use, Planned Use and Buying Influence," (mimeo), August–September 1982. Also see Ken M. Gettelman, Mark D. Albert, and Watson Nordquist (eds.), "Fundamentals of NC/CAM," in *1985 Guidebook, Modern Machine Shop,* (Cincinnati, Ohio, 1985), pp. 24–256.
5. Donald A. Hicks, *Automation Technology and Industrial Renewal: Adjustment Dynamics in the U.S. Metalworking Sector* (Washington, D.C.: American Enterprise Institute for Public Policy Research, 1986). For a discussion of the regional trends in the diffusion of a variety of automation production technologies, see J. Rees, R. Briggs, and R. Oakey, "The Adoption of New Technology in the American Machinery Industry," *Regional Studies,* Vol. 18.6, pp. 489–504.
6. B. Bluestone and B. Harrison, *The Deindustrialization of America* (New York: Basic Books, 1982), p. 224, likewise note that new plants are frequently smaller as they forsake secondary expansion in-place and expand instead to new locations far and near. See also R. Schmenner, "Some Firms Fight Ills of Bigness by Keeping Employee Units Small," *Wall Street Journal,* February 5, 1982, p. 1; and Robert G. Healy, *America's Industrial Future: An Environmental Perspective* (Washington, D.C.: The Conservation Foundation, 1982), pp. 31–32. The trend toward smaller plants has been observed for at least two decades now, as is evidenced in U.S. Department of Commerce, "Characteristics of 63 Modern Industrial Plants" (Washington, D.C.: U.S. Government Printing Office, 1966).
7. Daniel Bell, *The Coming of Post-Industrial Society* (New York: Basic Books, 1973), p. xv.
8. Michael Boretsky, *An Assessment of U.S. Competitiveness in High Technology Industries* (Washington, D.C.: International Trade Administration, U.S. Department of Commerce, February 1983), p. 34. In Boretsky's schema "technology-intensive" industries are defined as those having at least 10 percent, rather than 5 percent (the BLS criterion for "high-technology" industries), of employment being professional scientists and of R&D expenditures as a percentage of the GNP. For a related discussion,

see Charlotte Breckenridge, "High-Technology Employment: Growth in U.S. Regions, 1977–1981," Congressional Research Service, Library of Congress, HG 2401, December 16, 1983, Report No. 83-223 E, pp. 5–6.

9. The term "dematurity" is used to describe a post-Schumpeterian process that runs counter to that suggested by the notion of industrial life-cycles. See William J. Abernathy, Kim B. Clark, and Alan M. Kantrow, *Industrial Renaissance,* (New York: Basic Books, 1983). The term "neoindustrialization" is used by Frank P. Davidson with John Stuart Cox, in *Macro: A Clear Vision of How Science and Technology Will Shape Our Future,* (New York: W. Morrow, 1983).
10. Data provided by the U.S. Energy Information Administration, Department of Energy, and the Independent Petroleum Association of America. See also "Industrial Energy Consumption: 1958-81," In *1985 U.S. Industrial Outlook,* (Washington, D.C.: International Trade Administration, U.S. Department of Commerce, January 1985), pp. A-1 to A-7.
11. Data provided by the U.S. Department of Commerce. For a discussion of the ways in which the energy environment could well shift again in coming years, see Hans H. Landsberg, "A Time for Every Purpose: The Case of Energy." *Issues in Science and Technology,* Vol. 1, No. 2, Winter 1985, pp. 14–27.
12. See *Economic Review,* Federal Reserve Bank of Atlanta, "Growth Industries in the 1980s, Vol. 68, No. 4, April 1983, pp. 4–13.
13. William H. Branson, "The Myth of De-industrialization," in *Regulation: AEI Journal on Government and Society,* American Enterprise Institute for Public Policy Research, September–Octobr 1983, pp. 24 ff.
14. "America's Restructured Economy," *Business Week,* June 1, 1981, pp. 56–95. The identification and exploration of specially defined sectoral structures or development themes are increasingly common. See Marc U. Porat, *The Information Economy* (Washington, D.C.: Office of Telecommunications, 1977). For an exploration of the "new services economy," see Thomas M. Stanback, Jr., Peter J. Bearse, Thierry J. Noyelle, and Robert A. Karasek, *Services: The New Economy,* (Totowa, N.J.: Rowman & Allanheld, 1983). Finally, Peter F. Drucker, "Why America's Got So Many Jobs," *Wall Street Journal,* January 24, 1984, p. 32, has used the term "entrepreneurial" economy, which he defines as comprising new small business and high-technology subsectors.
15. For an application of the product-cycle paradigm in analyzing the restructuring of a regional economy, see Harold T. Gross, "Policy Options in Response to Industrial Life-Cycles: The Example of the Texas Gulf Coast Rice Farming Industry," unpublished Ph.D. dissertation, University of Texas at Dallas, May 1984.
16. Branson, "The Myth of Deindustrialization."
17. The labor productivity of high-technology industries in the 1970s was six times that of all U.S. business. See Pat Choate, "Jobs in the Future: The Impact of New Technology," paper presented at the Joint Conference of the American Enterprise Institute for Public Policy Research and the Konrad Adenauer Stiftung, "Technological Development, Causes, and Consequences: United States and West Germany," Bonn-Bad Godesburg, West Germany, November 26–27, 1984.
18. National Science Foundation, *Science Indicators, 1982* (Washington, D.C.: U.S. Government Printing Office, 1983), p. 21.
19. Richard W. Riche, Daniel E. Hecker, and John U. Burgan, "High Technology Today and Tomorrow: A Small Slice of the Employment Pie," *Monthly Labor Review,* Vol. 106, No. 11, November 1983, pp. 50–58. For a pessimistic prognosis of the labor effects of automation, see Gail Garfield Schwartz and William Neikirk, The Work Revolution: How High Tech Is Sweeping away Old Jobs and Industries and Creating New Ones in New Places, (New York: Rawson Associates, 1983).

20. Choate, "Jobs in the Future."
21. Bureau of Labor Statistics (BLS), *Productivity Chartbook* (Washington, D.C., 1981), Chart 22.
22. Joint Economic Committee, U.S. Congress, "Location of High Technology Firms and Regional Economic Development," (Washington, D.C.: U.S. Government Printing Office, June 1, 1982). Also see *Technology, Innovation, and Regional Economic Development* (Washington, D.C.: Office of Technology Assessment, July 1984).
23. Office of Technology Assessment, U.S. Congress, *U.S. Industrial Competitiveness: A Comparison of Steel, Electronics, and Automobiles* (Washington, D.C.: U.S. Government Printing Office, July 1981), p. 6. Also see National Academy of Engineering, *The Competitive Status of the U.S. Automobile Industry,* (Washington, D.C.: National Academy Press, 1982).
24. Robert Z. Lawrence, *Can America Compete?,* (Washington, D.C.: The Brookings Institution, 1984).
25. Bluestone and Harrison, *The Deindustrialization of America,* p. 6.
26. Ibid., p. 24.
27. N.D. Kondratiev, "The Long Waves in Economic Life," *Review,* No. 4, Spring 1979, pp. 519–562 (translation). Also see Joseph Schumpeter, *Business Cycles: A Theoretical Historical and Statistical Analysis of the Capitalist Process* (New York: McGraw-Hill, 1939), and Herbert Giersch, "The Age of Schumpeter," *American Economic Review,* May 1984, pp. 103–109.
28. Walter W. Rostow, *The World Economy: History and Prospect,* (Austin: University of Texas Press, 1978); Jay Forrester, "Innovation and Economic Change," *Futures,* No. 13, August 1981, pp. 323–331; Gerhard Mensch, *Stalemate in Technology* (Cambridge, Mass.: Ballinger, 1979). See also James W. Michaels, William Baldwin, and Lawrence Minard, "Echoes from a Siberian Prison Camp," *Forbes,* November 9, 1981, pp. 164–174.
29. Edwin Mansfield, "Long Waves and Technological Innovation," *American Economic Review,* May 1983, pp. 141–145; Nathan Rosenberg and Claudio R. Frischtak, "Long Waves and Economic Growth: A Critical Appraisal," *American Economic Review,* May 1983, pp. 146–151. For a critique of the long-wave perspective within the context of land policy, see Matthew Edel, "Land Policy, Economic Cycles, and Social Conflict," in I. Judith Innes deNeufville (id.), *The Land Use Policy Debate in the United States,* (New York: Pleaum Press, 1981, pp. 127–139.
30. Michaels et al., "'Echoes from a Siberian Prison Camp."
31. National Science Foundation, *Science Indicators, 1982.*
32. William H. Davidson, "Patterns of Factor-Saving Innovation in the Industrialized World," *European Economic Review,* Vol. 8, 1976, pp. 56, 214.
33. Nathaniel J. Mass and Peter M. Senge, "The Economic Long Wave: Implications for Industrial Recovery," *Economic Development Commentary* (Washington, D.C.: National Council for Urban Economic Development, Spring 1983), pp. 3–9.
34. Murray L. Weidenbaum and Michael J. Athey, "The 'Decline' of U.S. Manufacturing: Empirical Evidence and Policy Implications," Working Paper Number 87, Center for the Study of American Business, Washington University, St. Louis, May 1984. For a similar assessment, see Pat Choate, "Manufacturing: Meeting the Global Challenge," *Economic Development Commentary,* National Council for Urban Economic Development, Winter 1983, pp. 3–7; Marilyn Wilson, "Smokestack America: The Future Is Better Than You Think," *Dun's Business Month,* July 1983, pp. 30–35.
35. Statement by Raymond A. Hay, "Basic Industries in Trouble: Why... And Are There Solutions?" (Dallas; LTV Forum, 1983). p. 2.
36. Stanley J. Modic, "It's Time to Revitalize U.S. Industry," *Industry Week,* January 12, 1981, pp. 45–50.

37. Statement by N.E. Terleckyj, as cited in ibid.
38. Between 1972 and the commencement of the second of the back-to-back recessions in 1981, manufacturing jobs increased by 1,022,000 (5.3 percent), construction jobs by 287,000 (7.4 percent), and mining by 504,000 (80.3 percent), the latter an expression of the growing dependence on coal as a substitute for petroleum sources whose price and availability were increasingly judged to be unpredictable.
39. Hicks, *Automation Technology and Industrial Renewal.*
40. Private conversation with Harry Paxton, U.S. Steel, November 1984.
41. Raymond Vernon, "International Investment and International Trade in the Product Cycle," *Quarterly Journal of Economics,* Vol. 80, 1966, pp. 190–207. Also see Raymond Vernon, "The Product Cycle Hypothesis in a New International Environment," *Oxford Bulletin of Economics and Statistics,* 1980.
42. See especially John Rees, "Regional Industrial Shifts in the U.S. and the Internal Generation of Manufacturing in Growth Centers of the Southwest," in William C. Wheaton (ed.), *Interregional Movements and Regional Growth* (Washington, D.C.: The Urban Institute, 1979), pp. 51–73. Also see R.D. Norton and John Rees, "The Product Cycle and the Spatial Decentralization of American Manufacturing," *Regional Studies,* Vol. 13, 1979, pp. 141–151.
43. Schumpeter, *Capitalism, Socialism and Democracy,* pp. 83–84.
44. Data provided by the American Iron and Steel Institute. See also William T. Hogan, *World Steel in the 80s: A Case of Survival* (Lexington, Mass.: D.C. Heath, 1983), p. 119; and *American Metal Markets,* April 18, 1984, p. 7 (cited in Kent Jones, "Saving the Steel Industry," *Backgrounder,* (The Heritage Foundation, No. 54, May 21, 1984, p. 1).
45. Office of Technology Assessment, U.S. Congress, *U.S. Industrial Competitiveness: A Comparison of Steel, Electronics, and Automobiles,* (Washington, D.C.: U.S. Government Printing Office, July 1981), p. 75.
46. Ibid., p. 57.
47. Ibid.
48. With nearly four hundred pounds of aluminum inside—including forged aluminum-alloy pistons—and a fiberglass-reinforced plastic exterior, the Chevrolet Corvette shares the road with the new Pontiac Fiero, whose all-plastic body shrouds a steel chassis. Industry analysts predict that the mass-produced plastic automobile could well make its appearance in the next few years. See also "The Car of the Future Will Have New Skin and Bones," *Business Week,* July 29, 1985, pp. 50 ff.
49. Also see data cited in *Business Week,* June 13, 1983, p. 85.
50. Donald F. Barnett and Louis Schorsch, *Steel: Upheaval in a Basic Industry* (Cambridge, Mass.: Ballinger, 1983). See also the untitled statement by Robert W. Crandall in the LTV Forum, (Dallas, 1983), pp. 81–97. Even the mini-mill sector is not immune from the implications of overcapacity, low demand, and strong imports, however. See Mark Russell, "Steel Minimills Face Difficult Times: Sharp Competition Forces Companies to Change," *Wall Street Journal,* September 4, 1985, p. 6.
51. This section presents an analysis of data on the U.S. computer software industry (SIC 7372). (*Note:* SIC is an abbreviation for Standard Industrial Classification. The industry classification schema details more than 900 four-digit SIC codes established by the Office of Management and the Budget for use on government economic censuses and recordkeeping.) For a discussion of the development characteristics of the entire computer services industry (SIC 737), see Donald A. Hicks (with C.C. Pann), "Computer Software and Data Processing Services: Development Characteristics of A High-Technology Service Industry in Texas" (Austin, Tex.: Governor's Office of Economic Development and the Texas Computer Industry Council, July 1984).

52. Peter Hall, Ann R. Markusen, Richard Osborn, and Barbara Wachsman, "The Computer Software Sector," in Marc Weiss and Ann R. Markusen (eds.) *High Technology Industries and the Future of Employment,* 1983. The authors note that software understood as separate and distinct from hardware in the sense of the storage and processing capacity of a dedicated-use computer is only a quarter-century old. Software, as such, made its debut in the form of a primitive version of FORTRAN as recently as 1955. It was not until 1969 that the software industry as we understand it today was born. It was in that year that IBM, which had long provided programming capacity for its own computer equipment, was forced by the threat of antitrust action to spin off, or "unbundle," from its hardware production its systems and applications software production activities. This action prompted independent software vendors to develop and market their own products increasingly tailored to narrow and specific uses.
53. INPUT data cited in *Business Week,* February 27, 1984.
54. *Standard and Poor's Industry Surveys,* Vol. 2, July 1982, pp. 22–24.
55. Statement by Robert Leff, President, Softsel, April 1983.
56. The computer industry has traditionally been highly concentrated, though through the last decade the structure of the industry has broadened. By 1981, the computer industry was composed of approximately 314,000 employees in nearly a thousand (932) companies with a majority (53 percent) employing fewer than twenty people. Moreover, 44 percent of all shipments originated from the four largest companies and nearly two-thirds of all shipments originated from only four states (California, 30 percent; Massachusetts, 13 percent; New York, 11 percent; and Minnesota, 11 percent).
57. Since it is quite difficult to create a single comprehensive listing for use as a sampling frame in a survey of an industry subsector composed of usually tiny and unstable independent firms and sole proprietorships, a somewhat modified stratified sampling frame drawing on three largely independent sources was devised. One sample stratum was based on a Directory of Software Manufacturers published in a popular national publication targeted to personal computer users. A second stratum was built from lists of establishments in SIC 7372 in a dozen major metropolitan areas compiled by Contacts Influential, a major U.S. business listing company. (I appreciate the cooperation and assistance I received from the regional offices of the Contacts Influential organization located throughout the nation. Their support in kindly agreeing to permit my use of their listings in order to prepare a sampling frame for this stratum was truly invaluable.) The third stratum was a complete census of software establishments in the state of Texas made available by the Texas Industrial Commission in Austin. In all, surveys were sent to 2,085 software companies. Of these, 22.8 percent (476) were returned as undeliverable; these establishments were judged to be no longer in business—a common fate of new and small businesses that dominate this subsector. Of the 1,609 surveys that were delivered, 228 complete and usable surveys were received, which resulted in an overall survey response rate of 14.2 percent. The response rates of the respective sample strata varied considerably; in order, they were 18.2 percent, 45.2 percent, and 8.4 percent.

    The latter sampling frame based on the Texas-specific listing reflects the high business failure rate among such firms that accompanies the relatively high business formation rate overall in the Texas state economy. Independent evidence of the high degree of turnover in this industry subsector was provided by the Dallas Area Chamber of Commerce. Of 474 computer programming and software services (SIC 7372) establishments listed in the Dun's Marketing List in late 1983 as being located in the Dallas Primary Metropolitan Statistical Area, by early 1984, 174 (36.7 percent) were no longer listed with the Southwestern Bell Telephone Company. While the analysis reported in this section is based on data from the computer software and data processing services companies that were part of the Texas subsample exclusively, separate analyses of data from the other two strata reveal nearly complete comparability.

58. One-third (34.7 percent) of the companies report occupying 1,000 square feet or less; two-thirds (63.3 percent) occupy 3,000 square feet or less, and 75.5 percent occupy less than 5,000 square feet. This results in a space utilization of 362.4 square feet per employee.
59. Stanback et al., *Services: The New Economy,* pp. 82–83.

Chapter 4

1. See Alan R. Pred, *City Systems in Advanced Economies* (New York: John Wiley & Sons, 1977), for a treatment of the way in which the development of an urban system is responsive to the long-term industrial development of a national economy.
2. The use of the phrase "post-industrial urban system" may strike the reader as somewhat inconsistent given the way I have used the term "post-industrial" thus far. In previous chapters I have chosen to use the term "advanced industrial" rather than "post-industrial" to describe the development processes shaping our national economy in order to underscore the point that despite the renewed importance assigned the new services-based local and regional subeconomies, the national economy retains its essential goods production orientation. The phrase "advanced industrial development," then, highlights the fact that our industrialization has not ended, but rather continues to evolve, though in ways that are responsive to new economic and technological circumstances. The more consequential changes are those taking place *within* the goods and services sectors rather than any presumed shift in dominance between them. The term "post-industrial" has been reserved to describe the more subtle changes experienced within and among occupations, where and how work is done, and the ways in which the very notion of "work" has come to be redefined.

   In this chapter, the focus is on the spatial and locational aspects of advanced industrial development. Unlike the larger national economy, a so-called shift-to-services is much more pronouned within our nation's largest cities. In aggregate, our major industrial cities have gradually become the most clearly post-industrial places on the economic landscape. In them, manufacturing is either growing very slowly or most likely contracting. In concert with relatively rapid expansion of services in these same places, the role of manufacturing in our major industrial cities is being substantially diluted with the restructuring of their local economic bases. While early industrialization was originally responsible for expanding their roles from commercial to manufacturing centers in many cases, advanced industrial development is transforming them once again into centers for the provision and consumption of services primarily, and in selected cases, the manufacture of goods with high-technology content using the most advanced production technologies.
3. The need for reorienting economic base analysis to respond to the export capacity of advanced local-regional services economies is discussed in John M.L. Gruenstein and Sally Guerra, "Can Services Sustain a Regional Economy?" *Business Review,* Federal Reserve Bank of Philadelphia, July–August 1981, pp. 15–27. More background for the task of economic base analysis is available in Charles M. Tiebout, *The Community Economic Base Study* (New York: Committee for Economic Development, 1962). Also see Eva C. Galambos and Arthur F. Schreiber, *Economic Analysis for Local Government*

(Washington, D.C.: National League of Cities, November 1978).

4. Identifying the differences among urban areas is at least as informative as sketching out aggregate trends, but the task is not undertaken here. Major efforts at city classification and its techniques are found in Thierry J. Noyelle and Thomas M. Stanback, Jr., *The Economic Transformation of American Cities* (Totowa, N.J.: Rowman & Allanheld, 1983); Thomas M. Stanback, Jr., and Thierry J. Noyelle, *Cities in Transition: Changing Job Structure in Atlanta, Denver, Buffalo, Phoenix, Columbus, Nashville, and Charlotte* (Totowa, N.J.: Allanheld, Osmun, 1982); Brian J.L. Berry (ed.), *City Classification Handbook: Methods and Applications* (New York: Wiley, 1972); Beverley Duncan and Stanley Lieberson, *Metropolis and Region in Transition* (Beverly Hills, Calif.: Sage, 1970); Richard L. Forstall, "A New Social and Economic Grouping of Cities," *Municipal Year Book,* 1970, pp. 102–152; and Jeffrey Hadden and Edgar Borgatta, *American Cities: Their Social Characteristics,* (Chicago: Rand McNally, 1965).
5. E.A. Wood, "Industrial Development for a Community," *The Texas Planning Bulletin,* March 1938, pp. 3–4. Italics added.
6. John Long, *Population Deconcentration in the United States,* Special Demographic Analysis CDS–81–5 (Washington, D.C.: Bureau of the Census, November 1981).
7. The pace of population redistribution often picks up even in the trough of a business cycle downturn in regions most affected. Outmigration—and dampened inmigration—between April 1981 and July 1982, for example, claimed 1.8 percent of the population of the Great Lakes states of Michigan, Illinois, Indiana, Ohio, and Wisconsin, with a 2.1 percent loss in Michigan alone by July 1983.
8. See Long, *Population Deconcentration,* Table 6, p. 38.
9. Data from research by Kristin A. Hansen and Celia G. Boertlein, Bureau of the Census (cited in "Housing Costs Cut Mobility in U.S., *Dallas Morning News,* March 5, 1984, p. 4A).
10. See John L. Goodman, Jr., "Immobile Americans?" *Policy and Research Report,* The Urban Institute, Winter 1983, pp. 11–12.
11. See G.L. Houseman, *The Right of Mobility* (Port Washington, N.Y.: Kennikat Press, 1979).
12. Arguments for greater attention to issues surrounding labor mobility are offered in the U.S. Department of Housing and Urban Development, *The President's National Urban Policy Report, 1980* (Washington, D.C., 1980), pp. 13-9 to 13-10. However, these arguments in this official urban policy document are effectively neutralized by much stronger arguments in this report in favor of traditional local economic development efforts. Calls for a more explicit reliance on a labor mobility strategy are presented in Georges Vernez, Roger J. Vaughan, and Robert K. Yin, *Federal Activities in Urban Economic Development,* Report R-2372-EDA, (Santa Monica, Calif.: Rand Corporation, April 1979), pp. 139–141; R.D. Norton, *City Life-Cycles and American Urban Policy* (New York: Academic Press, 1979), pp. 26–28, 169–172; and the President's Commission for a National Agenda for the Eighties, *Urban America in the Eighties: Perspectives and Prospects* (Washington, D.C., 1981), pp. 57–62, 97–109.
13. Gordon Clark, *Interregional Migration: National Policy and Social Justice,* (Totowa, N.J.: Rowman & Allanheld, 1983).
14. According to official government data, this regional redistribution of manufacturing employment was accompanied by a regional convergence in income as a share of median income. Between 1940 and 1980, the Northeast and the South experienced the greatest relative shifts. Income declined from 131 percent to 106 percent of the U.S. median in the Northeast while increasing in the South from 66 percent to 91 percent. The West likewise experienced a less marked shift, with income declining from 115 percent to 108 percent, while incomes in the Midwest stayed relatively stable over this period, dropping only from 102 percent to 101 percent. See *Regional*

Growth: Historic Perspective (Washington, D.C.: Advisory Commission on Intergovernmental Relations, June 1980), pp. 9–12.

15. For a discussion of the increasing administrative centralization and industrial concentration amid the general deconcentration of production activities, see Barry Bluestone and Bennett Harrison, *The Deindustrialization of America: Plant Closings, Community Abandonment and the Dismantling of Basic Industry* (New York: Basic Books, 1982), pp. 118–129. Also see Robert Cohen, "The Internationalization of Capital and U.S. Cities," Ph.D. dissertation, New School for Social Research, 1979.
16. Brian J.L. Berry, *Growth Centers in the American Urban System,* 2 volumes (Cambridge, Mass.: Ballinger, 1973). Also see Alan R. Pred, "The Interurban Transmission of Growth in Advanced Economies: Empirical Findings versus Regional Planning Assumptions," *Regional Studies,* Vol. 10, pp. 151–171.
17. A.T. Thwaites, "Technological Change, Mobile Plants and Regional Development," *Regional Studies,* Vol. 12, pp. 445–461.
18. Niles Hansen, "The New International Division of Labor and Manufacturing Decentralization in the United States," *The Review of Regional Studies,* Vol. 9, No. 1, Spring 1979, pp. 1–11.
19. John Rees, "Regional Shifts in the U.S. and the Internal Generation of Manufacturing in Growth Centers of the Southwest," in W.C. Wheaton (ed.), *Interregional Movements and Regional Growth* (Washington, D.C.: The Urban Institute, 1979), pp. 51–73.
20. Statement by Calvin L. Beale at the Agricultural Outlook Conference, December 1982 (cited in Rochelle L. Stanfield, "Rapid Economic Growth of Rural Areas Brings Drawbacks along with Blessings," *National Journal,* September 24, 1983, pp. 1932–1937).
21. An overview of the countervailing costs and benefits of economic development efforts tied to luring branches of large multilocational firms versus the incubation and expansion of local small business is offered in Catherine Armington and Marjorie Odle, "Small Business—How Many Jobs?" *The Brookings Review,* Winter 1982, pp. 14–17.
22. The shift in the scale of urban economies from central-local to metropolitan-regional involves the assumption that the outer jurisdictional boundaries of the counties containing a central city and its suburbs are essentially the same as the "urban economy." In reality, however, there is undoubtedly considerable slippage between such conceptual and administrative boundaries. For a similar distinction at the next higher scale that asserts the primacy of the localized economies of "cities" and "city-regions" over the national economy, see Jane Jacobs, *Cities and the Wealth of Nations: Principles of Economic Life* (New York: Random House, 1984).
23. Edwin Mills, *Studies in the Structure of the Urban Economy* (Baltimore: The Johns Hopkins University Press, 1972).
24. Long, *Population Deconcentration,* Table 1, p. 9.
25. See Tables XII.1–XII.3 in Donald A. Hicks, "Urban and Economic Adjustment to the Postindustrial Era," in Donald A. Hicks and Norman J. Glickman (eds.), *Transition to the 21st Century: Prospects and Policies for Economic and Urban-Regional Transformation,* (Greenwich, Conn.: JAI Press, 1983), pp. 345–370. Ironically, the net contraction of central city populations around the country began to be evident just as the suburbanization process began to slow beginning in the 1950s, with the 1970s witnessing an even greater decline in the rate of suburbanization. On this point, see Katherine L. Bradbury, Anthony Downs, and Kenneth A. Small, *Urban Decline and the Future of American Cities* (Washington, D.C.: The Brookings Institution, 1982), p. 86, and Long, *Population Deconcentration,* p. 70.
26. While the role of transport technologies, such as the automobile and trolley systems, is important in understanding the origins of the suburbanization process, technology

does not function in a vacuum. The technologies that permitted people to express their residential preferences simultaneously enabled space and distance to become features of social stratification in the United States. Race-ethnic, social class, and later life-style enclaves have been the result. For a discussion of the way transport technology offered a spatial feature to U.S. class conflict, see Matthew Edel, "Land Policy, Economic Cycles and Social Conflict," in Judith Innes deNeufville (ed.), *The Land Use Policy Debate in the United States* (New York: Plenum Press, 1981), pp. 127–139.

27. Charlotte Fremon, "The Occupational Patterns in Urban Employment Change: 1965–1967," Urban Institute working paper. Washington, D.C., January 1970; Alan R. Pred, "The Location of High Value-Added Manufacturing," *Economic Geography* Vol. 41, No. 2, April 1965, pp. 108 ff. (cited in Stanback, et al., *Services: The New Economy,* p. 118). Moreover, in what was the most exhaustive examination of patterns of industrial development of its time, the Civic Development Committee of the National Electric Light Association invited the Metropolitan Life Insurance Company to study the relocation behavior of industry in North America during 1926–1927. Information from 2,084 communities representing approximately three-quarters of the United States and two-thirds of the Canadian total urban population was gathered and analyzed. Even then the greater importance of business establishments and retention over plant recruitment from elsewhere was evident. It was discovered that during 1926–1927 these communities gained a total of 10,000 new plants and 371,000 new jobs while simultaneously losing 5,908 plants. While relocations accounted for only 9.4 percent of the new plants, 82 percent of the plant losses were due to business failure and only 18 percent were due to relocation. Large cities of more than 50,000 accounted for fully 83 percent of plant losses and 70 percent of employment losses, while smaller cities generally gained outsized shares of these mobile plants. See James L. Madden, "Industry Grows but Seldom Moves," *Nation's Business,* August, 1929, pp. 45 ff.
28. See James Tobin, "Suburbanization and the Development of Motor Transportation," in Barry Schwartz (ed.), *The Changing Face of Suburbs* (Chicago: University of Chicago Press, 1976), pp. 95–111.
29. Edwin S. Mills, *Studies in the Structure of the Urban Economy* (Baltimore: The Johns Hopkins University Press, 1972), p. 47. Roger J. Vaughan, *The Urban Impacts of Federal Policies,* Vol. 2: Economic Development Report, R-2028-KF/RC (Santa Monica, Calif.: Rand Corporation, July 1977); Roger J. Vaughan and Mary E. Vogel, *The Urban Impacts of Federal Policies,* Vol. 4: Population and Residential Location, Report R-2205-KF/HEW (Santa Monica, Calif.: Rand Corporation, March 1979).
30. Gerald A. Carlino, *Economies of Scale in Manufacturing Location: Theory and Measurement* (Leiden: Martinus Nijhoff, 1978); Gerald A. Carlino, "Declining City Productivity and the Growth of Rural Regions," *Journal of Urban Economics,* Vol. 18, No. 1, July 1985, pp. 11–27.
31. Calculations from data reported in *American Federalism in the 1980s: Changes and Consequences* (Cambridge, Mass.: Roundtable on Governments, Lincoln Institute of Land Policy, August 1981), Appendix D, pp. 75–76; for the 1910–1963 period, see Edwin Mills, "Urban Density Functions," *Urban Studies,* Vol. 7, No. 1, February 1970, pp. 5–20.
32. Carlino, "Declining City Productivity."
33. Norton, *City Life Cycles,* pp. 6–10; see also Barry Schwartz (ed.), *The Changing Face of Suburbs* (Chicago: University of Chicago Press, 1976).
34. Arthur P. Solomon, "The Emerging Metropolis," in Arthur P. Solomon (ed.), *The Prospective City,* 1980, pp. 3–25. See also Georges Vernez, Roger J. Vaughan, and Robert K. Yin, *Federal Activities in Urban Economic Development,* Report R-2372-EDA (Santa Monica, Calif.: Rand Corporation, April 1979), p. v.

35. Long, *Population Deconcentration,* Table 14, p. 65.
36. Bureau of the Census, Current Population Report, Series P-20, No. 331, "Geographical Mobility: March 1975–March 1978" (1978): Table 1; Calvin Beale, "The Recent Shift of the United States Population to Nonmetropolitan Areas, 1970–1975," *International Regional Science Review,* Vol. 2, 1977, pp. 113–122; also see: "The Population Turnaround in Rural-Small Town America," in W.P. Browne and D.F. Hadwiger (eds.), *Rural Policy Problems: Changing Dimensions* (Lexington, Mass.: Lexington Books, 1982); Kevin McCarthy and Peter A. Morrison, "The Changing Demographic and Economic Structure of Nonmetropolitan Areas in the United States," *International Regional Science Review,* Vol. 3, 1977, pp. 123–142.
37. Calculations based on data cited in Stanfield "Rapid Economic Growth of Rural Areas," p. 1933.
38. Long, *Population Deconcentration,* pp. 5, 87.
39. Daniel R. Vining, Jr., Robert L. Pallone, Han Lin Li, and Chung Hsin Yang, "Population Dispersal from Core Regions: A Description and Tentative Explanation of the Patterns in 21 Countries," in Hicks and Glickman, *Transition to the 21st Century,* pp. 81–111; Daniel R. Vining, Jr., and Thomas Kontuly, "Population Dispersal for Major Metropolitan Regions: An International Comparison," *International Regional Science Review,* Vol. 3, No. 1, 1978, pp. 49–73; Daniel R. Vining, Jr., "Migration Between the Core and the Periphery," *Scientific American,* Vol. 247, No. 6, December 1982, pp. 45–53.
40. William Alonso, "The Population Factor and Urban Structure," in Solomon, *The Prospective City,* pp. 34–35; Peter A. Morrison, Roger J. Vaughan, Georges Vernez, and Barbara R. Williams, *Recent Contributions to the Urban Policy Debate,* Report R-2394-RC (Santa Monica, Calif.: Rand Corporation, March 1979), pp. 6–16.
41. Carlino, "Declining City Productivity."
42. Martin Holdrich, "Prospects for Metropolitan Growth," *American Demographics,* April 1984, pp. 33–37; John Long's calculations yield a decline from 74.2 percent to 73.3 percent based on data before the addition of 36 new SMSAs on July 1, 1981.
43. Peter A. Morrison and Allan Abrahamse, "Is Population Decentralization Lengthening Commuting Distances?" Report N-1934-NICHD (Santa Monica, Calif.: Rand Corporation, December 1982).
44. Gladys K. Bowles and Calvin L. Beale, "Commuting Patterns of Nonmetropolitan Household Head, 1975," (Washington, D.C.: Economics, Statistics, and Cooperative Service, U.S. Department of Agriculture; Athens: Institute for Behavioral Research, University of Georgia, 1980).
45. Data cited in Stanfield, "Rapid Economic Growth of Rural Areas," p. 1933.
46. The economies of selected nonmetropolitan counties have not only diversified, but they have also succeeded in capturing an outsized proportion of the high-technology employment created in the past decade and a half. See Ted K. Bradshaw and Edward J. Blakely, *Rural Policy Problems: Changing Dimensions* (Lexington, Mass.: Lexington Books, 1982).
47. Gerald Carlino, "New Employment Growth Trends: The U.S. and the Third District," in *Business Review,* Federal Reserve Bank of Philadelphia, September–October 1983, p. 13.
48. Holdrich, "Prospects for Metropolitan Growth," pp. 36–37.
49. Robert M. Ady, "High-Technology Plants: Different Criteria for the Best Location," *Economic Development Commentary,* Council of Urban Economic Development, Winter 1983, pp. 8–10; See also Manny Ellenis, "Six Major Trends Affecting Site Selection Decisions to the Year 2000," *Dun's Business Month,* November 1983, pp. 116–130.
50. Bahar B. Norris, "Nonmetropolitan Manufacturing Growth in the Post-World War II United States: A Test of Industrial Location Theory," unpublished Ph.D. disserta-

tion, University of Texas at Dallas, August 1984.

51. Entropy indices for U.S. high-technology industries illustrate that despite the notable degree of geographical concentration both within and among industrial groupings, between 1972 and 1977 there was a widespread tendency for both high-technology plants and employment to disperse in patterns detectable at the county level. See Amy K. Glasmeier, Peter G. Hall, and Ann R. Markusen, "Recent Evidence on High-Technology Industries' Spatial Tendencies: A Preliminary Investigation," Appendix C in Office of Technology Assessment, *Technology, Innovation and Regional Economic Development* (Washington, D.C., 1984), pp. 145–167.
52. This section heading is borrowed from a phrase used by Wilbur R. Thompson in hearings before the Subcommittee on the City of the Committee on Banking, Finance and Urban Affairs, U.S. House of Representatives, Ninety-Sixth Congress, Second Session, held on September 16–17, 1980, *Urban Revitalization and Industrial Policy,* p. 175.
53. August Losch, *Economics of Location* (translated from the 2nd rev. ed. by William H. Woglom) (New Haven: Yale University Press, 1954), pp. 326–327. Italics added.
54. Jean Gottman, "Urban Centrality and the Interweaving of Quaternary Activities," *Ekistiks,* Vol. 29, 1970, pp. 322–331; see also Chapter 2, "The Economy and Cities," in Royce Hansen, *Rethinking Urban Policy: Urban Development in An Advanced Economy,* (Washington, D.C.: National Academy Press, 1983), pp. 11–37; *Business Week,* "Service Industries Prop up the Old Factory Cities," July 5, 1982, pp. 100 ff.
55. Multipliers tied to service industries typically are generally the highest of all industries in metropolitan area economies. See Anthony H. Pascal and Aaron S. Gurwitz, *Systematic Planning for Local Economic Development,* Report R-2932-HUD (Santa Monica, Calif.: Rand Corporation, forthcoming). The special adjustment problems of regional economies tied to declining industries long associated with such high multipliers for which service industry growth cannot readily compensate are revealed in a study of the evolution of the Cleveland economy led by the decline of durable goods manufacturing. (Aaron S. Gurwitz and G. Thomas Kingsley, *The Cleveland Metropolitan Economy* [Santa Monica, Calif.: The Rand Corporation, March 1982]). However, in a study of the prospects for the diversification of the Houston economy on the heels of the decline in the energy industry, the income and employment multipliers for the service jobs which are expanding indicate that they are likely not to compensate for the energy-manufacturing jobs that are being lost. Bernard L. Weinstein, Donald R. Hoyte, N.A. Adamson, Harold T. Gross, and John Rees, "Structural Change in the Oil Industry and Its Impact on the Gulf Coast Economy," John Gray Institute, Lamar University, Beaumont, Texas, April 1985.
56. George Peterson, "Effects of Economic Cycles," *Policy and Research Report,* The Urban Institute, Vol. 13, No. 2, Summer 1983, pp. 1–3.
57. John D. Kasarda, "The Implications of Contemporary Redistribution Trends for National Urban Policy," in Hicks and Glickman, *Transition to the 21st Century,* pp. 17–47. See also James M. Simmie, "Beyond the Industrial City," *Journal of the American Planning Association,* Vol. 49, Winter 1981, pp. 58–59.
58. Robert O'Connell, Andrew Silton, and Roger Vaughan, "Telecommunications: The New Urban Infrastructure," *Economic Development Commentary,* Council on Urban Economic Development, Spring 1982, pp. 3–8; George Beckerman, "The Teleport: Telecommunication and Urban Growth," *Commentary,* Council on Urban Economic Development, Spring 1983, pp. 10–15.
59. Robert Guenther, "Office Construction Boom Persists Despite Record Rate for Vacancies," *Wall Street Journal,* November 15, 1984, p. 37. For further background on space usage trends across industries and over time, see *Estimating Land and Floor Area Implicit in Employment Projections* (Philadelphia: IDE Associates, July 1972), and

Thomas Muller, *Economic Impacts of Land Development: Employment, Housing, and Property Values* (Washington, D.C.: The Urban Institute, 1976), p. 25.

60. This section is based on an analysis reported in Donald A. Hicks and Joel H. James, "The City as a 'Textured' Business Climate: Establishment and Employment Contraction and Expansion in the City of Dallas Between 1977 and 1982," Center for Policy Studies, University of Texas at Dallas, November 1983.
61. Data accessibility influenced the choice of Dallas to illustrate in greater detail this recycling and transformation process. As a Southern city, Dallas experienced explosive metropolitan growth relatively recently. Between 1969 and 1979, the South alone continued to report a positive relationship between total employment growth and place size. And the gap in manufacturing growth rates between metropolitan and nonmetropolitan areas was considerably narrower in the South than in any other region. Consequently, in contrast to large and older industrial cities in other regions, the restructuring of the Dallas economy proceeded in the face of modest local and dramatic regional growth. In this sense, cities of Dallas's vintage have grown into their advanced urban forms rather than contracting or retreating into them.
62. During 1970–1980 the population density of Dallas, the nation's seventh largest city, declined from 3,179 to 2,715 persons per square mile. Of the nation's ten largest cities, population densities increased in only four (Los Angeles, Houston, San Diego, and Phoenix) while it declined at faster rates than Dallas in two (Philadelphia and Detroit) and at the same rate or slower in three (New York, Chicago, and Baltimore).
63. Recent race-residence trends in Dallas illustrate that the recycling of industrial-era cities need not necessarily lead to increased racial polarization or segregation. As measured by the index of dissimilarity (ID), Dallas was one of the least residentially segregated cities by race in 1940 with an ID of 80.2, which indicates that approximately 80.2 percent of the nonwhites that would have to relocate to be equally distributed with whites within some territorial unit such as a city block or census tract. (Charleston, South Carolina, recalled by one student of segregation as the birthplace of the Confederacy, was the *least* segregated of the nation's largest 109 cities: ID = 60.1. This indicates that in the South, residential segregation did not begin to accompany other forms of social segregation until the scale of Southern cities began to expand in the 1950s.) During 1940–1960, residential segregation increased dramatically in Dallas, with ID values rising from 88.4 in 1950 to 94.6 in 1960. However, residential segregation by race began to ebb slowly during the 1960s and then rapidly during the 1970s. By 1970, Dallas's ID had declined to 92.7, followed by a far steeper decline to 76.7 by 1980. This pattern is superimposed on a long-term decline in the proportion of black Dallas residents from 21.2 percent in 1900, to 15.1 percent in 1920, to 13.1 percent in 1950. Thereafter, that proportion has increased dramatically to 24.9 percent in 1970 and 29.4 percent in 1980. While residential segregation declined in the twenty-eight largest U.S. cities during the 1970s—with average IDs dropping from 87 to 81 measured at the block level—there is no evidence that this trend was a function of changes in either absolute size of the black population or its share of city population. See Karl Taeuber, *Racial Residential Segregation, 1980* (Washington, D.C.: Citizens Committee for Civil Rights, March 1983).
64. Over the past decade, there has been a clear pattern of declining vacancies in the office market of the city of Dallas until 1978–1979. Since then, and as a result of an unprecedented boom in office construction particularly in the Dallas Central Business District, overall absorption (millions of square feet increase in occupancy) has failed to keep pace with overall vacancy (millions of square feet unleased). The size of this huge gap indicating the extent to which the city's office market has been overbuilt is considerably understated by the tendency for developers to lure tenants from one site to another, thus assuring that while the latest project may be substantially leased

even before completion, nearby buildings that are only slightly older register the effects of such corporate "musical chairs" by high vacancy rates.

65. Donald A. Hicks and Joel H. James, "The City of Dallas as a 'Textured' Business Climate: Establishment and Employment Contraction and Expansion in the City of Dallas Between 1977 and 1982," Center for Policy Studies, University of Texas at Dallas, November 1983.
66. Anticipated skill shortages in the Dallas area through 1988 include plumber trainees, accountants, electronics engineers and programmers, engineer scientific programmers, skilled craftsmen, systems analysts, electrical engineers, electronic technicians, and mechanical engineers. Human Resources Committee, Dallas Chamber of Commerce, *Forecast of Dallas Area Critical Labor Needs: 1984 through 1988* (Dallas, December 1984). Nonetheless, while there has been considerable concern over the skill mismatches between the specialized skill demands of new central city economies and the often weak skill pools of central city residents resulting in high structural unemployment in central cities, the evidence of such mismatch is far from compelling. See Charlotte Fremon, "The Occupational Patterns in Urban Employment Change: 1965–67," Urban Institute working paper (Washington, D.C.: January 1970), p. 51, and cited in Arthur P. Solomon, "The Emerging Metropolis," in A.P. Solomon (ed.), *The Prospective City* (Cambridge, Mass.: The MIT Press, 1980), p. 12.
67. Bertrand Renaud, "Structural Changes in Advanced Economies and Their Impact on Cities in the 1980s" (Washington, D.C.: Urban Development Department, The World Bank, March 1982), p. 18.
68. For a compendium of theories of urban decline, see Katherine L. Bradbury, Anthony Downs, and Kenneth A. Small, "Explaining Urban Decline: Theories and a Proposal for Testing" (Washington, D.C.: The Brookings Institution, March 1979). While my discussion does not dwell on the possibility that urban "decline" is an inevitable accompaniment of advanced urban development, a reasonable case can be made for this view. Bradbury et al. offer extensive evidence of the fact that U.S. cities' growth and decline are governed by a wide variety of factors. Consequently, understanding that the causes of suburbanization (and urban decline) are so many and diverse actually strengthens, rather than weakens, the conclusion that both are inevitable. When so many different factors are capable of registering the same general effects—including declining population, income and even employment—the conclusion that urban decline is inevitable does not appear to be unwarranted. See Bradbury et al., *Urban Decline* pp. 139, 164.
69. This perspective is amply illustrated by Peter Libassi and Victor Hausner, "Revitalizing Central City Investment" (Columbus, Ohio: Academy for Contemporary Problems, June 1977).
70. Development Choices for the Eighties, *Background Information Summary* (Washington, D.C.: Urban Land Institute, March 24, 1980).
71. See Dennis E. Gale, *Neighborhood Revitalization and the Postindustrial City* (Lexington, Mass.: Lexington Books, 1984); Michael H. Schill and Richard P. Nathan, *Revitalizing America's Cities: Neighborhood Reinvestment and Displacement* (Albany: State University of New York Press, 1983).
72. Bradbury et al., *Urban Decline,* pp. 29–31.
73. Ibid., Tables 3.2–3.4, pp. 36–41.
74. Martin Carnoy and Derek Shearer, *Economic Democracy* (White Plains, N.Y.: M.E. Sharpe, 1980). Quote from annotation by Philip Shapira, "Economic Development Analysis and Planning in Advanced Industrialized Economies: A Bibliography" (Chicago: Council of Planning Librarians, May 1983), p. viii.
75. See especially Committee on National Urban Policy, *The Evolution of National Urban*

Policy 1970–1980: Lessons from the Past (Washington, D.C.: National Academy Press, 1982).

76. Federal obligations to both places and political jurisdictions exceeded those targeted to people directly during the 1970s, despite the doubling of aid in the latter category during 1976–1978. Georges Vernez, "Federal Response to Urban Economic Development Problems," in Peter A. Morrison, Roger J. Vaughan, Georges Vernez, and Barbara R. Williams (eds.), *Recent Contributions to the Urban Policy Debate,* Report R–2394–RC (Santa Monica, Calif.: Rand Corporation, March 1979), Table IV.1, p. 27. For an analysis of federal urban economic development activities by type of recipient (i.e., place, jurisdiction, people, or market failure), see Georges Vernez, Roger J. Vaughan, and Robert K. Yin, *Federal Activities in Urban Economic Development* Report R–2372–EDA (Santa Monica, Calif.: Rand Corporation, April 1979). An argument for spatially sensitive policies to restrict capital mobility and thereby control the pace at which a new services–based urban economy emerges is offered by Thierry J. Noyelle, "The Service Era: Focusing Public Policy on People and Places," *Economic Development Commentary,* Summer 1984, pp. 12–17.
77. For an illustration of this presumed "contradiction between capital and community," see quote by John Friedmann cited in Bluestone and Harrison, *The Deindustrialization of America,* p. 20.
78. The classic analysis of the ways in which the crucial sociological dimensions—how people perceive space and ultimately use it—of "community" no longer overlap more traditional and simplistic geospatial conceptions of the term are offered by Roland L. Warren, *The Community in America* (Chicago: Rand McNally, 1978). For a discussion of the difficulties associated with distinguishing territorial and nonterritorial conceptions of "community," see Jack P. Gibbs, "Types of Urban Units," in Kent P. Schwirian, *Contemporary Topics in Urban Sociology* (Morristown, N.J.: General Learning Press, 1977), pp. 255–259.
79. As Bradbury et al. (in *Urban Decline,* Table 5-3, p. 98) have illustrated, among the wide variety of hypotheses that have been tested to account for urban decline, those factors that have *pulled* households and firms out of central cities appear to be more potent than factors that might *push* them out. The role of the positive attractiveness of suburban areas to households and firms seeking greater economic opportunity emerges as far more important than the role of negative characteristics of central cities from which both might wish to escape.

Chapter 5

1. B.A. Evans, "More Smokestacks for the Southwest," *Dallas,* November 1925.
2. See Amos H. Hawley, *Urban Society: An Ecological Approach* (New York: Ronald Press, 1971).
3. Christopher Freeman, Jr., John Clark, and Luc Soete, *Unemployment and Technological Innovation* (Westport, Conn.: Greenwood Press, 1982).
4. *Technology, Innovation, and Regional Economic Development,* OTA–STI–238 (Washington, D.C.: U.S. Congress, Office of Technology Assessment, July 1984), p. 5. The reader interested in the development of the high-technology sector should see Lynn E. Browne, "High Technology and Business Services," *New England Economic Review,* Federal Reserve Bank of Boston, July/August 1983; Lynn E. Browne, "Can High Tech Save the Great Lakes States?" *New England Economic Review,* Federal Reserve Bank of Boston, November–December 1983; Clark E. Chastain, "So You

Want to Attract High-Tech Firms," Part 1, *Colorado Business Review,* October–November 1983; John S. Hekman, "The Future of High Technology Industry in New England," *New England Economic Review,* January–February 1980; Federal Reserve Bank of Boston, "High Technology and Regional Economic Development," April 1984; Raymond Radosevich and Lee Zink, "Efficacy of the High Technology Development Strategy in the Southwest," *The Southwestern Review,* Spring 1983, pp. 19–32; William H. Miller, "The Phony War between High Tech and Low Tech," *Industry Week,* October 3, 1983; "The High Technology Revolution's Impact on Economic Development," *Nation's Business,* November 1983; Paul Oosterhuis, "High Technology Industries and Tax Policy in the 1980s," *National Journal,* January 1, 1983, pp. 46–49; Cheryl Russell, "High-Tech Hoopla," *American Demographics,* June 1983; John F. Schnell, "An Introduction to High Tech," *Illinois Business Review,* Vol. 40, No. 3, June 1983; Courtenay Slater, "Getting High on Technology," *American Demographics,* November 1983. Stuart Zipper, "How Does High Tech Love Texas?" *Texas Business,* August 1983; Charlotte Breckenridge, "High Technology Employment: Growth in U.S. Regions, 1977–1981," (Washington, D.C.: Congressional Research Service, Library of Congress, December 16, 1983); California Commission on Industrial Innovation, Final Report, *Winning Technologies: A New Industrial Strategy for California and the Nation* (Sacramento, September 1982); Office of the Comptroller, State of Texas, "Texas Maintains Its Share of High Technology," *Fiscal Notes,* December 1983; Vimla Aggarwal and Sumer C. Aggarwal, "High Technology: Economic Hope or Hoax," *Business Horizons,* January–February, 1984; *High Technology: Public Policies for the 1980s,* A National Journal Issues Book (Washington, D.C., 1983).

5. Joint Economic Committee, "Location of High Technology Firms and Regional Economic Development," Washington, D.C.: U.S. Congress, June 1, 1982, p. 2.
6. Richard W. Riche, Daniel E. Hecker, and John U. Burgan, "High Technology Today and Tomorrow: A Small Slice of the Employment Pie," *Monthly Labor Review,* November 1983, pp. 50–59.
7. Moreover, Glasmeier et al. have shown that high-technolgy development has had its greatest impacts in business establishment and employment growth on smaller localities that seldom attract public attention. These include Santa Cruz (California), Oxnard (California), Lakeland (Florida), Lubbock (Texas), and Savannah (Georgia) among the top gainers during 1972–1977. See Amy K. Glasmeier, Peter G. Hall, and Ann R. Markusen, "Recent Evidence on High-Technology Industries' Spatial Tendencies: A Preliminary Investigation," in *Technology, Innovation, and Regional Economic Development,* pp. 154–155. For an overview of the literature on technology and regional development, see Edward J. Malecki, "Technology and Regional Development: A Survey," *International Regional Science Review,* Vol. 8, No. 2, October 1983, pp. 89–125.
8. While the distribution and growth of the high-technology sector does not appear to be influenced by a host of traditional location factors, little progress has been made on what the new location factors might be. For a discussion of this, see Glasmeier et al., "Recent Evidence on High-Technology Industries' Spatial Tendencies," pp. 160, 165.
9. John Rees, "Regional Industrial Shifts in the U.S. and the Internal Generation of Manufacturing Growth Centers in the Southwest," in W. Wheaton (ed.), *Interregional Movements and Regional Growth,* (Washington, D.C.: Urban Institute, 1979), and R.D. Norton and John Rees, "The Product Cycle and the Spatial Decentralization of American Manufacturing," *Regional Studies,* Vol. 13, 1979, pp. 141–151.
10. Brian J.L. Berry, "Hierarchical Diffusion: The Basis of Development Filtering and Spread in a System of Growth centers," in N. Hansen (ed.), *Growth Centers in Regional Economic Development* (New York: Free Press, 1972).

11. See Donald A. Hicks, "Advanced Industrial Development in Texas," in Anthony Champagne, Edward Harpham, and Glenn Robinson, *Texas Politics and Policy* (forthcoming), and M. Ray Perryman, "Texas Economic Forecast: The Perryman Report 1984" (Dallas, 1984). For an analysis of the Texas high-technology sector, see Susan Goodman and Victor L. Arnold, "High Technology in Texas," *Texas Business Review,* November–December 1983, pp. 290–295.
12. *Texas: Past and Future: A Survey* (Austin: Texas 2000 Commission, Office of the Governor, 1981), p. 24; and *Texas Trends* (Austin: Texas 2000 Commission, Office of the Governor, August 1980).
13. Office of Technology Assessment, *Technology, Innovation and Regional Economic Development,* Chapter 2. See also Bob McKay, "Location of High Technology Industry in Texas: A Critical Assessment" (Austin: Texas Economic Development Commission, June 1984).
14. A historical succession of the classic contributions would include A. Weber, *Theory: The Location of Industries,* trans. C.J. Friedrich (Chicago: University of Chicago Press, 1929); E.M. Hoover, *Location Theory and the Shoe and Leather Industries* (Cambridge, Mass.: Harvard University Press, 1937); W. Isard, *Location and Space Economy* (Cambridge, Mass.: MIT Press, 1956); E.S. Mills, "An Aggregative Model of Resource Allocation in a Metropolitan Area," *American Economic Review,* Papers LVII, 1967, pp. 197–210; H.W. Richardson, *Regional Growth Theory* (New York: Halsted Press, 1973).
15. See John Rees, "Regional Economic Decentralization Processes in the United States and Their Policy Implications," in D.A. Hicks and N.J. Glickman (eds.), *Transition to the 21st Century: Prospects and Policies for Economic and Urban-Regional Transformation* (Greenwich, Conn.: JAI Press, 1983), pp. 241–278; also see Bahar B. Norris, "Nonmetropolitan Manufacturing Growth in the Post–World War II United States: A Test of Industrial Location Theory," unpublished doctoral dissertation, University of Texas at Dallas, August 1984.
16. Gerald A. Carlino, "Declining City Productivity and the Growth of Rural Regions: A Test of Alternative Explanations," Working Paper No. 83-8, Federal Reserve Bank of Philadelphia, 1983.
17. Peter Hall et al. suggest that the software industry dates only to 1969. See Peter Hall, Ann R. Markusen, Richard Osborn, and Barbara Wachsman, "The Computer Software Sector," in Marc Weiss and Ann R. Markusen (eds.), *High Technology Industries and the Future of Employment (1983).* This section is based on Donald A. Hicks (with Czi C. Pann), "Computer Software and Data Processing Services: Development Characteristics of a High-Technology Service Industry in Texas," Center for Policy Studies, University of Texas at Dallas, July 1984.
18. Marc U. Porat, *The Information Economy,* (Washington, D.C.: Office of Telecommunications, 1981).
19. "Office Equipment Systems and Services, Current Analysis, *Standard and Poor's Industry Surveys* Vol. 150, No. 10, Sec. 1, March 11, 1982, pp. 1–43.
20. Riche et al., "High Technology Today and Tomorrow."
21. Data reported in *Software: A New Industry,* (Paris: Committee for Information, Computer, and Communications Policy, Organization for Economic Cooperation and Development, 1984), p. 191.
22. Data provided by the Texas Comptroller of Public Accounts and Texas Employment Commission, as reported in *Fiscal Notes,* Office of the Comptroller, Austin, Texas, December 1983.
23. The data reported in this section are from a nationwide survey of 2,085 computer software (SIC 7372) firms I conducted in 1983. The analysis reported here is from the Texas subset of that larger national data base.

24. Thomas M. Stanback, Jr., Peter J. Bearse, Thierry J. Noyelle, and Robert A. Karasek, *Services: The New Economy* (Totowa, N.J.: Allanheld & Rowman, 1983), p. 110.
25. See Catherine Armington, Candee Harris, and Marjorie Odle, "Formation and Growth in High-Technology Firms: A Regional Assessment," Appendix B in *Technology, Innovation, and Regional Economic Development,* p. 142.
26. An estimated 20.0 percent of the firms in the overall high-technology sector of the D-FW regional economy similarly report an international export orientation.
27. For a discussion of the dynamics of business location decisionmaking from a corporate perspective, see H.A. Stafford, *Principles of Industrial Facility Location* (Atlanta: Conway Publications, 1980).
28. The importance of proximity to an airport likewise emerged in the work of Glasmeier, et al. "Recent Evidence of High Technologies' Spatial Tendencies," pp. 163–165. While Glasmeier et al. discovered that high-technology industries are not significantly attracted to locations that can offer their employees relatively low housing prices or access to skill upgrading through local major universities, such findings appear to be consistent with those applicable to a wider range of steering factors discussed here.
29. "[I]t is not surprising that the technical skills variable remains the most important distinguishing factor in explaining high-technology development." See Armington et al., "Formation and Growth," p. 134.
30. Data on the subsample of Texas computer software firms indicate a space/employee ratio of 371.6 square feet per employee. The ratios for a succession of birth cohorts of Texas software companies clearly reveal the limited space requirements of relatively new and small firms within the sector. These ratios are:

Birth Cohort	*Space/Employee Ratio*
1950s	442.9 sq. ft./employee
1960s	329.6 sq. ft./employee
1970–1974	399.4 sq. ft./employee
1975–1979	378.6 sq. ft./employee
1980–1982	276.4 sq. ft./employee

31. While access to an appropriately skilled labor force has consistently emerged as a significant locational factor governing high-technology industrial growth, the relationship between high-technology plants and major universities has been shown to be only moderate at best. See Glasmeier, Hall, and Markusen, "Recent Evidence on High-Technology Industries' Spatial Tendencies: A Preliminary Investigation," p. 161; and Joint Economic Committee, U.S. Congress, *Location of High Technology Firms and Regional Economic Development* (Washington, D.C.: June 1, 1982), Table III.5, p. 23, and Table III.6, p. 25. Further discussion of these results is provided by John Rees, "High-Technology Location and Regional Development: The Theoretical Base," Appendix A, Office of Technology Assessment, *Technology, Innovation, and Regional Economic Development,* pp. 102–105. Also see A.S. Vavra and Larry S. Hill, "Site Selection Factors: Communities vs. Industry," unpublished manuscript, Texas A&M University, Fall 1983, p. 4 (cited in Bob McKay, "Location of High Technology Industry in Texas: A Critical Assessment," Texas Economic Development Commission, Austin, Texas, June 1984, p. 5).
32. It is reasonable to suggest that continuing deficiencies in the ability of key metropolitan areas around Texas to meet their rapidly growing needs for skilled workers in computer-related fields—and the needs of these workers for continuing education from locally accessible institutions—may well blunt the capacity of the state—and the Dallas area in particular—to retain its leadership in this industry. These findings may reveal a relatively unique historical relationship between industry and universities

in Texas. Even through their interdependence ultimately may be far greater than either party has acknowledged historically, more traditional cultural patterns may well continue to operate to insulate each of these sectors from being aware of the importance of the other. Arguably, in other places outside Texas where industry–university relationships have evolved differently, the pattern of findings might be quite different.

33. The business listing used for deriving a sampling frame included only incorporated business; therefore, for reasons related both to legal status and employment size, the stratum of "mom and pop" software producers and related service establishments is likely underrepresented. A more comprehensive census of such software production and service settings that extended to unincorporated establishments undoubtedly would have uncovered a more substantial proportion of firms in such residential settings.
34. This section is based on Donald A. Hicks and William H. Stolberg, "The High-Technology Sector: Growth and Development in the Dallas-Fort Worth Regional Economy, 1964–1984," Center for Applied Research, University of Texas at Dallas, October 1984.
35. Berry, "Hierarchical Diffusion," p. 18.
36. See H. Giersch, "The Age of Schumpeter," *American Economic Review,* May 1984, pp. 103–106. See also Wilbur Thompson, "Internal and External Factors in the Development of Urban Economies," in H. Perloff and L. Wingo (eds.), *Issues in Urban Economics* (Baltimore: The Johns Hopkins Press, 1968); A. Pred, "Diffusion, Organizational Spatial Structure and City–System Development," *Economic Geography,* Vol. 51, 1975, pp. 252–268; R.D. Norton, *City Life-Cycles and American Urban Policy* (New York: Academic Press, 1979), pp. 129–130; R. Struyk and F. James, *Intrametropolitan Industrial Location* (Lexington, Mass.: Lexington Books, 1975).
37. See U.S. Department of Housing and Urban Development, *The President's National Urban Policy Report, 1980* (Washington, D.C.: 1980), p. 3–11. More recent research on the impacts of economic cycles on cities has concluded that such cycles have their greatest effect on business formations and failures, predominantly a suburban phenomenon today. The implication is that the historical role of the central city as the incubator of new businesses has been surrendered. See George E. Peterson et al., "Effects of Economic Cycles," *Policy and Research Report,* The Urban Institute, Vol. 13, No. 2, Summer 1983, pp. 1–3.
38. E.A. Wood, "Industrial Development for a Community," *Texas Planning Bulletin,* March 1938, pp. 3–8.
39. Anne Bagamery, "No Policy Is Good Policy," *Forbes,* June 18, 1984, p. 140.
40. For a brief discussion of the concentrative effect of defense spending on high-technology development, see Glasmeier et al., "Recent Evidence on High-Technology Industries' Spatial Tendencies," p. 148.
41. For a brief introduction to the concept of "export employment" and the limits of its interpretation, see Eva C. Galambos and Arthur Schreiber, *Economic Analysis for Local Government* (Washington, D.C.: National League of Cities, November 1978), pp. 15 ff.
42. For discussion of the shift-to-services within an urban-regional context, see Royce Hanson (ed.), *Rethinking Urban Policy: Urban Development in an Advanced Economy,* Committee on National Urban Policy, National Research Council (Washington, D.C.: National Academy Press, 1983). For a review of why a region's service sector is unwisely assumed to be subordinate to the goods-producing sector, see John M.L. Gruenstein and Sally Guerra, "Can Services Sustain a Regional Econommy?" *Business Review,* Federal Reserve Bank of Philadelphia, July–August 1981, pp. 15–27.
43. Stanback et al., *Services: The New Economy,* p. 1.
44. The classification of high-technology sector employment follows Riche et al., "High

Technology Today and Tomorrow," Table 1, p. 52.

45. For a discussion of the differences between economic growth and development, see Peter Libassi and Victor Hausner, "Revitalizing Central City Investment" (Columbus, Ohio: Academy for Contemporary Problems, June 1977).
46. Donald A. Hicks (with Czi C. Pann), "Computer Software and Data Processing Services: Development Characteristics of a High-Technology Service Industry in Texas," Center for Policy Studies, University of Texas at Dallas, July 1984.
47. In an effort to supplement the exploratory work of the Dallas Mayor's Joint Task Force on High Technology, the Dallas Area Chamber of Commerce began compiling a census of high-technology firms in the six county area anchored by Dallas County. By July 1984, this census included approximately 800 firms. The data set used in this section permits analyses of trends at the establishment level as well as sectoral analysis.
48. On this point, see also Armington et al., "Formation and Growth in High-Technology Firms," p. 129.
49. Ibid., p. 142. See also Office of the U.S. Trade Representative, *U.S. National Study on Trade in Services* (Washington, D.C., December 1983).
50. These data are reported in F. Karl Willenbrock, "Human Resource Needs for High Technology Industry," *High Technology Public Policies for the 1980s* (Washington, D.C.: National Journal, 1983), pp. 83–87. Estimates of the size and composition of such skill pools vary considerably, however. In 1983 the National Science Foundation estimated the number of doctoral scientists—that is, employed scientists and engineers—in 1982 to be 3.3 million. Cited in *Site Selection Handbook/85,* 1985, p. 629.
51. For a similar analysis of the dispersion-concentration patterns of U.S. high-technology plants and employment between 1972 and 1977 at the county level of analysis, see Glasmeier et al., "Recent Evidence of High Technologies' Spatial Tendencies," pp. 146 ff. The Dallas-Fort Worth metropolitan area was created by combining the two separate metropolitan areas surrounding the cities of Dallas and Fort Worth, respectively, in the early 1970s. In this chapter, the D-FW region is defined in such a way as to equate to the eleven-county metropolitan designation that was used until early 1984, at which time it was scaled back to only nine counties. While Dallas County (Dallas) and Tarrant County (Fort Worth) are coanchors for this region, the following analysis assumes that the former is economically dominant within the region.
52. For an analysis of why new vintage manufacturing has been able to filter into areas at the periphery of metropolitan economies and in the rural and nonmetropolitan places between them, see Norris, "Nonmetropolitan Manufacturing Growth."
53. A review of alternative theoretical perspectives on systems of urban cities is found in Thierry J. Noyelle and Thomas M. Stanback, Jr., *The Economic Transformation of American Cities* (Totowa, N.J.: Allanheld and Rowman, 1983), pp. 27–50.
54. From a similar perspective, Goddard has indicated that the ability of technology to "decouple" specific economic activities from specific locations may be more important than whether or not this constitutes a centralizing or decentralizing dynamic. See John B. Goddard, "The Impact of New Information Technology on Urban and Regional Structure in Europe," paper presented at Landtronics: Anglo/American conference, London, England, June 1985.

Chapter 6

1. Bureau of Labor Statistics *Productivity Chartbook, 1981,* p. 11. See also Battelle Memorial Institute, *Agriculture 2000: A Look at the Future* (Columbus, Ohio: Battelle Press, 1983).

2. The competition to shape the perceptions of agriculturally related developments in the minds of the American public is illustrated by the following selected headlines in the national press. Neal Peirce, "U.S. Agricultural Doomsday Looms," *Washington Post,* February 22, 1982; B. Bruce-Briggs, "Antineomalthusian," *The New York Times Book Review,* September 13, 1981, pp. 9 ff. (a book review of Julian L. Simon, *The Ultimate Resource* [Princeton, N.J.: Princeton University Press, 1981]); Ann Crittenden, "Demand Outpaces World Food Supply," *New York Times,* August 16, 1981, pp. 1 ff.; Mary Thornton, "Food Exports Threatened by Loss of Farmland in the U.S.," *Washington Star,* November 26, 1979; Martha M. Hamilton, "Shortage of U.S. Farmland Predicted," *Washington Post,* January 17, 1981, pp. B1–B2; Seth King, "New Plans in the Works to Save the Good Earth," *New York Times,* October 11, 1981, p. 8E; Frank J. Popper, "Land Crunch?" *New York Times,* March 29, 1981; Julian L. Simon, "And Now the Good News: Life on Earth Is Improving," *The Washington Post,* July 13, 1980, pp. E1–E2.
3. Peter Wolf, *Land in America: Its Value, Use and Control* (New York: Pantheon, 1981), p. 340.
4. Office of the U.S. Trade Representative, *Twenty-Sixth Annual Report of the President of the United States on the Trade Agreements Program, 1981–82* (Washington, D.C., 1983), p. 16.
5. U.S. Bureau of the Census, *Statistical Abstract of the United States, 1982–83* (103rd ed.) (Washington, D.C., 1982), p. 676.
6. Data cited in Office of the U.S. Trade Representative, *Annual Report of the President of the United States on the Trade Agreements Program, 1983,* Twenty-seventh Issue (Washington, D.C., April 1084), p. 13.
7. *Statistical Abstract of the United States, 1982–83,* p. 678.
8. Peter Drucker, "Demographics and American Economic Policy," in M. Wachter and S. Wachter (eds.), *Toward a New U.S. Industrial Policy?* (Philadelphia: University of Pennsylvania Press, 1981, p. 237).
9. *Bureau of Economic Analysis, 1980 OBERS-Regional Projections,* Vol. 3 (Washington, D.C.: U.S. Department of Commerce, 1982), Table 4. Robert Z. Lawrence, *Can America Compete?* (Washington, D.C.: The Brookings Institution, 1984).
10. Data cited in *National Journal,* April 16, 1983, p. 820.
11. U.S. Bureau of the Census, jointly with U.S. Department of Agriculture, *Current Population Reports,* Series P-27, No. 56, *Farm Population of the United States: 1982,* (Washington, D.C.: U.S. Government Printing Office, 1983), p. 1.
12. *Statistical Abstract of the United States, 1982–83,* p. 649.
13. Office of the U.S. Trade Representative *Twenty-Sixth Annual Report of the President of the United States, 1981–82,* p. 44.
14. U.S. Department of Housing and Urban Development, *The President's National Urban Policy Report, 1980* (Washington, D.C., 1980), Chap. 1.
15. *Statistical Abstract of the United States, 1982–83,* p. 651.
16. Bluestone and Harrison, *The Deindustrialization of America;* Committee on National Urban Policy, Commission on Behavioral and Social Sciences and Education, National Research Council, *Rethinking Urban Policy: Urban Development in An Advanced Economy* (Washington, D.C.: National Academy Press, 1983), p. 14.
17. *Statistical Abstract of the United States, 1982–83,* p. 655.
18. Ibid., p. 652.
19. Donald A. Hicks, "'Plastic' Land in the Post-Industrial Era," *Cato Journal,* Vol. 2, No. 2, Fall 1982, pp. 437–468; D.A. Hicks, "National Urban Land Policy: Facing the Inevitability of City and Regional Evolution," in George Lefcoe (ed.), *Urban Land Policy for the 1980s: The Message for State and Local Government* (Lexington, Mass.: Lexington Books, 1983), pp. 21–39.

20. *Statistical Abstract of the United States, 1982–83,* p. 653.
21. Data cited in Robert B. Delano, "The U.S. Farm of the Future," *Dallas Morning News,* May 23, 1983. Not surprisingly, trends in land prices have not been uniform across the nation. In Texas, for instance, the inflation-adjusted value of rural land has risen continually through the 1980s, after lagging behind comparable areas in the Midwest during the 1970s. Intrastate variation in demand for rural land highlights the importance of both net returns from agricultural activity and the price of nonagricultural activity in nearby metropolitan economies on land values. See Ivan W. Schmedemann and Don Holtkamp, "Land Prices in Texas 1980–82," *Trend: Texas Real Estate News and Developments* (Texas Real Estate Research Center Texas A&M University, February 1983), pp. 1–3.
22. *Statistical Abstract of the United States, 1982–83,* p. 656.
23. Ibid., p. 660.
24. U.S. Department of Agriculture, *Farm Real Estate Market Developments* (Washington, D.C., July 1982).
25. Data cited in *National Journal,* June 11, 1983, p. 1284.
26. See *Texas: Past and Future: A Survey* (Austin: Texas 2000 Commission, Office of the Governor, 1981); *Texas Trends* (Austin: Texas 2000 Commission, Office of the Governor, August 1980).
27. Michael Wilson, "Cornfield vs. Concrete: The Concern over Farmland Conversion," *The National Future Farmer,* February–March 1982, pp. 18 ff. (citing 1977 Soil Conservation Service data.)
28. Cornucopia Project Report, 1981. Concern over soil erosion is certainly not new. The Soil Conservation Service has been promoting the use by farmers of ameliorative techniques to prevent the loss of topsoil through erosion for a half-century. Over the decades, new strategies for organizing and working the land have included terracing, crop rotation, and contour plowing, as well as the more recent technique of "no-till" farming, which allows topsoil to be less vulnerable to water and wind erosion.
29. Harold Faber, "New York Forests Expand in Acreage," *New York Times,* May 1982.
30. National Agricultural Lands Study, *Where Have the Farmlands Gone?* (Washington, D.C.: U.S. Government Printing Office, 1979).
31. John Long, *Population Deconcentration in the United States,* Special Demographic Analysis CDS-81-5 (Washington, D.C.: Bureau of the Census, 1981).
32. Glenn V. Fuguitt and Paul R. Voss, *Growth and Change in Rural America* (Washington, D.C.: Urban Land Institute, 1979).
33. Martha M. Hamilton, "Shortage of U.S. Farmland Predicted," *Washington Post,* January 17, 1981, pp. B1–B2.
34. Wilson, "Cornfield vs. Concrete."
35. Frank Schnidman, "Agricultural Land Preservation: Serious Land Policy Concern or Latest 'Public Interest' Ploy?" Monograph 81–1 (Cambridge, Mass.: Lincoln Institute of Land Policy, 1981).
36. Julian L. Simon and Seymour Sudman, *International Regional Science Review,* 1982.
37. Julian L. Simon, "Are We Losing Ground?" *Illinois Business Review,* Vol. 37, April 1980, pp. 1–6.
38. Pierre R. Crosson and Anthony T. Stout, cited in *Business Week,* "Soil Erosion May Not Be the Farmer's Nemesis after All," July 30, 1984, pp. 16 ff. See also Pierre R. Crosson and Sterling Brubaker, *Resource and Environmental Effects of U.S. Agriculture* (Baltimore: The Johns Hopkins University Press, Resources for the Future, 1982).
39. Daniel R. Vining, Sr., Thomas Plaut, and Kenneth Biesi, "Urban Encroachment on Prime Agricultural Land in the United States," *International Regional Science Review,* Vol. 2, No. 2, 1977, pp. 143–156; B.L. Dillman and C.F. Cousins, "Urban Encroach-

ment on Prime Agricultural Land," *International Regional Science Review,* Vol. 7, No. 3, 1982, pp. 285–292.

40. John Fraser Hart, "Urban Encroachment on Rural Areas," *The Geographical Review,* Vol. 66, January 1976, pp. 1–17; Clifton B. Luttrell, "Our 'Shrinking' Farmland: Mirage or Potential Crisis" (Federal Reserve Bank of St. Louis, October 1980, pp. 11–18).
41. Julian L. Simon, *The Ultimate Resource,* (Princeton, N.J.: Princeton University Press, 1981).
42. Schnidman, "Agricultural Land Preservation."
43. "Should Water Subsidies Go Down the Drain?" *Business Week,* March 5, 1984, pp. 104 ff.
44. Quote by William L. Brown in Philip Shabecoff, "A Million Species Are Endangered," *New York Times,* November 22, 1981, p. 8.
45. Janet Guyon, "Is the World Ready to Be Nourished on Tobacco? Some Investors Hope So but Run into Obstacles," *Wall Street Journal,* May 13, 1983, p. 42.
46. Ann Crittenden, "The Gene Machine Hits the Farm," *New York Times,* June 28, 1981, pp. 1F ff.
47. Office of the U.S. Trade Representative, 1982, p. 17.
48. *Statistical Abstract of the United States, 1982–83,* p. 657.
49. Perry L. Adkisson, "Future of World Agriculture," *Dallas Morning News,* January 4, 1983, p. 11A.
50. *Statistical Abstract of the United States, 1982–83,* p. 657.
51. W. Gene Wilson and Gene D. Sullivan, "The Advent of Biotechnology: Implications for Southeastern Agriculture," *Economic Review,* Federal Reserve Bank of Atlanta, March 1984, pp. 42–50.
52. Thomas A. Hiatt, "Rapid Commercialization of Gene-Splicing Predicted," *Dallas Morning News,* December 19, 1982, p. 37AA.
53. Hilary H. Smith, "Texas Agricultural Productivity: Is Research the Remedy?" *Economic Review,* Federal Reserve Bank of Dallas, November 1983, pp. 15–28; Office of Technology Assessment, U.S. Congress, *An Assessment of the United States Food and Agricultural System,* Report OTA–F–155 (Washington, D.C.: U.S. Government Printing Office, 1981).
54. Statement by Carl Haub and Robert Fox, quoted in Daniel Rapaport, "Double or Nothing?" *National Journal,* October 8, 1983, p. 2073.
55. Crittenden, "The Gene Machine Hits the Farm."
56. *Global 2000 Report to the President,* Vols. I, II and III (Washington, D.C., 1980); Paul R. Ehrlich, *The Population Bomb* (New York: Ballantine Books, 1968).
57. Julian L. Simon, "Are We Losing Our Farmland?" *The Public Interest,* Vol. 67, 1982, pp. 49–62; also see Herman Kahn, William Brown, and Leon Martel, *The Next 200 Years* (New York: Morrow, 1976); Wilfred Beckerman, *Two Cheers for the Affluent Society: A Spirited Defense of Economic Growth* (New York: St. Martins Press, 1975); John R. Maddox, *The Doomsday Syndrome* (New York: McGraw Hill, 1972).
58. Lester C. Thurow, "Farms: A Policy Success," *Newsweek,* May 16, 1983, p. 78.
59. Bruce Gardner, "Agriculture's Revealing—and Painful—Lesson for Industrial Policy," *The Backgrounder,* Heritage Foundation, January 3, 1984.

Chapter 7

1. This chapter is based in part on Donald A. Hicks, "The Property Tax in a New Industrial Era," in C. Lowell Harriss (ed.), *The Property Tax and Local Finance* (New

York: The Academy of Political Science, 1983), pp. 208–221.

2. Clarence Heer, "Taxation and Public Finance," *Report of the President's Research Committee on Social Trends* (Washington, D.C., 1933), 1331–1390.
3. This mix of topics is the subject of the selections in Donald A. Hicks and Norman J. Glickman (eds.), *Transition to the 21st Century: Prospects and Policies for Economic and Urban-Regional Transformation* (Greenwich, Conn.: JAI Press, 1983).
4. The reader should see Woolery's discussions of new kinds of property and new rights in property in Arlo Woolery, "Alternative Methods of Taxing Property," in Harriss, *The Property Tax and Local Finance,* pp. 180–188.
5. For a more detailed treatment of the loosening fit between the local property tax and the tax base reworked by local economic development trends, see Dick Netzer, "Does the Property Tax Have a Future?" in Harriss, *The Property Tax and Local Finance,* pp. 222–236.
6. Tax Foundation, *Monthly Tax Features,* Vol. 27, No. 6, July 1983, p. 2; and *Monthly Tax Features,* Vol. 29, No. 6, June/July 1985, p. 3.
7. For a general overview of land and the property tax, see Peter Wolf, *Land in America: Its Value, Use and Control* (New York: Pantheon Books, 1981), pp. 101–137.
8. Tax Foundation, "Localities Becoming More Self-Supporting, *Monthly Tax Features,* Vol. 28, No. 5, May 1984, pp. 3–4; also Tax Foundation, "Localities Outpace States and Washington in Raising Own-Source Tax/Nontax Revenues," *Monthly Tax Features,* Vol. 29, No. 3, March 1985, pp. 5–6.
9. Anthony H. Pascal, Mark D. Menchik, Jan M. Chaiken, Phyllis L. Ellickson, Warren E. Walker, Dennis N. DeTray, and Arthur E. Wise, *Fiscal Containment of Local and State Government* (Santa Monica, Calif.: Rand Corporation, September 1979), R-2494-FF/RC; and Anthony H. Pascal and Mark D. Menchik, "Fiscal Containment: Who Gains? Who Loses?" (Santa Monica, Calif.: Rand Corporation, September 1979), R-2494/1-FF/RC. As of November, 1983 according to ACIR 24 states had raised the property tax rates their municipal government could impose (cited in *National Journal,* August 11, 1984, p. 1515).
10. Bureau of the Census, *Statistical Abstract of the United States,* 1982–1983 (Washington, D.C.), p. 303.
11. Ibid., p. 77.
12. Tax Foundation, "Property Taxes Show First Decline Since World War II: Most of Drop Occurs in California Following 'Prop 13'," *Monthly Tax Features,* Vol. 25, March 1981, p. 4.
13. Tax Foundation, "U.S. Family $427 Poorer in '81 than '71: Federal Taxes and Inflation Take Big Bite," *Monthly Tax Features,* Vol. 25, June–July 1981, p. 4.
14. Tax Foundation, "Localities Becoming More Self-Supporting," *Monthly Tax Features,* pp. 3–4.
15. For a distinctly different emphasis on the enduring local-regional scale of urban economies, see Jane Jacobs, *Cities and the Wealth of Nations* (New York: Random House, 1984).
16. Melvin M. Webber, "The Urban Place and the Nonplace Urban Realm;" in Melvin Webber (ed.), *Explanations into Urban Structure* (Philadelphia: University of Pennsylvania Press, 1964), pp. 79–153; also, "the Post-City Age," Daedalus, Vol. 97, No. 4, Fall 1968, pp. 1091–1110; and "Order in Diversity: Community without Propinquity," in Lowdon Wingo (ed.), *Cities and Space: The Future Use of Urban Land* (Baltimore: The John Hopkins University Press, 1963). L. Wingo and Carol C. Webber, "Culture, Territoriality and the Elastic Mile," in H. Wentworth Eldredge (ed.), *Taming Megalopolis,* Vol. 1 (New York: Anchor Books, 1967).
17. Tax Foundation, "$1.3 Trillion Spending in FY 83 by All Levels of Government in U.S.," *Monthly Tax Features,* Vol. 27, No. 6, July 1983, pp. 1 ff.

18. 1982 census and government surveys reported by Commerce Clearinghouse News Bureau of Chicago (cited in *Dallas Magazine,* August 1984, p. 124).
19. Alfred L. Malabre, Jr., "Record Surpluses at the State and Local Levels Ease Worry over Washington's Budget Deficit," *Wall Street Journal,* December 6, 1983, pp. 58 ff.
20. E.A. Ross (Leland Stanford University professor) 1896 quote, cited in W. Gerald Holmes, "Common Sense Is the Best Subsidy," *Nation's Business,* October 1930, pp. 74 ff.
21. See Michael Kieschnik, *Taxes and Growth: Business Incentives and Economic Development* (Washington, D.C.: Council of State Planning Agencies, 1981); Bennett Harrison and Susan Kanter, "The Political Economy of States' Job Creation and Business Incentives," *Journal of the American Institute of Planners,* Vol. 44, October 1978, pp. 424–435. For a discussion of the taxes as location factors, see Bahar B. Norris, "Nonmetropolitan Manufacturing Growth in the Post-World War II United States: A Test of Industrial Location Theory," unpublished Ph.D. dissertation, University of Texas at Dallas, August 1984, pp. 23–25. The ineffectiveness of industrial development efforts aimed at luring plants from elsewhere is recorded as early as 1926–1927 in James L. Madden, "Industry Grows but Seldom Moves," *Nation's Business,* August 1929, pp. 45 ff.
22. Joint Economic Committee, U.S. Congress, *State and Local Economic Development Strategy: A 'Supply Side' Perspective* (Washington, D.C.: U.S. Government Printing Office, 1981).
23. See Catherine Armington, Candee Harris, and Marjorie Odle, "Formation and Growth in High-Technology Firms: A Regional Assessment, Appendix B, Office of Technology Assessment, *Technology Innovation, and Regional Economic Development* (Washington, D.C., 1984), p. 108–143.
24. Roger J. Vaughan, *State Taxation and Economic Developmtn* (Washington, D.C.: The Council of State Planning Agencies, 1979); Bernard L. Weinstein, "Tax Incentives for Growth," *Society,* Vol. 14, No. 3, 1977, pp. 73–75.
25. Joint Economic Committee, U.S. Congress, *Location of High Technology Firms and Regional Economic Development* (Washington, D.C.: U.S. Government Printing Office, 1982). This study also reported that a recent reduction of property taxes in Massachusetts appeared to be associated—if indirectly—with the recent upturn in the state economy with its newly found dependence on an expanding high-technology manufacturing base. See also Armington et al. "Formation and Growth in High-Technology Firms," pp.134, 138, 142. The same finding emerged in Donald A. Hicks, "Computer Software and Data Processing Services: Development Characteristics of a High-Technology Service Industry in Texas," (Austin, Texas: Governor's Office of Economic Development and Texas Computer Industry Council, July 1984).
26. E. Mills described W. Alonso's urban land market dynamics as follows:

    > Other things being equal, goods and services are produced downtown if their production functions permit substitution of capital and labor for land. If not, they are produced in suburbs or, as in the case of agriculture, outside urban areas altogether. Furthermore, goods and services produced both downtown and in the suburbs are produced with higher ratios of nonland-to-land inputs downtown than in the suburbs. Therefore, understanding how the urban economy ticks is mainly a matter of understanding how markets combine land with other inputs in varying proportions of different places to produce goods and services.

    E. Hills, *Urban Economics,* 2nd ed. (Glenview, Ill.: Scott, Foresman, 1980), p. 56; cited in John F. McDonald, "The Substitution of Land for Other Inputs in Urban Areas," *Papers of the Regional Science Association,* Vol. 48, 1981, pp. 39–52.
27. See Neal R. Peirce, "The Tax Bill Has Bad News for Urban Areas," *National Journal,* June 8, 1985, p. 1370; also, "Tax Reform Already has Developers Spooked," *Business Week,* June 24, 1985, p. 42.

28. For a dramatic example of the ways in which new telecommunications technologies can substantially reduce the need for office space and hasten the dispersal of centralized work forces, see John Urquhart, "Telephone Companies Adapt to Changing Needs for Space," The *Wall Street Journal,* Nov. 13, 1985, p. 35. The potential impact of communication-based office automation on central business district high-rise office development was noted early by R.D. Norton, *City-Life Cycles and American Urban Policy* (New York: Academic Press, 1979), p. 172.
29. *The Report of the President's Commission on Housing* (Washington, D.C., 1982), pp. 71 ff.; Ira S. Lowry, "Managing the Existing Housing Stock: Prospects and Problems," (Santa Monica, Calif.: Rand Corporation, February 1982).
30 Bureau of the Census,*Statistical Abstract of the United States, 1982–83,* Table 1351, p. 752.
31. Ibid., Table 60, p. 43.
32. Data provided by the National Association of Realtors.
33. What future, if any, the property tax might have has been the focus of explicit analysis for many years. For the conditions under which two analysts can foresee the demise of the property tax, see George W. Mitchell, "Is This Where We Came In?" *Proceedings of the Forty-Ninth Annual Conference* (National Tax Association, 1956), p. 494; and Dick Netzer, "Does the Property Tax Have a Future?" In Harriss, *The Property Tax and Local Finance,* p.231.
34. Barry A. Currier, "An Analysis of Differential Taxation as a Method of Maintaining Agricultural and Open Space Land Uses," *University of Florida Law Review,* Vol. 30, No. 5, Fall 1978, pp. 821–842.
35. John D. Kasarda, "The Implications of Contemporary Redistribution Trends for National Urban Policy," in Donald A. Hicks and Norman J. Glickman (eds.) *Transition to the 21st Century: Prospects and Policies for Economic and Urban-Regional Transformation* (Greenwich, Conn.: JAI Press, 1983).
36. E. Savas, *Privatizing the Public Sector: How to Shrink Government* (Chatham, N.J.: Chatham House, 1982); Donald Fisk, Herbert Keisling, and Thomas Muller, *Private Provision of Public Services: An Overview* (Washington, D.C.: The Urban Institute, 1978).
37. Figures provided by the Tax Foundation indicate that states and localities have increased their dependence on user fees and nontax revenues significantly in past decades, from amounts equivalent to 20 percent total tax revenue collected in 1960 to 34 percent in 1980. For explorations of the transition by localities from tax to such nontax financing alternatives, see Anthony Pascal, *A Guide to Installing Equitable Beneficiary-Based Finance in Local Government* (Santa Monica, Calif.: Rand Corporation, June 1984). "Exploring Benefit-Based Financing for Local Government Services: Must User Charges Harm the Disadvantaged?" (Santa Monica, Calif.: Rand Corporation, February 1984). For an account of how cities undergoing industrial restructuring amid population decline can avoid crippling fiscal mismatches, see Terry Nichols Clark and Lorna Crowley Ferguson, *City Money: Political Processes, Fiscal Strain, and Retrenchment* (New York: Columbia University Press, 1983).

Index

A

B

D

E

F

G

H

I

N

O